Java
多线程编程核心技术
（第2版）

Java Multi-thread Programming
(Second Edition)

高洪岩 著

机械工业出版社
China Machine Press

图书在版编目（CIP）数据

Java 多线程编程核心技术 / 高洪岩著 . —2 版 . —北京：机械工业出版社，2019.1（2021.10 重印）

（Java 核心技术系列）

ISBN 978-7-111-61490-6

I. J…　II. 高…　III. JAVA 语言 – 程序设计　IV. TP312.8

中国版本图书馆 CIP 数据核字（2018）第 274886 号

Java 多线程编程核心技术（第 2 版）

出版发行：机械工业出版社（北京市西城区百万庄大街 22 号　邮政编码：100037）	
责任编辑：赵亮宇	责任校对：殷　虹
印　　刷：北京捷迅佳彩印刷有限公司	版　　次：2021 年 10 月第 2 版第 5 次印刷
开　　本：186mm×240mm　1/16	印　　张：28.25
书　　号：ISBN 978-7-111-61490-6	定　　价：99.00 元

凡购本书，如有缺页、倒页、脱页，由本社发行部调换

客服热线：（010）88379426　88361066　　　　投稿热线：（010）88379604

购书热线：（010）68326294　　　　　　　　　　读者信箱：hzjsj@hzbook.com

版权所有·侵权必究

封底无防伪标均为盗版

本书法律顾问：北京大成律师事务所　韩光 / 邹晓东

Preface 前　言

本书是国内首本整本系统、完整地介绍 Java 多线程技术的书籍，作为笔者，我要感谢大家的支持与厚爱。

本书第 1 版在出版后获得了广大 Java 程序员与学习者的关注，技术论坛、博客、公众号等平台大量涌现出针对 Java 多线程技术的讨论与分享。能为国内 IT 知识的建设贡献微薄之力是让我最欣慰的。

有些读者在第一时间就根据书中的知识总结了学习笔记，并在博客中进行分享，笔者非常赞赏这种传播知识的精神。知识就要分享，知识就要传播，这样才能共同进步。

第 2 版与第 1 版的区别

本书第 1 版上市后收到了大量的读者反馈，我对每一个建议都细心地进行整理，力求在第 2 版中得以完善。

第 2 版在第 1 版的基础上着重加强了 8 点更新：

1）大量知识点重排，更有利于阅读与理解；

2）更新了读者提出的共性问题并进行集中讲解；

3）丰富 Thread.java 类 API 的案例，使其更具有实用性；

4）对线程的信息进行监控实时采样；

5）强化了 volatile 语义、多线程核心 synchronized 的案例；

6）力求知识点连贯，方便深度学习与理解，增加原子与线程安全的内容；

7）深入浅出地介绍代码重排特性；

8）细化工具类 ThreaLocal 和 InheritableThreadLocal 的源代码分析与原理。

由于篇幅有限，有关线程池的知识请参考笔者的另一本书——《 Java 并发编程：核心方法与框架》，那本书中有针对 Java 并发编程技术的讲解。在向分布式领域进军时还需要用到 NIO 和 Socket 技术，故推荐笔者的拙作《NIO 与 Socket 编程技术指南》，希望可以给读者带来一些帮助。

本书秉承大道至简的主导思想，只介绍 Java 多线程开发中最值得关注的内容，希望抛砖引玉，以个人的一些想法和见解，为读者拓展出更深入、更全面的思路。

本书特色

在撰写本书的过程中，我尽量少用"啰唆"的文字，全部以 Demo 式案例来讲解技术点的实现，使读者看到代码及运行结果后就可以知道项目要解决的是什么问题，类似于网络中博客的风格，让读者用最短的时间学习知识点，明白知识点如何应用，以及在使用时要避免什么，使读者能够快速学习知识并解决问题。

读者对象

- Java 程序员；
- 系统架构师；
- Java 多线程开发者；
- Java 并发开发者；
- 大数据开发者；
- 其他对多线程技术感兴趣的人员。

如何阅读本书

本书本着实用、易懂的学习原则，利用 7 章来介绍 Java 多线程相关的技术。

第 1 章讲解了 Java 多线程的基础，包括 Thread 类的核心 API 的使用。

第 2 章讲解了在多线程中对并发访问的控制，主要是 synchronized 的使用。由于此关键字在使用上非常灵活，所以该章用很多案例来说明它的使用，为读者学习同步知识打好坚实的基础。

第 3 章讲解了线程之间的通信与交互细节。该章主要介绍 wait()、notifyAll() 和 notify() 方法的使用，使线程间能够互相通信，合作完成任务。该章还介绍了 ThreadLocal 类的使用。学习完该章，读者就能在 Thread 多线程中进行数据的传递了。

第 4 章讲解了 Lock 对象。因为 synchronized 关键字使用起来比较麻烦，所以 Java 5 提供了 Lock 对象，更好地实现了并发访问时的同步处理，包括读写锁等。

第 5 章讲解了 Timer 定时器类，其内部原理是使用多线程技术。定时器在执行计划任务时是很重要的，在进行 Android 开发时也会深入使用。

第 6 章讲解的单例模式虽然很简单，但如果遇到多线程将会变得非常麻烦。如何在多线程中解决这么棘手的问题呢？本章会全面给出解决方案。

第 7 章对前面章节遗漏的技术空白点进行补充，通过案例使多线程的知识体系更加完整，尽量做到不出现技术空白点。

交流和支持

由于笔者水平有限，加上编写时间仓促，书中难免会出现一些疏漏或者不准确的地方，恳请读者批评指正，期待能够得到你们的真挚反馈，在技术之路上互勉共进。

联系笔者的邮箱是 279377921@qq.com。

致谢

在本书出版的过程中，感谢公司领导和同事的大力支持，感谢家人给予我充足的时间来撰写稿件，感谢出生 3 个多月的儿子高晟京，看到你，我有了更多动力，最后感谢在此稿件上耗费大量精力的高婧雅编辑与她的同事们，是你们的鼓励和帮助，引导我顺利完成了本书。

高洪岩

前言

第1章　Java 多线程技能

1.1 进程和多线程概述 ... 1
1.2 使用多线程 ... 5
1.2.1 继承 Thread 类 ... 5
1.2.2 使用常见命令分析线程的信息 ... 8
1.2.3 线程随机性的展现 ... 11
1.2.4 执行 start() 的顺序不代表执行 run() 的顺序 ... 12
1.2.5 实现 Runnable 接口 ... 13
1.2.6 使用 Runnable 接口实现多线程的优点 ... 14
1.2.7 实现 Runnable 接口与继承 Thread 类的内部流程 ... 16
1.2.8 实例变量共享造成的非线程安全问题与解决方案 ... 17
1.2.9 Servlet 技术造成的非线程安全问题与解决方案 ... 21
1.2.10 留意 i-- 与 System.out.println() 出现的非线程安全问题 ... 24
1.3 currentThread() 方法 ... 26
1.4 isAlive() 方法 ... 29
1.5 sleep(long millis) 方法 ... 31
1.6 sleep(long millis, int nanos) 方法 ... 33
1.7 StackTraceElement[] getStackTrace() 方法 ... 33
1.8 static void dumpStack() 方法 ... 35
1.9 static Map<Thread, StackTraceElement[]> getAllStackTraces() 方法 ... 36
1.10 getId() 方法 ... 38
1.11 停止线程 ... 38
1.11.1 停止不了的线程 ... 39
1.11.2 判断线程是否为停止状态 ... 41
1.11.3 能停止的线程——异常法 ... 43
1.11.4 在 sleep 状态下停止线程 ... 47
1.11.5 用 stop() 方法暴力停止线程 ... 49
1.11.6 stop() 方法与 java.lang.ThreadDeath 异常 ... 51
1.11.7 使用 stop() 释放锁给数据造成不一致的结果 ... 52
1.11.8 使用 "return;" 语句停止线程的缺点与解决方案 ... 54
1.12 暂停线程 ... 57
1.12.1 suspend() 方法与 resume() 方法的使用 ... 57
1.12.2 suspend() 方法与 resume() 方法的缺点——独占 ... 58

1.12.3　suspend() 方法与 resume() 方法的缺点——数据不完整 …………62
1.13　yield() 方法 …………………………63
1.14　线程的优先级 ………………………64
　　1.14.1　线程优先级的继承特性 ………65
　　1.14.2　优先级的规律性 ………………66
　　1.14.3　优先级的随机性 ………………68
　　1.14.4　优先级对线程运行速度的影响 …70
1.15　守护线程 ……………………………71
1.16　本章小结 ……………………………73

第 2 章　对象及变量的并发访问 …………74

2.1　synchronized 同步方法 ………………74
　　2.1.1　方法内的变量为线程安全 ……74
　　2.1.2　实例变量非线程安全问题与解决方案 ………………………77
　　2.1.3　同步 synchronized 在字节码指令中的原理 ……………………80
　　2.1.4　多个对象多个锁 …………………81
　　2.1.5　将 synchronized 方法与对象作为锁 ………………………84
　　2.1.6　脏读 ……………………………89
　　2.1.7　synchronized 锁重入 ……………91
　　2.1.8　锁重入支持继承的环境 …………93
　　2.1.9　出现异常，锁自动释放 …………94
　　2.1.10　重写方法不使用 synchronized …96
　　2.1.11　public static boolean holdsLock (Object obj) 方法的使用 …99
2.2　synchronized 同步语句块 ……………99
　　2.2.1　synchronized 方法的弊端 ………99
　　2.2.2　synchronized 同步代码块的使用 ……………………………102
　　2.2.3　用同步代码块解决同步方法的弊端 ……………………………104
　　2.2.4　一半异步，一半同步 ……………105
　　2.2.5　synchronized 代码块间的同步性 …108
　　2.2.6　println() 方法也是同步的 ………110
　　2.2.7　验证同步 synchronized(this) 代码块是锁定当前对象的 ………110
　　2.2.8　将任意对象作为锁 ………………113
　　2.2.9　多个锁就是异步执行 ……………116
　　2.2.10　验证方法被调用是随机的 ……118
　　2.2.11　不同步导致的逻辑错误及其解决方法 ……………………121
　　2.2.12　细化验证 3 个结论 ……………124
　　2.2.13　类 Class 的单例性 ………………129
　　2.2.14　静态同步 synchronized 方法与 synchronized(class) 代码块 ……130
　　2.2.15　同步 syn static 方法可以对类的所有对象实例起作用 …………135
　　2.2.16　同步 syn(class) 代码块可以对类的所有对象实例起作用 ……137
　　2.2.17　String 常量池特性与同步相关的问题与解决方案 ……………138
　　2.2.18　同步 synchronized 方法无限等待问题与解决方案 …………141
　　2.2.19　多线程的死锁 …………………143
　　2.2.20　内置类与静态内置类 …………146
　　2.2.21　内置类与同步：实验 1 ………149
　　2.2.22　内置类与同步：实验 2 ………151
　　2.2.23　锁对象改变导致异步执行 ……153
　　2.2.24　锁对象不改变依然同步执行 …156
　　2.2.25　同步写法案例比较 ……………158
2.3　volatile 关键字 ………………………159
　　2.3.1　可见性的测试 …………………159
　　2.3.2　原子性的测试 …………………168
　　2.3.3　禁止代码重排序的测试 ………176
2.4　本章小结 ……………………………187

第 3 章　线程间通信 ········· 188

- 3.1 wait/notify 机制 ········· 188
  - 3.1.1 不使用 wait/notify 机制实现线程间通信 ········· 188
  - 3.1.2 wait/notify 机制 ········· 191
  - 3.1.3 wait/notify 机制的原理 ········· 192
  - 3.1.4 wait() 方法的基本使用 ········· 192
  - 3.1.5 完整实现 wait/notify 机制 ········· 194
  - 3.1.6 使用 wait/notify 机制实现 list.size() 等于 5 时的线程销毁 ········· 195
  - 3.1.7 对业务代码进行封装 ········· 198
  - 3.1.8 线程状态的切换 ········· 201
  - 3.1.9 wait() 方法：立即释放锁 ········· 202
  - 3.1.10 sleep() 方法：不释放锁 ········· 203
  - 3.1.11 notify() 方法：不立即释放锁 ········· 204
  - 3.1.12 interrupt() 方法遇到 wait() 方法 ········· 206
  - 3.1.13 notify() 方法：只通知一个线程 ········· 208
  - 3.1.14 notifyAll() 方法：通知所有线程 ········· 211
  - 3.1.15 wait(long) 方法的基本使用 ········· 212
  - 3.1.16 wait(long) 方法自动向下运行需要重新持有锁 ········· 214
  - 3.1.17 通知过早问题与解决方法 ········· 217
  - 3.1.18 wait 条件发生变化与使用 while 的必要性 ········· 220
  - 3.1.19 生产者/消费者模式的实现 ········· 224
  - 3.1.20 通过管道进行线程间通信——字节流 ········· 250
  - 3.1.21 通过管道进行线程间通信——字符流 ········· 253
  - 3.1.22 实现 wait/notify 的交叉备份 ········· 256
- 3.2 join() 方法的使用 ········· 259
  - 3.2.1 学习 join() 方法前的铺垫 ········· 259
  - 3.2.2 join() 方法和 interrupt() 方法出现异常 ········· 261
  - 3.2.3 join(long) 方法的使用 ········· 263
  - 3.2.4 join(long) 方法与 sleep(long) 方法的区别 ········· 264
  - 3.2.5 join() 方法后面的代码提前运行——出现意外 ········· 268
  - 3.2.6 join() 方法后面的代码提前运行——解释意外 ········· 270
  - 3.2.7 join(long millis, int nanos) 方法的使用 ········· 273
- 3.3 类 ThreadLocal 的使用 ········· 273
  - 3.3.1 get() 方法与 null ········· 274
  - 3.3.2 类 ThreadLocal 存取数据流程分析 ········· 275
  - 3.3.3 验证线程变量的隔离性 ········· 277
  - 3.3.4 解决 get() 方法返回 null 的问题 ········· 282
  - 3.3.5 验证重写 initialValue() 方法的隔离性 ········· 283
- 3.4 类 InheritableThreadLocal 的使用 ········· 284
  - 3.4.1 类 ThreadLocal 不能实现值继承 ········· 285
  - 3.4.2 使用 InheritableThreadLocal 体现值继承特性 ········· 286
  - 3.4.3 值继承特性在源代码中的执行流程 ········· 288
  - 3.4.4 父线程有最新的值，子线程仍是旧值 ········· 291
  - 3.4.5 子线程有最新的值，父线程仍是旧值 ········· 293
  - 3.4.6 子线程可以感应对象属性值的变化 ········· 294
  - 3.4.7 重写 childValue() 方法实现对继承的值进行加工 ········· 297
- 3.5 本章小结 ········· 298

第 4 章　Lock 对象的使用 299

4.1 使用 ReentrantLock 类 299
4.1.1 使用 ReentrantLock 实现同步 299
4.1.2 验证多代码块间的同步性 301
4.1.3 await() 方法的错误用法与更正 304
4.1.4 使用 await() 和 signal() 实现 wait/notify 机制 307
4.1.5 await() 方法暂停线程运行的原理 309
4.1.6 通知部分线程——错误用法 312
4.1.7 通知部分线程——正确用法 314
4.1.8 实现生产者/消费者模式一对一交替输出 317
4.1.9 实现生产者/消费者模式多对多交替输出 319
4.1.10 公平锁与非公平锁 321
4.1.11 public int getHoldCount() 方法的使用 324
4.1.12 public final int getQueueLength() 方法的使用 325
4.1.13 public int getWaitQueueLength (Condition condition) 方法的使用 327
4.1.14 public final boolean hasQueuedThread (Thread thread) 方法的使用 328
4.1.15 public final boolean hasQueuedThreads() 方法的使用 329
4.1.16 public boolean hasWaiters (Condition condition) 方法的使用 331
4.1.17 public final boolean isFair() 方法的使用 332
4.1.18 public boolean isHeldByCurrentThread() 方法的使用 333
4.1.19 public boolean isLocked() 方法的使用 334
4.1.20 public void lockInterruptibly() 方法的使用 335
4.1.21 public boolean tryLock() 方法的使用 336
4.1.22 public boolean tryLock (long timeout, TimeUnit unit) 方法的使用 338
4.1.23 public boolean await (long time, TimeUnit unit) 方法的使用 339
4.1.24 public long awaitNanos (long nanosTimeout) 方法的使用 341
4.1.25 public boolean awaitUntil (Date deadline) 方法的使用 342
4.1.26 public void awaitUninterruptibly() 方法的使用 344
4.1.27 实现线程按顺序执行业务 346
4.2 使用 ReentrantReadWriteLock 类 349
4.2.1 ReentrantLock 类的缺点 349
4.2.2 ReentrantReadWriteLock 类的使用——读读共享 351
4.2.3 ReentrantReadWriteLock 类的使用——写写互斥 352
4.2.4 ReentrantReadWriteLock 类的使用——读写互斥 352
4.2.5 ReentrantReadWriteLock 类的使用——写读互斥 354
4.3 本章小结 355

第 5 章　定时器 Timer 356
5.1 定时器 Timer 的使用 356

5.1.1 schedule(TimerTask task, Date time) 方法的测试 ········· 356
5.1.2 schedule(TimerTask task, Date firstTime, long period) 方法的测试 ········· 366
5.1.3 schedule(TimerTask task, long delay) 方法的测试 ········· 374
5.1.4 schedule(TimerTask task, long delay, long period) 方法的测试 ··· 374
5.1.5 scheduleAtFixedRate (TimerTask task, Date firstTime, long period) 方法的测试 ········· 375
5.2 本章小结 ········· 384

第 6 章 单例模式与多线程 ········· 385
6.1 立即加载 / 饿汉模式 ········· 385
6.2 延迟加载 / 懒汉模式 ········· 387
　6.2.1 延迟加载 / 懒汉模式解析 ········· 387
　6.2.2 延迟加载 / 懒汉模式的缺点 ········· 388
　6.2.3 延迟加载 / 懒汉模式的解决方案 ········· 390
6.3 使用静态内置类实现单例模式 ········· 399
6.4 序列化与反序列化的单例模式实现 ········· 400
6.5 使用 static 代码块实现单例模式 ········· 402
6.6 使用 enum 枚举数据类型实现单例模式 ········· 404
6.7 完善使用 enum 枚举数据类型实现单例模式 ········· 405
6.8 本章小结 ········· 407

第 7 章 拾遗增补 ········· 408
7.1 线程的状态 ········· 408
　7.1.1 验证 NEW、RUNNABLE 和 TERMINATED ········· 410
　7.1.2 验证 TIMED_WAITING ········· 411
　7.1.3 验证 BLOCKED ········· 412
　7.1.4 验证 WAITING ········· 414
7.2 线程组 ········· 415
　7.2.1 线程对象关联线程组：一级关联 ········· 416
　7.2.2 线程对象关联线程组：多级关联 ········· 417
　7.2.3 线程组自动归属特性 ········· 418
　7.2.4 获取根线程组 ········· 419
　7.2.5 线程组中加线程组 ········· 420
　7.2.6 组内的线程批量停止 ········· 421
　7.2.7 递归取得与非递归取得组内对象 ········· 422
7.3 Thread.activeCount() 方法的使用 ········· 423
7.4 Thread.enumerate(Thread tarray[]) 方法的使用 ········· 423
7.5 再次实现线程执行有序性 ········· 424
7.6 SimpleDateFormat 非线程安全 ········· 426
　7.6.1 出现异常 ········· 426
　7.6.2 解决异常的方法 1 ········· 428
　7.6.3 解决异常的方法 2 ········· 430
7.7 线程中出现异常的处理 ········· 431
　7.7.1 线程出现异常的默认行为 ········· 431
　7.7.2 使用 setUncaughtException-Handler() 方法进行异常处理 ········· 432
　7.7.3 使用 setDefaultUncaughtExceptionHandler() 方法进行异常处理 ········· 433
7.8 线程组内处理异常 ········· 434
7.9 线程异常处理的优先性 ········· 437
7.10 本章小结 ········· 442

第 1 章 Chapter 1

Java 多线程技能

作为本书的第 1 章，重点是让读者快速进入 Java 多线程的学习，所以本章主要介绍 Thread 类的核心方法。Thread 类的核心方法较多，读者应该着重掌握如下技术点：
- 线程的启动；
- 如何使线程暂停；
- 如何使线程停止；
- 线程的优先级；
- 线程安全相关的问题。

以上内容也是本章学习的重点与思路，掌握这些内容是进入 Java 多线程学习的必经之路。

1.1 进程和多线程概述

本书主要介绍在 Java 语言中使用的多线程技术，但讲到多线程技术时不得不提及"进程"这个概念，"百度百科"对"进程"的解释如图 1-1 所示。

图 1-1 进程的定义

初看这段文字十分抽象，难以理解，那么再来看如图 1-2 所示的内容。

图 1-2　Windows 7 系统中的进程列表

难道一个正在操作系统中运行的 exe 程序可以理解成一个"进程"？没错！

通过查看"Windows 任务管理器"窗口中的列表，完全可以将运行在内存中的 exe 文件理解成进程——进程是受操作系统管理的基本运行单元。

程序是指令序列，这些指令可以让 CPU 完成指定的任务。*.java 程序经编译后形成 *.class 文件，在 Windows 中启动一个 JVM 虚拟机相当于创建了一个进程，在虚拟机中加载 class 文件并运行，在 class 文件中通过执行创建新线程的代码来执行具体的任务。创建测试用的代码如下：

```java
public class Test1 {
    public static void main(String[] args) {
        try {
            Thread.sleep(Integer.MAX_VALUE);
        } catch (InterruptedException e) {
            e.printStackTrace();
        }
    }
}
```

在没有运行这个类之前，任务管理器中以"j"开头的进程列表如图 1-3 所示。

第 1 章　Java 多线程技能　❖　3

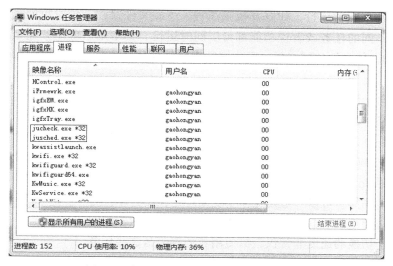

图 1-3　任务管理器中以 "j" 开头的进程

Test1 类重复运行 3 次后的进程列表如图 1-4 所示。可以看到，在任务管理器中创建了 3 个 javaw.exe 进程，说明每执行一次 main() 方法就创建一个进程，其本质上就是 JVM 虚拟机进程。

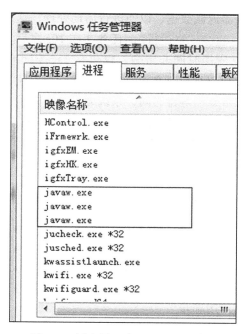

图 1-4　创建了 3 个 javaw.exe 进程

那什么是线程呢？线程可以理解为在进程中独立运行的子任务，例如，QQ.exe 运行时，很多的子任务也在同时运行,如好友视频线程、下载文件线程、传输数据线程、发送表情线程等,这些不同的任务或者说功能都可以同时运行,其中每一项任务完全可以理解成是"线程"在工作,传文件、听音乐、发送图片表情等这些功能都有对应的线程在后台默默地运行。

进程负责向操作系统申请资源。在一个进程中,多个线程可以共享进程中相同的内存或文件资源。

使用多线程有什么优点呢?其实如果有使用"多任务操作系统"的经验,如 Windows 系列,大家应该都有这样的体会:使用多任务操作系统 Windows,可以大幅利用 CPU 的空闲时间来处理其他任务,例如,可以一边让操作系统处理正在用打印机打印的数据,一边使用 Word 编辑文档。CPU 在这些任务中不停地进行切换,由于切换的速度非常快,给使用者的感受是这些任务在同时运行,所以使用多线程技术可以在同一时间内执行更多不同的任务。

为了更加有效地理解多线程的优势,下面先来看如图 1-5 所示的单任务运行环境。

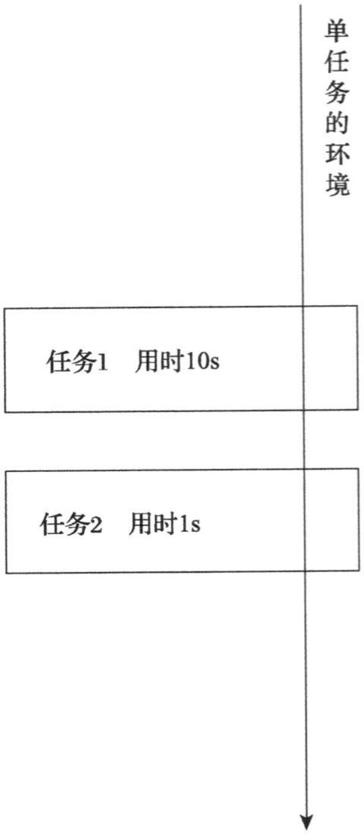

在图 1-5 中,任务 1 和任务 2 是两个完全独立、不相关的任务。任务 1 在等待远程服务器返回数据,以便进行后期处理,这时 CPU 一直呈等待状态,一直在"空运行"。任务 2 在 10s 之后被运行,虽然执行完任务 2 所用时间非常短,仅仅是 1s,但也必须等任务 1 运行结束后才可以运行任务 2,本程序运行在单任务环境中,所以任务 2 有非常长的等待时间,系统运行效率大幅降低。单任务的特点就是排队执行,即同步,就像在 cmd 中输入一条命令后,必须等待这条命令执行完才可以执行下一条命令。在同一时间只能执行一个任务,CPU 利用率大幅降低,这就是单任务运行环境的缺点。

图 1-5 单任务运行环境

多任务运行环境如图 1-6 所示。

在图 1-6 中,CPU 完全可以在任务 1 和任务 2 之间来回切换,使任务 2 不必等到 10s 之后再运行,系统和 CPU 的运行效率大大提升,这就是为什么要使用多线程技术、为什么要学习多线程。多任务的特点是在同一时间可以执行多个任务,这也是多线程技术的优点。使用多线程也就是在使用异步。

在通常情况下,单任务和多任务的实现与操作系统有关。例如,在一台计算机上使用同一个 CPU,安装 DOS 磁盘操作系统只能实现单任务运行环境,而

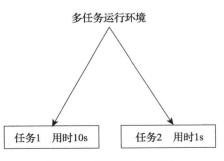

图 1-6 多任务运行环境

安装 Windows 操作系统则可以实现多任务运行环境。

在什么场景下使用多线程技术？笔者总结了两点。

1）阻塞。一旦系统中出现了阻塞现象，则可以根据实际情况来使用多线程技术提高运行效率。

2）依赖。业务分为两个执行过程，分别是 A 和 B。当 A 业务发生阻塞情况时，B 业务的执行不依赖 A 业务的执行结果，这时可以使用多线程技术来提高运行效率；如果 B 业务的执行依赖 A 业务的执行结果，则可以不使用多线程技术，按顺序进行业务的执行。

在实际的开发应用中，不要为了使用多线程而使用多线程，要根据实际场景决定。

> **注意** 多线程是异步的，所以千万不要把 Eclipse 代码的顺序当作线程执行的顺序，线程被调用的时机是随机的。

1.2 使用多线程

想学习一个技术就要"接近"它，所以本节首先通过一个示例来接触一下线程。

一个进程正在运行时至少会有一个线程在运行，这种情况在 Java 中也是存在的，这些线程在后台默默地执行，例如，调用 public static void main() 方法的线程就是这样的，而且它由 JVM 创建。

创建示例项目 callMainMethodMainThread，并创建 Test.java 类，代码如下：

```java
package test;

public class Test {

    public static void main(String[] args) {
        System.out.println(Thread.currentThread().getName());
    }

}
```

程序运行结果如图 1-7 所示。

在控制台输出的 main 其实就是一个名称为 main 的线程在执行 main() 方法中的代码。另外，需要说明一下，在控制台输出的 main 和 main 方法没有任何关系，它们仅仅是名字相同而已。

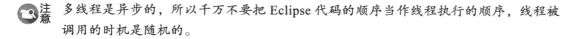

图 1-7 程序运行结果（主线程 main 出现）

1.2.1 继承 Thread 类

Java 的 JDK 开发包已经自带了对多线程技术的支持，通过它可以方便地进行多线程编程。实现多线程编程主要有两种方式：一种是继承

Thread 类，另一种是实现 Runnable 接口。

在学习如何创建新的线程前，先来看看 Thread 类的声明结构：

```
public class Thread implements Runnable
```

从上面的源代码中可以发现，Thread 类实现了 Runnable 接口，它们之间具有多态关系，多态结构的示例代码如下：

```
Runnable run1 = new Thread();
Runnable run2 = new MyThread();
Thread t1 = new MyThread();
```

其实使用继承 Thread 类的方式创建新线程时，最大的局限是不支持多继承，因为 Java 语言的特点是单根继承，所以为了支持多继承，完全可以实现 Runnable 接口，即一边实现一边继承，但这两种方式创建线程的功能是一样的，没有本质的区别。

本节主要介绍第一种方式。创建名称为 t1 的 Java 项目，创建一个自定义的线程类 MyThread.java，此类继承自 Thread，并且重写 run() 方法。在 run() 方法中添加线程要执行的任务代码如下：

```
package com.mythread.www;

public class MyThread extends Thread {
@Override
    public void run() {
        super.run();
        System.out.println("MyThread");
    }
}
```

运行类代码如下：

```
package test;

import com.mythread.www.MyThread;

public class Run {

    public static void main(String[] args) {
        MyThread mythread = new MyThread();
        mythread.start();//耗时大
        System.out.println("运行结束！");//耗时小
    }

}
```

上面代码使用 start() 方法来启动一个线程，线程启动后会自动调用线程对象中的 run() 方法，run() 方法里面的代码就是线程对象要执行的任务，是线程执行任务的入口。

程序运行结果如图 1-8 所示。

从图 1-8 的程序运行结果来看，MyThread.java 类中的 run() 方法的执行时间相对于输出"运行结束！"的执行时间晚，因为 start() 方法的执行比较耗时，这也增加了先输出"运行结束！"字符串的概率。start() 方法耗时的原因是执行了多个步骤，步骤如下：

图 1-8　程序运行结果

1）通过 JVM 告诉操作系统创建 Thread。

2）操作系统开辟内存并使用 Windows SDK 中的 createThread() 函数创建 Thread 线程对象。

3）操作系统对 Thread 对象进行调度，以确定执行时机。

4）Thread 在操作系统中被成功执行。

以上 4 步完整地执行后所消耗的时间一定大于输出"运行结束！"字符串的时间。另外，main 线程执行 start() 方法时不必等待 4 步都执行完毕，而是立即继续执行 start() 方法后面的代码，这 4 步会与输出"运行结束！"的代码一同执行，由于输出"运行结束！"耗时比较少，所以在大多数的情况下，先输出"运行结束！"，后输出"MyThread"。

但在这里，还是有非常非常小的、非常渺茫的机会能输出如下运行结果：

```
MyThread
运行结束！
```

输出上面的结果说明执行完整的 start() 方法的 4 步后，才执行输出"运行结束！"字符串的代码，这也说明线程执行的顺序具有随机性。然而由于输出这种结果的机会很小，使用手动的方式来重复执行" Run as "->" Java Application "难以重现，这时可以人为地制造这种输出结果，即在执行输出"运行结束！"代码之前先执行代码 Thread.sleep(300)，让 run() 方法有充足的时间来先输出"MyThread"，后输出"运行结束！"，示例代码如下：

```
package test;

import com.mythread.www.MyThread;

public class Run2 {
    public static void main(String[] args) throws InterruptedException {
        MyThread mythread = new MyThread();
        mythread.start();
        Thread.sleep(200);
        System.out.println("运行结束！");
    }
}
```

在使用多线程技术时，代码的运行结果与代码的执行顺序或调用顺序是无关的。另外，线程是一个子任务，CPU 以不确定的方式，或者说是以随机的时间来调用线程中的 run() 方法，所以先输出"运行结束！"和先输出"MyThread"具有不确定性。

注意　如果多次调用 start() 方法，则出现异常 Exception in thread "main" java.lang. Illegal-ThreadStateException。

1.2.2　使用常见命令分析线程的信息

可以在运行的进程中创建线程，如果想查看这些线程的状态与信息，则可采用 3 种常见命令，它们分别是 jps+jstack.exe、jmc.exe 和 jvisualvm.exe，它们在 jdk\bin 文件夹中。

创建测试用的程序并运行，代码如下：

```
package test.run;

public class Run3 {
    public static void main(String[] args) throws InterruptedException {
        for (int i = 0; i < 5; i++) {
            new Thread() {
                public void run() {
                    try {
                        Thread.sleep(500000);
                    } catch (InterruptedException e) {
                        e.printStackTrace();
                    }
                };
            }.start();
        }
    }
}
```

1）采用第 1 种方式查看线程的状态——使用 jps+jstack.exe 命令。在 cmd 中输入 jps 命令查看 Java 进程，其中进程 id 是 13824 的进程就是当前运行类 Run3 对应的 Java 虚拟机进程，然后使用 jstack 命令查看该进程下线程的状态，命令如下：

```
C:\>cd jdk1.8.0_161
C:\jdk1.8.0_161>cd bin
C:\jdk1.8.0_161\bin>jps
13824 Run3
8328 Jps
C:\jdk1.8.0_161\bin>jstack -l 13824
```

按 Enter 键后就可以看到线程的状态。

2）采用第 2 种方式查看线程的状态——使用 jmc.exe 命令。双击 jmc.exe 命令出现如图 1-9 所示的欢迎界面。

关闭欢迎界面后双击 Run3 进程，再双击"MBean 服务器"，然后单击"线程"选项卡，出现如图 1-10 所示的界面。

图 1-9 jmc.exe 命令的欢迎界面

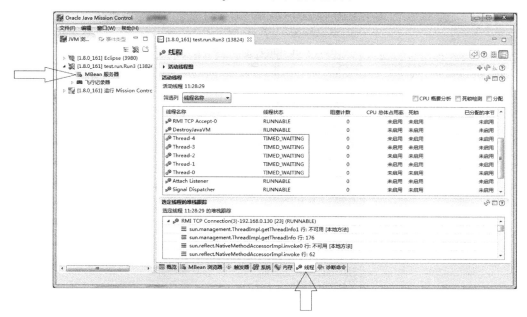

图 1-10 线程列表

在线程列表中可以看到 5 个线程的名称与状态。

3）采用第 3 种方式查看线程的状态——使用 jvisualvm.exe 命令。双击 jvisualvm.exe 命令，出现如图 1-11 所示的界面。

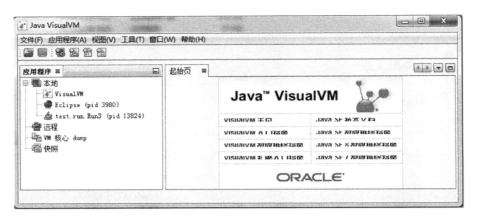

图 1-11　jvisualvm.exe 命令的主界面

双击 Run3 进程，再单击"线程"选项卡就看到了 5 个线程，如图 1-12 所示。

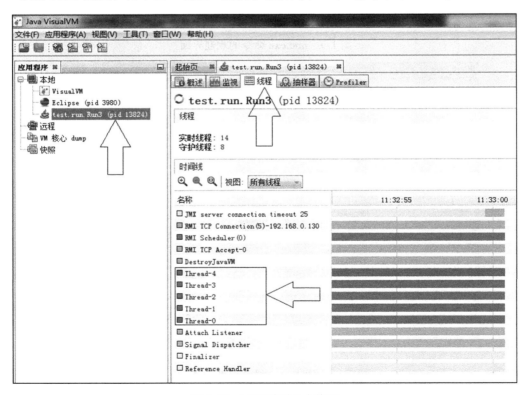

图 1-12　可以看到 5 个线程

但采用 jvisualvm.exe 命令看不到线程运行的状态，所以推荐使用 jmc.exe 命令来分析线程对象的相关信息。

1.2.3 线程随机性的展现

前面介绍过线程的调用是随机的，但这一点并没有在代码中得以体现，都是理论的内容，所以本节将在名称为 randomThread 的 Java 项目中演示线程的随机性。

创建自定义线程类 MyThread.java 的代码如下：

```java
package mythread;

public class MyThread extends Thread {
    @Override
    public void run() {
        for (int i = 0; i < 10000; i++) {
            System.out.println("run=" + Thread.currentThread().getName());
        }
    }
}
```

再创建运行类 Test.java 代码：

```java
package test;

import mythread.MyThread;

public class Test {
    public static void main(String[] args) {
        MyThread thread = new MyThread();
        thread.setName("myThread");
        thread.start();

        for (int i = 0; i < 10000; i++) {
            System.out.println("main=" + Thread.currentThread().getName());
        }
    }
}
```

Thread.java 类中的 start() 方法通知 "线程规划器" ——此线程已经准备就绪，准备调用线程对象的 run() 方法。这个过程其实就是让系统安排一个时间来调用 Thread 中的 run() 方法，即让线程执行具体的任务，具有随机顺序执行的效果。

如果调用代码 "thread.run();" 而不是 "thread.start();"，其实就不是异步执行了，而是同步执行，那么此线程对象并不交给 "线程规划器" 来进行处理，而是由 main 主线程来调用 run() 方法，也就是必须等 run() 方法中的代码执行完毕后才可以执行后面的代码。

以异步方式运行的效果如图 1-13 所示。

多线程随机输出的原因是 CPU 将时间片分给不同的线程，线程获得时间片后就执行任务，所以这些线程在交替地执行并输出，导致输出

```
run=myThread
run=myThread
run=myThread
main=main
main=main
main=main
main=main
main=main
main=main
run=myThread
run=myThread
run=myThread
run=myThread
run=myThread
run=myThread
run=myThread
run=myThread
run=myThread
```

图 1-13 随机被执行的线程

结果呈现乱序的效果。时间片即 CPU 分配给各个程序的时间。每个线程被分配一个时间片,在当前的时间片内 CPU 去执行线程中的任务。需要注意的是,CPU 在不同的线程上进行切换是需要耗时的,所以并不是创建的线程越多,软件运行效率就越高,相反,线程数过多反而会降低软件的执行效率。

1.2.4 执行 start() 的顺序不代表执行 run() 的顺序

注意,执行 start() 方法的顺序不代表线程启动的顺序。创建测试用的项目名称为 z,MyThread.java 类代码如下:

```java
package extthread;

public class MyThread extends Thread {

    private int i;

    public MyThread(int i) {
        super();
        this.i = i;
    }
    @Override
    public void run() {
        System.out.println(i);
    }

}
```

运行类 Test.java 代码如下:

```java
package test;

import extthread.MyThread;

public class Test {

    public static void main(String[] args) {
        MyThread t11 = new MyThread(1);
        MyThread t12 = new MyThread(2);
        MyThread t13 = new MyThread(3);
        MyThread t14 = new MyThread(4);
        MyThread t15 = new MyThread(5);
        MyThread t16 = new MyThread(6);
        MyThread t17 = new MyThread(7);
        MyThread t18 = new MyThread(8);
        MyThread t19 = new MyThread(9);
        MyThread t110 = new MyThread(10);
        MyThread t111 = new MyThread(11);
        MyThread t112 = new MyThread(12);
        MyThread t113 = new MyThread(13);
```

```
        t11.start();
        t12.start();
        t13.start();
        t14.start();
        t15.start();
        t16.start();
        t17.start();
        t18.start();
        t19.start();
        t110.start();
        t111.start();
        t112.start();
        t113.start();
    }

}
```

程序运行结果如图 1-14 所示。

使用代码：

```
MyThread thread = new MyThread();
thread.start();
```

启动一个线程后，JVM 直接调用 MyThread.java 类中的 run() 方法。

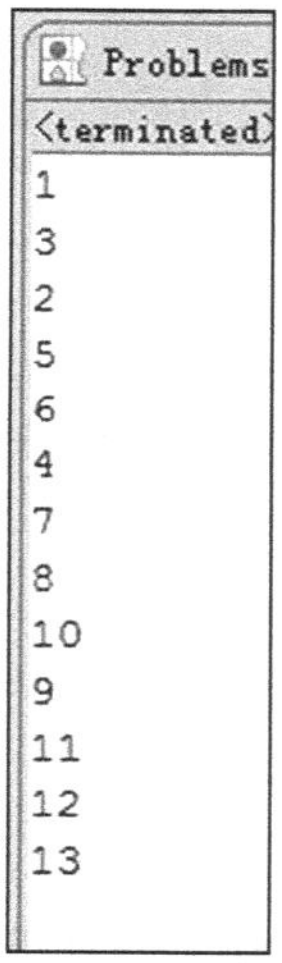

图 1-14　线程启动顺序与 start() 执行顺序无关

1.2.5　实现 Runnable 接口

如果想创建的线程类已经有一个父类了，就不能再继承自 Thread 类，因为 Java 不支持多继承，所以需要实现 Runnable 接口来解决这样的情况。

创建项目 t2，继续创建一个实现 Runnable 接口的 MyRunnable 类，代码如下：

```
package myrunnable;

public class MyRunnable implements Runnable {
@Override
    public void run() {
        System.out.println("运行中!");
    }
}
```

如何使用这个 MyRunnable.java 类呢？这就要看一下 Thread.java 的构造函数了，如图 1-15 所示。

在 Thread.java 类的 8 个构造函数中，有 5 个可以传递 Runnable 接口。为了说明构造函数支持传入一个 Runnable 接口的对象，运行如下类代码：

构造方法摘要
Thread()   　　分配新的 Thread 对象。
Thread(Runnable target)   　　分配新的 Thread 对象。
Thread(Runnable target, String name)   　　分配新的 Thread 对象。
Thread(String name)   　　分配新的 Thread 对象。
Thread(ThreadGroup group, Runnable target)   　　分配新的 Thread 对象。
Thread(ThreadGroup group, Runnable target, String name)   　　分配新的 Thread 对象，以便将 target 作为其运行对象，将指定的 name 作为其名称，并作为 group 所引用的线程组的一员。
Thread(ThreadGroup group, Runnable target, String name, long stackSize)   　　分配新的 Thread 对象，以便将 target 作为其运行对象，将指定的 name 作为其名称，作为 group 所引用的线程组的一员，并具有指定的*堆栈大小*。
Thread(ThreadGroup group, String name)   　　分配新的 Thread 对象。

图 1-15　Thread.Java 的构造函数

```
public class Run {
    public static void main(String[] args) {
        Runnable runnable=new MyRunnable();
        Thread thread=new Thread(runnable);
        thread.start();
        System.out.println("运行结束!");
    }
}
```

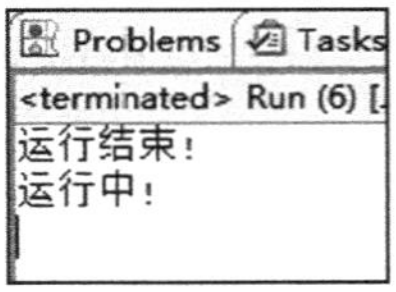

程序运行结果如图 1-16 所示。

图 1-16　程序运行结果

图 1-16 所示的输出结果没有什么特殊之处。

1.2.6　使用 Runnable 接口实现多线程的优点

使用继承 Thread 类的方式来开发多线程应用程序在设计上是有局限性的，因为 Java 是单根继承，不支持多继承，所以为了改变这种限制，可以使用实现 Runnable 接口的方式来实现多线程技术。下面来看使用 Runnable 接口必要性的演示代码。

创建测试用的项目 moreExtends，首先来看业务 A 类，代码如下：

```
package service;

public class AServer {
    public void a_save_method() {
        System.out.println("a中的保存数据方法被执行");
    }
}
```

再来看业务 B 类，代码如下：

```
package service;
```

```
public class BServer1 extends AServer,Thread
{
    public void b_save_method() {
        System.out.println("b中的保存数据方法被执行");
    }
}
```

BServer1.java 类不支持在 extends 关键字后写多个类名,即 Java 并不支持多继承的写法,所以在代码"public class BServer1 extends AServer, Thread"处出现如下异常信息:

```
Syntax error on token "extends", delete this token
```

这时就有使用 Runnable 接口的必要性了,创建新的业务 B 类,代码如下:

```
package service;

public class BServer2 extends AServer implements Runnable {
    public void b_save_method() {
        System.out.println("b中的保存数据方法被执行");
    }

    @Override
    public void run() {
        b_save_method();
    }
}
```

程序不再出现异常,通过实现 Runnable 接口,可间接地实现"多继承"的效果。

另外,需要说明的是 Thread.java 类也实现了 Runnable 接口,如图 1-17 所示。

```
115  class Thread implements Runnable {
116      /* Make sure registerNatives is the first thing <clinit> does. */
117      private static native void registerNatives();
118      static {
119          registerNatives();
120      }
```

图 1-17 Thread.Java 类实现 Runnable 接口

这意味着构造函数 Thread(Runnable target) 不仅可以传入 Runnable 接口的对象,而且可以传入一个 Thread 类的对象,这样做完全可以将一个 Thread 对象中的 run() 方法交由其他线程进行调用,示例代码如下:

```
public class Test {
    public static void main(String[] args) {
        MyThread thread = new MyThread();
        //MyThread是Thread的子类,而Thread是Runnable实现类
        //所以MyThread也相当于Runnable的实现类
        Thread t = new Thread(thread);
```

```
        t.start();
    }
}
```

在非多继承的情况下，使用继承 Thread 类和实现 Runnable 接口两种方式在取得程序运行的结果上并没有什么太大的区别，一旦出现"多继承"的情况，则采用实现 Runnable 接口的方式来处理多线程的问题是很有必要的。

1.2.7 实现 Runnable 接口与继承 Thread 类的内部流程

使用如下代码以实现 Runnable 接口法启动一个线程的执行过程和在前面章节使用继承 Thread 类启动一个线程时的执行过程是不一样的：

```
MyRunnable run = new MyRunnable();
Thread t = new Thread(run);
t.start();
```

JVM 直接调用的是 Thread.java 类的 run() 方法，该方法源代码如下：

```
@Override
public void run() {
    if (target != null) {
        target.run();
    }
}
```

在方法中判断 target 变量是否为 null，不为 null 则执行 target 对象的 run() 方法，target 存储的对象就是前面声明的 MyRunnable run 对象，对 Thread 构造方法传入 Runnable 对象，再结合 if 判断就可以执行 Runnable 对象的 run() 方法了。变量 target 是在 init() 方法中进行赋值初始化的，核心源代码如下：

```
private void init(ThreadGroup g, Runnable target, String name,
        long stackSize, AccessControlContext acc,
        boolean inheritThreadLocals) {
    ......
    this.target = target;
    ......
}
```

而方法 init() 是在 Thread.java 构造方法中被调用的，源代码如下：

```
public Thread(Runnable target) {
    init(null, target, "Thread-" + nextThreadNum(), 0);
}
```

通过分析 JDK 源代码可以发现，实现 Runnable 接口法在执行过程上相比继承 Thread 类法稍微复杂一些。

1.2.8 实例变量共享造成的非线程安全问题与解决方案

自定义线程类中的实例变量针对其他线程可以有共享与不共享之分,这在多个线程之间交互时是很重要的技术点。

1. 不共享数据的情况

不共享数据的情况如图 1-18 所示。

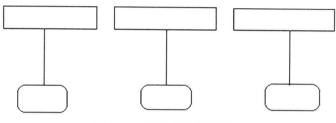

图 1-18 不共享数据的情况

下面通过一个示例来看一下不共享数据的情况。

创建实验用的 Java 项目,名称为 t3,MyThread.java 类代码如下:

```java
public class MyThread extends Thread {
    private int count = 5;

    public MyThread(String name) {
        super();
        this.setName(name);//设置线程名称
    }

    @Override
    public void run() {
        super.run();
        while (count > 0) {
            count--;
            System.out.println("由 " + this.currentThread().getName()
                    + " 计算,count=" + count);
        }
    }
}
```

运行类 Run.java 代码如下:

```java
public class Run {
    public static void main(String[] args) {
        MyThread a=new MyThread("A");
        MyThread b=new MyThread("B");
        MyThread c=new MyThread("C");
        a.start();
        b.start();
```

```
            c.start();
        }
}
```

程序运行结果如图 1-19 所示。

由图 1-19 可以看到,该程序一共创建了 3 个线程,每个线程都有各自的 count 变量,自己减少自己的 count 变量的值,这样的情况就是变量不共享,此示例并不存在多个线程访问同一个实例变量的情况。

如果想实现 3 个线程共同去对一个 count 变量进行减法操作,则代码该如何设计呢?

2. 共享数据的情况

共享数据的情况如图 1-20 所示。

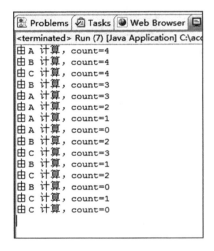

图 1-19　不共享数据情况的示例

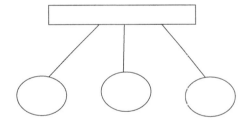

图 1-20　共享数据的情况

共享数据的情况就是多个线程可以访问同一个变量,例如,在实现投票功能的软件时,多个线程同时处理同一个人的票数。

下面通过一个示例来看下共享数据的情况。

创建 t4 测试项目,MyThread.java 类代码如下:

```java
public class MyThread extends Thread {

    private int count=5;

    @Override
    public void run() {
        super.run();
            count--;
            //此示例不要用while语句,会造成其他线程得不到运行的机会
            //因为第一个执行while语句的线程会将count值减到0
```

```
            //一直由一个线程进行减法运算
            System.out.println("由 "+this.currentThread().getName()+" 计算,
                count="+count);
        }
    }
```

运行类 Run.java 代码如下:

```java
public class Run {
    public static void main(String[] args) {
        MyThread mythread=new MyThread();

        Thread a=new Thread(mythread,"A");
        Thread b=new Thread(mythread,"B");
        Thread c=new Thread(mythread,"C");
        Thread d=new Thread(mythread,"D");
        Thread e=new Thread(mythread,"E");
        a.start();
        b.start();
        c.start();
        d.start();
        e.start();
    }
}
```

程序运行结果如图 1-21 所示。

从图 1-21 可以看到,A 线程和 B 线程输出的 count 值都是 3,说明 A 和 B 同时对 count 进行处理,产生了"非线程安全"问题。而我们想要得到输出的结果却不是重复的,应该是依次递减的。

出现非线程安全的情况是因为在某些 JVM 中,count-- 的操作要分解成如下 3 步,(执行这 3 个步骤的过程中会被其他线程所打断):

1) 取得原有 count 值。
2) 计算 count-1。
3) 对 count 进行赋值。

在这 3 个步骤中,如果有多个线程同时访问,那么很大概率出现非线程安全问题,得出重复值的步骤如图 1-22 所示。

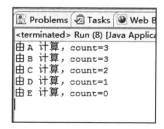

图 1-21 共享变量值重复了,出现线程安全问题

int count=5;

时间 线程	1	2	3	4	5	6	7	8
A	5			4		4		
B			5		4		4	

图 1-22 得出重复值的步骤

A 线程和 B 线程对 count 执行减 1 计算后得出相同值 4 的过程如下。

1）在时间单位为 1 处，A 线程取得 count 变量的值 5。
2）在时间单位为 2 处，B 线程取得 count 变量的值 5。
3）在时间单位为 3 处，A 线程执行 count-- 计算，将计算后的值 4 存储到临时变量中。
4）在时间单位为 4 处，B 线程执行 count-- 计算，将计算后的值 4 也存储到临时变量中。
5）在时间单位为 5 处，A 线程将临时变量中的值 4 赋值给 count。
6）在时间单位为 6 处，B 线程将临时变量中的值 4 也赋值给 count。
7）最终结果就是 A 线程和 B 线程得到相同的计算结果 4，非线程安全问题出现了。

其实这个示例就是典型的销售场景，5 个销售员，每个销售员卖出一个货品后不可以得出相同的剩余数量，必须在每个销售员卖完一个货品后，其他销售员才可以在新的剩余物品数上继续减 1 操作，这时就需要使多个线程之间进行同步操作，即用按顺序排队的方式进行减 1 操作，更改代码如下：

```java
public class MyThread extends Thread {
    private int count=5;
    @Override
    synchronized public void run() {
        super.run();
            count--;
            System.out.println("由 "+this.currentThread().getName()+" 计算，
                count="+count);
    }
}
```

重新运行程序后，便不会出现值一样的情况了，如图 1-23 所示。

通过在 run() 方法前加入 synchronized 关键字，使多个线程在执行 run() 方法时，以排队的方式进行处理。一个线程调用 run() 方法前，先判断 run() 方法有没有被上锁，如果 run() 方法被上锁，则说明其他线程正在调用 run() 方法，必须等其他线程对 run() 方法调用结束后，该线程才可以执行 run() 方法，这样也就实现了排队调用 run() 方法的目的，从而实现按

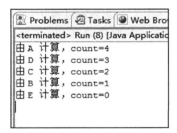

图 1-23 方法调用被同步

顺序对 count 变量减 1 的效果。synchronized 可以对任意对象及方法加锁，而加锁的这段代码称为"互斥区"或"临界区"。

当一个线程想要执行同步方法里面的代码时，线程会首先尝试去申请这把锁，如果能够申请到这把锁，那么这个线程就会执行 synchronized 里面的代码。如果不能申请到这把锁，那么这个线程就会不断尝试去申请这把锁，直到申请到为止，而且多个线程会同时去争抢这把锁。

1.2.9　Servlet 技术造成的非线程安全问题与解决方案

非线程安全问题主要指多个线程对同一个对象中的同一个实例变量进行操作时会出现值被更改、值不同步的情况，进而影响程序执行流程。下面通过一个示例来学习如何解决因 Servlet 技术造成的非线程安全问题。

创建 t4_threadsafe 项目，以实现非线程安全的环境。LoginServlet.java 代码如下：

```java
package controller;

//本类模拟成一个Servlet组件
public class LoginServlet {

    private static String usernameRef;
    private static String passwordRef;

    public static void doPost(String username, String password) {
        try {
            usernameRef = username;
            if (username.equals("a")) {
                Thread.sleep(5000);
            }
            passwordRef = password;

            System.out.println("username=" + usernameRef + " password="
                    + password);
        } catch (InterruptedException e) {
            // TODO Auto-generated catch block
            e.printStackTrace();
        }
    }
}
```

线程 ALogin.java 代码如下：

```java
package extthread;

import controller.LoginServlet;

public class ALogin extends Thread {
    @Override
    public void run() {
        LoginServlet.doPost("a", "aa");
    }
}
```

线程 BLogin.java 代码如下：

```java
package extthread;
```

```
import controller.LoginServlet;

public class BLogin extends Thread {
    @Override
    public void run() {
        LoginServlet.doPost("b", "bb");
    }
}
```

运行类 Run.java 代码如下:

```
public class Run {

    public static void main(String[] args) {
        ALogin a = new ALogin();
        a.start();
        BLogin b = new BLogin();
        b.start();
    }

}
```

程序运行结果如图 1-24 所示。

运行结果是错误的,在研究问题的原因之前,首先要知道两个线程向同一个对象的 public static void doPost(String username, String password) 方法传递参数时,方法的参数值不会被覆盖,方法的参数值是绑定到当前执行线程上的。

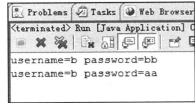

图 1-24 线程非安全

执行错误结果的过程如下。

1)在执行 main() 方法时,执行的结构顺序如下:

```
ALogin a = new ALogin();
a.start();
BLogin b = new BLogin();
b.start();
```

这样的代码被顺序执行时,很大概率是 ALogin 线程先执行,BLogin 线程后执行,因为 ALogin 线程是首先执行 start() 方法的,并且在执行 a.start() 之后又执行 BLogin b = new BLogin(),实例化代码需要耗时,更增大了 ALogin 线程先执行的概率。

但如果 BLogin 线程的确得到了先执行的机会,那么运行结果有可能出现两种:

运行结果 1:

```
b bb
a aa
```

运行结果 2:

```
a bb
a aa
```

2）ALogin 线程执行了 public static void doPost(String username, String password) 方法，对 username 和 password 传入值 a 和 aa。

3）ALogin 线程执行 usernameRef = username 语句，将值 a 赋值给 usernameRef。

4）ALogin 线程执行 if (username.equals("a")) 代码符合条件，执行 Thread.sleep(5000) 停止运行 5s。

5）BLogin 线程也执行 public static void doPost(String username, String password) 方法，对 username 和 password 传入值 b 和 bb。

6）由于 LoginServlet.java 是单例的，只存在 1 份 usernameRef 和 passwordRef 变量，所以 ALogin 线程对 usernameRef 赋的 a 值被 BLogin 线程的值 b 所覆盖，usernameRef 值变成 b。

7）BLogin 线程执行 if (username.equals("a")) 代码不符合条件，不执行 Thread.sleep(5000)，而继续执行后面的赋值语句，将 passwordRef 值变成 bb。

8）BLogin 线程执行输出语句，输出 b 和 bb 的值。

9）5s 之后，ALogin 线程继续向下运行，参数 password 的值 aa 是绑定到当前线程的，所以不会被 BLogin 线程的 bb 值所覆盖，将 ALogin 线程 password 的值 aa 赋值给变量 passwordRef，而 usernameRef 还是 BLogin 线程赋的值 b。

10）ALogin 线程执行输出语句，输出 b 和 aa 的值。

以上就是对运行过程的分析。

需要注意的是，如果代码改成：

```
ALogin a = new ALogin();
BLogin b = new BLogin();
a.start();
b.start();
```

则 BLogin 线程先执行的比重会加大，并且输出如下两种结果的概率较大。

运行结果 1：

```
a bb
a aa
```

运行结果 2：

```
b bb
b aa
```

解决"非线程安全"问题可以使用 synchronized 关键字，更改代码如下：

```
synchronized public static void doPost(String username, String password) {
    try {
        usernameRef = username;
        if (username.equals("a")) {
            Thread.sleep(5000);
        }
        passwordRef = password;
```

```
            System.out.println("username=" + usernameRef + " password="
                    + password);
        } catch (InterruptedException e) {
            // TODO Auto-generated catch block
            e.printStackTrace();
        }
    }
}
```

程序运行结果如图 1-25 所示。

在 Web 开发中，Servlet 对象本身就是单例的，所以为了不出现非线程安全问题，建议不要在 Servlet 中出现实例变量。

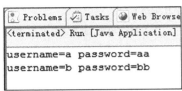

图 1-25　排队进入方法线程安全了

1.2.10　留意 i-- 与 System.out.println() 出现的非线程安全问题

在前面章节中，解决非线程安全问题使用的是 synchronized 关键字，本节将通过案例去细化 println() 方法与 i-- 联合使用时"有可能"出现的另外一种异常情况，并说明其原因。

创建名称为 sameNum 的项目，自定义线程 MyThread.java 代码如下：

```java
package extthread;

public class MyThread extends Thread {

    private int i = 5;

    @Override
    public void run() {
        System.out.println("i=" + (i--) + " threadName="
                + Thread.currentThread().getName());
        //注意：代码i--;单独一行运行
        //被改成在当前项目中的println()方法中直接进行输出
    }

}
```

运行类 Run.java 代码如下：

```java
package test;

import extthread.MyThread;

public class Run {

    public static void main(String[] args) {

        MyThread run = new MyThread();
```

```
        Thread t1 = new Thread(run);
        Thread t2 = new Thread(run);
        Thread t3 = new Thread(run);
        Thread t4 = new Thread(run);
        Thread t5 = new Thread(run);

        t1.start();
        t2.start();
        t3.start();
        t4.start();
        t5.start();

    }

}
```

程序运行后还是会出现非线程安全问题,如图 1-26 所示。

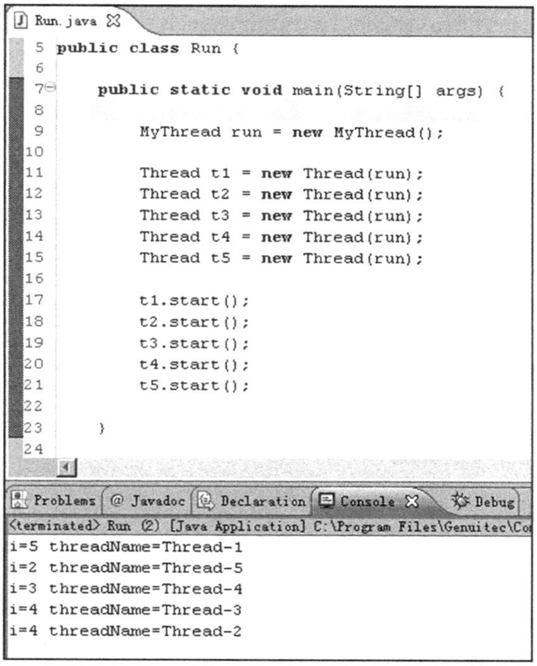

图 1-26 非线程安全问题继续出现

本实验的测试目的是要说明:虽然 println() 方法在内部是同步的,但 i-- 的操作是在进入 println() 之前发生的,所以发生非线程安全问题仍有一定的概率,如图 1-27 所示。所以,为了防止发生非线程安全问题,推荐使用同步方法。

```
PrintStream.class
    public void println(String x) {
    synchronized (this) {
        print(x);
        newLine();
    }
    }
```

图 1-27　println() 方法在内部是同步的

1.3　currentThread() 方法

currentThread() 方法可返回代码段正在被哪个线程调用。下面通过一个示例进行说明。
创建 t6 项目，创建 Run1.java 类代码如下：

```
public class Run1 {
    public static void main(String[] args) {
        System.out.println(Thread.currentThread().getName());
    }
}
```

程序运行结果如图 1-28 所示。

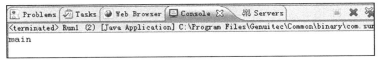

图 1-28　Run1.java 运行结果

该运行结果说明 main() 方法是被名为 main 的线程调用的。
继续实验，创建 MyThread.java 类，代码如下：

```
public class MyThread extends Thread {

    public MyThread() {
        System.out.println("构造方法的打印：" + Thread.currentThread().getName());
    }

    @Override
    public void run() {
        System.out.println("run方法的打印：" + Thread.currentThread().getName());
    }

}
```

运行类 Run2.java 代码如下：

```
public class Run2 {
    public static void main(String[] args) {
```

```
        MyThread mythread = new MyThread();
        mythread.start();
        // mythread.run();
    }
}
```

程序运行结果如图 1-29 所示。

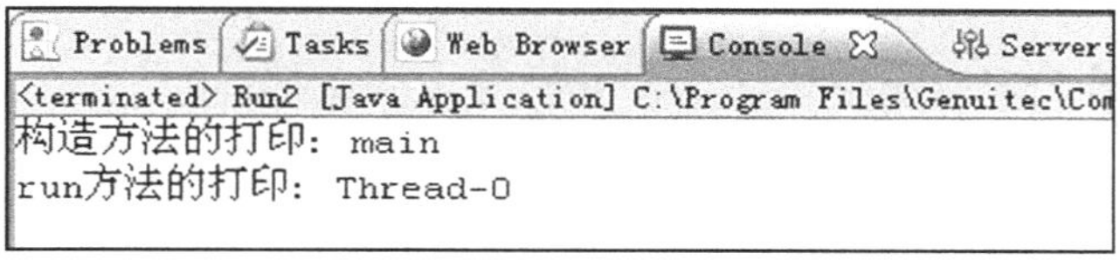

图 1-29 Run2.java 运行结果

从图 1-29 的运行结果可以发现，MyThread.java 类的构造函数是被 main 线程调用的，而 run() 方法是被名称为 Thread-0 的线程调用的，run() 方法是自动调用的方法。

文件 Run2.java 代码更改如下：

```
public class Run2 {
    public static void main(String[] args) {
        MyThread mythread = new MyThread();
        // mythread.start();
        mythread.run();
    }
}
```

程序运行结果如图 1-30 所示。

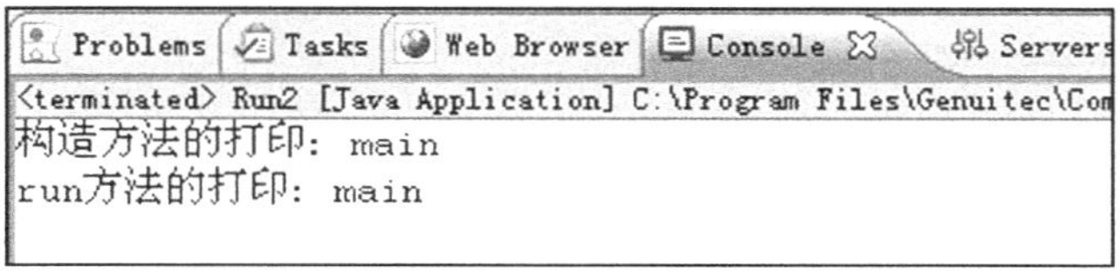

图 1-30 结果显示均被 main 主线程所调用

执行方法 run() 和 start() 还是有一些区别的。

1）my.run();：立即执行 run() 方法，不启动新的线程。

2）my.start();：执行 run() 方法时机不确定，启动新的线程。

再来测试一个比较复杂的情况，创建测试用的项目 currentThreadExt，创建 Java 文件 CountOperate.java，代码如下：

```
package mythread;

public class CountOperate extends Thread {

    public CountOperate() {
```

```java
        System.out.println("CountOperate---begin");
        System.out.println("Thread.currentThread().getName()="
            + Thread.currentThread().getName());
        System.out.println("this.getName()=" + this.getName());
        System.out.println("CountOperate---end");
    }

    @Override
    public void run() {
        System.out.println("run---begin");
        System.out.println("Thread.currentThread().getName()="
            + Thread.currentThread().getName());
        System.out.println("this.getName()=" + this.getName());
        System.out.println("run---end");
    }

}
```

创建 Run.java 文件,代码如下:

```java
package test;

import mythread.CountOperate;

public class Run {

    public static void main(String[] args) {
        CountOperate c = new CountOperate();
        Thread t1 = new Thread(c);
        t1.setName("A");
        t1.start();
    }

}
```

程序运行结果如下:

```
CountOperate---begin
Thread.currentThread().getName()=main
this.getName()=Thread-0
CountOperate---end
run---begin
Thread.currentThread().getName()=A
this.getName()=Thread-0
run---end
```

代码 this.getName() 代表 MyThread 对象的 name 名称,由于 MyThread 对象的 name 名称从未改变,所以默认为 Thread-0。

1.4 isAlive() 方法

isAlive() 方法的功能是判断当前的线程是否存活。
新建项目 t7，类文件 MyThread.java 代码如下：

```java
public class MyThread extends Thread {
    @Override
    public void run() {
        System.out.println("run=" + this.isAlive());
    }
}
```

运行 Run.java 代码如下：

```java
public class Run {
    public static void main(String[] args) {
        MyThread mythread = new MyThread();
        System.out.println("begin ==" + mythread.isAlive());
        mythread.start();
        System.out.println("end ==" + mythread.isAlive());
    }
}
```

程序运行结果如图 1-31 所示。

isAlive() 方法的作用是测试线程是否处于活动状态。那什么是活动状态呢？线程已经启动且尚未终止的状态即活动状态。如果线程处于正在运行或准备开始运行的状态，就认为线程是"存活"的。

需要说明的是，对于代码：

```java
System.out.println("end ==" + mythread.isAlive());
```

虽然其输出的值是 true，但此值是不确定的。输出 true 值是因为 mythread 线程还未执行完毕，如果代码更改如下：

图 1-31 程序运行结果

```java
public static void main(String[] args) throws InterruptedException {
    MyThread mythread = new MyThread();
    System.out.println("begin ==" + mythread.isAlive());
    mythread.start();
    Thread.sleep(1000);
    System.out.println("end ==" + mythread.isAlive());
}
```

则代码：

```java
System.out.println("end ==" + mythread.isAlive());
```

输出的结果为 false，因为 mythread 对象已经在 1s 之内执行完毕。

需要注意的是，main 主线程执行 Thread.sleep(1000) 方法会使 main 主线程停止 1s，而不是将 mythread 线程停止 1s。

另外，在使用 isAlive() 方法时，如果将线程对象以构造参数的方式传递给 Thread 对象进行 start() 启动，则运行的结果和前面示例是有差异的，造成这种差异的原因是 Thread.currentThread() 和 this 的差异，下面测试一下这个实验。

创建测试用的 isaliveOtherTest 项目，创建 CountOperate.java 文件，代码如下：

```java
package mythread;

public class CountOperate extends Thread {

    public CountOperate() {
        System.out.println("CountOperate---begin");

        System.out.println("Thread.currentThread().getName()="
                + Thread.currentThread().getName());
        System.out.println("Thread.currentThread().isAlive()="
                + Thread.currentThread().isAlive());

        System.out.println("this.getName()=" + this.getName());
        System.out.println("this.isAlive()=" + this.isAlive());

        System.out.println("CountOperate---end");
    }

    @Override
    public void run() {
        System.out.println("run---begin");

        System.out.println("Thread.currentThread().getName()="
                + Thread.currentThread().getName());
        System.out.println("Thread.currentThread().isAlive()="
                + Thread.currentThread().isAlive());

        System.out.println("this.getName()=" + this.getName());
        System.out.println("this.isAlive()=" + this.isAlive());

        System.out.println("run---end");
    }

}
```

创建 Run.java 文件，代码如下：

```java
package test;

import mythread.CountOperate;

public class Run {
    public static void main(String[] args) {
        CountOperate c = new CountOperate();
```

```
        Thread t1 = new Thread(c);
        System.out.println("main begin t1 isAlive=" + t1.isAlive());
        t1.setName("A");
        t1.start();
        System.out.println("main end t1 isAlive=" + t1.isAlive());
    }

}
```

程序运行结果如下：

```
CountOperate---begin
Thread.currentThread().getName()=main
Thread.currentThread().isAlive()=true
this.getName()=Thread-0
this.isAlive()=false
CountOperate---end
main begin t1 isAlive=false
main end t1 isAlive=true
run---begin
Thread.currentThread().getName()=A
Thread.currentThread().isAlive()=true
this.getName()=Thread-0
this.isAlive()=false
run---end
```

注意，关键字 this 代表 this 所在类的对象。

1.5 sleep(long millis) 方法

sleep() 方法的作用是在指定的时间（毫秒）内让当前"正在执行的线程"休眠（暂停执行），这个"正在执行的线程"是指 this.currentThread() 返回的线程。

下面通过一个示例进行说明。创建项目 t8 项目，类 MyThread1.java 代码如下：

```
public class MyThread1 extends Thread {
    @Override
    public void run() {
        try {
            System.out.println("run threadName="
                    + this.currentThread().getName() + " begin");
            Thread.sleep(2000);
            System.out.println("run threadName="
                    + this.currentThread().getName() + " end");
        } catch (InterruptedException e) {
            // TODO Auto-generated catch block
            e.printStackTrace();
        }
    }
}
```

如果调用 sleep() 方法所在的类是 Thread，则执行代码：

```
Thread.sleep(3000);
this.sleep(3000);
```

效果是一样的。

如果调用 sleep() 方法所在的类不是 Thread，则必须使用代码：

```
Thread.sleep(3000);
```

实现暂停的功能。

运行类 Run1.java 代码如下：

```java
public class Run1 {
    public static void main(String[] args) {
        MyThread1 mythread = new MyThread1();
        System.out.println("begin =" + System.currentTimeMillis());
        mythread.run();
        System.out.println("end   =" + System.currentTimeMillis());
    }
}
```

直接调用 run() 方法，程序运行结果如图 1-32 所示。

继续实验，创建 MyThread2.java 代码如下：

```java
public class MyThread2 extends Thread {
    @Override
    public void run() {
        try {
            System.out.println("run threadName="
                    + this.currentThread().getName() + " begin ="
                    + System.currentTimeMillis());
            Thread.sleep(2000);
            System.out.println("run threadName="
                    + this.currentThread().getName() + " end   ="
                    + System.currentTimeMillis());
        } catch (InterruptedException e) {
            // TODO Auto-generated catch block
            e.printStackTrace();
        }
    }
}
```

图 1-32　将 main 线程暂停 2s 的运行结果

创建 Run2.java 代码如下：

```java
public class Run2 {
    public static void main(String[] args) {
        MyThread2 mythread = new MyThread2();
        System.out.println("begin =" + System.currentTimeMillis());
        mythread.start();
        System.out.println("end   =" + System.currentTimeMillis());
```

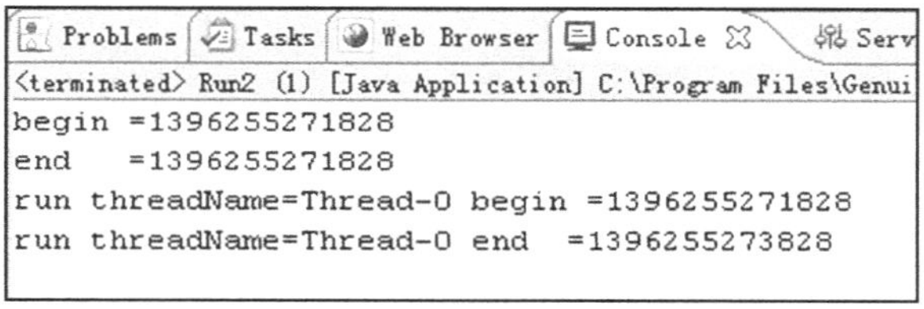

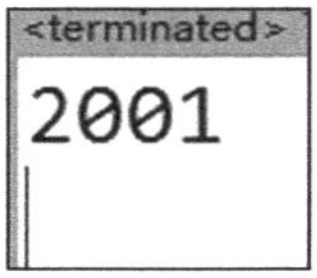

        }
}
```

使用 start() 方法启动线程，程序运行结果如图 1-33 所示。

图 1-33　程序运行结果

由于 main 线程与 MyThread2 线程是异步执行的，所以首先输出的信息为 begin 和 end，而 MyThread2 线程后运行的，在最后两行间隔了 2s 输出 run … begin 和 run … end 相关的信息。

## 1.6　sleep(long millis, int nanos) 方法

sleep(long millis, int nanos) 方法的作用是在指定的毫秒数加指定的纳秒数内让当前正在执行的线程休眠（暂停执行），此操作受到系统计时器和调度程序的精度和准确性的影响。

创建测试用的代码如下：

```java
public class Test1 {
    public static void main(String[] args) throws InterruptedException {
        long beginTime = System.currentTimeMillis();
        Thread.currentThread().sleep(2000, 999999);
        long endTime = System.currentTimeMillis();
        System.out.println(endTime - beginTime);
    }
}
```

程序运行结果如图 1-34 所示。

图 1-34　将 main 线程暂停 2001ms 的运行结果

1.7　StackTraceElement[] getStackTrace() 方法

StackTraceElement[] getStackTrace() 方法的作用是返回一个表示该线程堆栈跟踪元素数组。如果该线程尚未启动或已经终止，则该方法将返回一个零长度数组。如果返回的数组不是零长度的，则其第一个元素代表堆栈顶，它是该数组中最新的方法调用。最后一个元素代表堆栈底，是该数组中最旧的方法调用。

创建测试用的代码如下：

```java
package test1;
```
```

```java
public class Test1 {

    public void a() {
        b();
    }

    public void b() {
        c();
    }

    public void c() {
        d();
    }

    public void d() {
        e();
    }

    public void e() {
        StackTraceElement[] array = Thread.currentThread().getStackTrace();
        if (array != null) {
            for (int i = 0; i < array.length; i++) {
                StackTraceElement eachElement = array[i];
                System.out.println("className=" + eachElement.getClassName() + " "
                        + "methodName=" + eachElement.getMethodName() + " fileName="
                        + eachElement.getFileName() + " lineNumber="+ eachElement.
                        getLineNumber());
            }
        }
    }

    public static void main(String[] args) {
        Test1 test1 = new Test1();
        test1.a();
    }
}
```

程序运行结果如下：

```
className=java.lang.Thread methodName=getStackTrace fileName=Thread.java
    lineNumber=1559
className=test1.Test1 methodName=e fileName=Test1.java lineNumber=22
className=test1.Test1 methodName=d fileName=Test1.java lineNumber=18
className=test1.Test1 methodName=c fileName=Test1.java lineNumber=14
className=test1.Test1 methodName=b fileName=Test1.java lineNumber=10
className=test1.Test1 methodName=a fileName=Test1.java lineNumber=6
className=test1.Test1 methodName=main fileName=Test1.java lineNumber=36
```

在控制台中输出了当前线程的堆栈跟踪信息。

1.8 static void dumpStack() 方法

static void dumpStack() 方法的作用是将当前线程的堆栈跟踪信息输出至标准错误流。该方法仅用于调试。

创建测试用的代码如下：

```java
package test6;

public class Test1 {

    public void a() {
        b();
    }

    public void b() {
        c();
    }

    public void c() {
        d();
    }

    public void d() {
        e();
    }

    public void e() {
        int age = 0;
        age = 100;
        if (age == 100) {
            Thread.dumpStack();
        }
    }

    public static void main(String[] args) {
        Test1 test1 = new Test1();
        test1.a();
    }
}
```

程序运行结果如图 1-35 所示。

```
java.lang.Exception: Stack trace
        at java.lang.Thread.dumpStack(Thread.java:1336)
        at test6.Test1.e(Test1.java:25)
        at test6.Test1.d(Test1.java:18)
        at test6.Test1.c(Test1.java:14)
        at test6.Test1.b(Test1.java:10)
        at test6.Test1.a(Test1.java:6)
        at test6.Test1.main(Test1.java:31)
```

图 1-35 程序运行结果

1.9 static Map<Thread, StackTraceElement[]> getAllStackTraces() 方法

static Map<Thread, StackTraceElement[]> getAllStackTraces() 方法的作用是返回所有活动线程的堆栈跟踪的一个映射。映射键是线程，而每个映射值都是一个 StackTraceElement 数组，该数组表示相应 Thread 的堆栈转储。返回的堆栈跟踪的格式都是针对 getStackTrace 方法指定的。在调用该方法的同时，线程可能也在执行。每个线程的堆栈跟踪仅代表一个快照，并且每个堆栈跟踪都可以在不同时间获得。如果虚拟机没有线程的堆栈跟踪信息，则映射值中将返回一个零长度数组。

创建测试用的代码如下：

```java
package test8;

import java.util.Iterator;
import java.util.Map;

public class Test1 {

    public void a() {
        b();
    }

    public void b() {
        c();
    }

    public void c() {
        d();
    }

    public void d() {
        e();
    }

    public void e() {
        Map<Thread, StackTraceElement[]> map =
            Thread.currentThread().getAllStackTraces();
        if (map != null && map.size() != 0) {
            Iterator keyIterator = map.keySet().iterator();
            while (keyIterator.hasNext()) {
                Thread eachThread = (Thread) keyIterator.next();
                StackTraceElement[] array = map.get(eachThread);
                System.out.println("------每个线程的基本信息");
                System.out.println("   线程名称: " + eachThread.getName());
                System.out.println(" StackTraceElement[].length=" + array.length);
                System.out.println("   线程的状态: " + eachThread.getState());
                if (array.length != 0) {
                    System.out.println(" 输出StackTraceElement[]数组具体信息: ");
```

```java
                for (int i = 0; i < array.length; i++) {
                    StackTraceElement eachElement = array[i];
                    System.out.println("    " + eachElement.getClassName()
                            + " " + eachElement.getMethodName() + " "+
                            eachElement.getFileName() + " " + eachElement.
                            getLineNumber());
                }
            } else {
                System.out.println("    没有StackTraceElement[]信息,因为线程
                        " + eachThread.getName() + "中的StackTraceElement[].
                        length==0");
            }
            System.out.println();
            System.out.println();
        }
    }
}

public static void main(String[] args) {
    Test1 test1 = new Test1();
    test1.a();
}
}
```

程序运行结果如下:

------每个线程的基本信息
    线程名称: Signal Dispatcher
StackTraceElement[].length=0
    线程的状态: RUNNABLE
    没有StackTraceElement[]信息,因为线程Signal Dispatcher中的StackTraceElement[].
        length==0

------每个线程的基本信息
    线程名称: main
StackTraceElement[].length=8
    线程的状态: RUNNABLE
    输出StackTraceElement[]数组具体信息:
        java.lang.Thread dumpThreads Thread.java -2
        java.lang.Thread getAllStackTraces Thread.java 1610
        test8.Test1 e Test1.java 25
        test8.Test1 d Test1.java 21
        test8.Test1 c Test1.java 17
        test8.Test1 b Test1.java 13
        test8.Test1 a Test1.java 9
        test8.Test1 main Test1.java 54

------每个线程的基本信息
    线程名称: Attach Listener
StackTraceElement[].length=0
    线程的状态: RUNNABLE

没有StackTraceElement[]信息，因为线程Attach Listener中的StackTraceElement[].
        length==0

------每个线程的基本信息
    线程名称：Finalizer
StackTraceElement[].length=4
    线程的状态：WAITING
    输出StackTraceElement[]数组具体信息：
        java.lang.Object wait Object.java -2
        java.lang.ref.ReferenceQueue remove ReferenceQueue.java 143
        java.lang.ref.ReferenceQueue remove ReferenceQueue.java 164
        java.lang.ref.Finalizer$FinalizerThread run Finalizer.java 209

------每个线程的基本信息
    线程名称：Reference Handler
StackTraceElement[].length=4
    线程的状态：WAITING
    输出StackTraceElement[]数组具体信息：
        java.lang.Object wait Object.java -2
        java.lang.Object wait Object.java 502
        java.lang.ref.Reference tryHandlePending Reference.java 191
        java.lang.ref.Reference$ReferenceHandler run Reference.java 153

1.10　getId() 方法

getId() 方法用于取得线程的唯一标识。

创建测试用的项目 runThread，创建 Test.java 类，代码如下：

```
package test;

public class Test {
    public static void main(String[] args) {
        Thread runThread = Thread.currentThread();
        System.out.println(runThread.getName() + " " + runThread.getId());
    }
}
```

程序运行结果如图 1-36 所示。

从运行结果来看，当前执行代码的线程名称为 main，线程 id 值为 1。

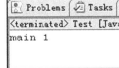

图 1-36　程序运行结果

1.11　停止线程

停止线程是多线程开发的一个很重要的技术点，掌握此技术可以对线程的停止进行有效的处理，停止线程在 Java 语言中并不像 break 语句那样干脆，还需要一些技巧性的处理。

使用 Java 内置支持多线程的类去设计多线程应用是很常见的事情，然而多线程给开发人员带来了一些新的挑战，如果处理不好就会导致超出预期的行为及难于定位的错误。

本节将讨论如何更好地停止一个线程。停止一个线程意味着在线程处理完任务之前停止正在做的操作，也就是放弃当前的操作，虽然这看起来非常简单，但是必须做好防范措施，以便达到预期的效果。停止一个线程可以使用 Thread.stop() 方法，但不推荐使用此方法，虽然它确实可以停止一个正在运行的线程，但是这个方法是不安全的，而且是被弃用作废的，意味着在将来的 Java 版本中，这个方法将不可用或不被支持。

大多数情况下，停止一个线程使用 Thread.interrupt() 方法，但这个方法不会终止一个正在运行的线程，还需要加入一个判断才可以完成线程的停止。关于此知识点，后面有专门的章节对其进行介绍，这里不再赘述。

在 Java 中有 3 种方法可以使正在运行的线程终止运行：

1）使用退出标志使线程正常退出。

2）使用 stop() 方法强行终止线程，但是这个方法不推荐使用，因为 stop() 和 suspend()、resume() 一样，都是作废过期的方法，使用它们可能发生不可预料的结果。

3）使用 interrupt() 方法中断线程。

这 3 种方法会在后面的章节进行介绍。

1.11.1　停止不了的线程

本示例将调用 interrupt() 方法来停止线程，但 interrupt() 方法的使用效果并不像 for+break 语句那样，马上就停止循环。调用 interrupt() 方法仅仅是在当前线程中做了一个停止的标记，并不是真正停止线程。

创建名称为 t11 的项目，文件 MyThread.java 代码如下：

```java
public class MyThread extends Thread {
    @Override
    public void run() {
        super.run();
        for (int i = 0; i < 500000; i++) {
            System.out.println("i=" + (i + 1));
        }
    }
}
```

运行类 Run.java 代码如下：

```java
package test;

import exthread.MyThread;

public class Run {
```

```java
public static void main(String[] args) throws InterruptedException {
    MyThread thread = new MyThread();
    thread.start();
    Thread.sleep(2000);
    thread.interrupt();
    System.out.println("zzzzzzzz");
}
}
```

程序运行结果如图 1-37 所示。

把 Eclipse 软件中的控制台的日志复制到文本编辑器软件中，显示的行数是 500001 行，如图 1-38 所示。

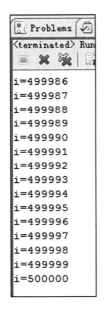

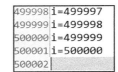

图 1-37　程序运行结果（正常　　　图 1-38　确认是 500001 行且
　　　　　循环输出 50 万次）　　　　　　　　　线程未停止

在 331063 行处输出了 zzzzzzzz，说明 sleep 时间为 2s 时，for 语句执行了 331063 次循环，日志如下：

```
i=331062
i=331063
zzzzzzzz
i=331064
i=331065
```

从运行结果来看，调用 interrupt() 方法并没有将线程停止，那如何停止线程呢？

1.11.2 判断线程是否为停止状态

在学习如何停止线程之前,先来看一下如何判断线程的状态已经是停止的。在 Java 的 SDK 中,Thread.java 类提供了两个判断方法。

1) public static boolean interrupted():测试 currentThread() 是否已经中断。

2) public boolean this.isInterrupted():测试 this 关键字所在类的对象是否已经中断。

其中,interrupted() 方法的声明如图 1-39 所示。isInterrupted() 方法的声明如图 1-40 所示。

```
interrupted
public static boolean interrupted()
```

图 1-39 interrupted() 方法的声明

```
isInterrupted
public boolean isInterrupted()
```

图 1-40 isInterrupted() 方法的声明

这两个方法可以判断线程是否是停止状态,那么这两个方法有什么区别呢?先来看看方法 this.interrupted() 方法的解释:测试当前线程是否已经中断,当前线程是指运行 this.interrupted() 方法的线程。为了对此方法有更深入的了解,创建项目,名称为 t12,类 MyThread.java 代码如下:

```java
public class MyThread extends Thread {
    @Override
    public void run() {
        super.run();
        for (int i = 0; i < 500000; i++) {
            System.out.println("i=" + (i + 1));
        }
    }
}
```

类 Run.java 代码如下:

```java
public class Run {
    public static void main(String[] args) {
        try {
            MyThread thread = new MyThread();
            thread.start();
            Thread.sleep(1000);
            thread.interrupt();
            //Thread.currentThread().interrupt();
            System.out.println("是否停止1? ="+thread.interrupted());
            System.out.println("是否停止2? ="+thread.interrupted());
        } catch (InterruptedException e) {
            System.out.println("main catch");
            e.printStackTrace();
        }
```

```
            System.out.println("end!");
        }
    }
```

程序运行结果如图 1-41 所示。

Run.java 类虽然通过在 thread 对象上调用代码：

```
thread.interrupt();
```

来停止 thread 对象所代表的线程，在后面又使用代码：

```
System.out.println("是否停止1? ="+thread.interrupted());
System.out.println("是否停止2? ="+thread.interrupted());
```

来判断 thread 对象所代表的线程是否停止，但从控制台输出的结果来看，线程并未停止，这也证明了 interrupted() 方法的解释：测试当前线程是否已经中断。这个"当前线程"是 main，从未中断过，所以输出的结果是 2 个 false。

图 1-41 程序运行结果

> **注意**：测试代码中使用 thread.interrupted() 来判断 currentThread() 是否被中断，也可以使用代码 Thread.interrupted() 判断，因为在 Thread.java 类中调用静态 static 方法时，大多数是针对 currentThread() 线程进行操作的。

如何使 main 线程有中断效果呢？创建 Run2.java，代码如下：

```java
public class Run2 {
    public static void main(String[] args) {
        Thread.currentThread().interrupt();
        System.out.println("是否停止1? =" + Thread.interrupted());
        System.out.println("是否停止2? =" + Thread.interrupted());
        System.out.println("end!");
    }
}
```

程序运行结果如图 1-42 所示。

从上述结果来看，interrupted() 方法的确判断了当前线程是否是停止状态，但为什么第二个布尔值是 false 呢？查看一下 interrupted() 方法在官方帮助文档中的解释：

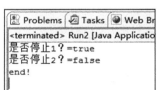

图 1-42 主线程 main 已是停止状态的运行结果

> 测试当前线程是否已经中断。线程的中断状态由该方法清除。换句话说，如果连续两次调用该方法，则第二次调用将返回 false（在第一次调用已清除了其中断状态之后，且第二次调用检验完中断状态前，当前线程再次中断的情况除外）。

文档已经解释得很详细，interrupted() 方法具有清除状态的功能，所以第二次调用 interrupted() 方法返回的值是 false。

isInterrupted() 方法的声明如下：

```
public boolean isInterrupted()
```

从声明中可以看出 isInterrupted() 方法不是 static() 方法，作用于调用这个方法的对象。

继续创建 Run3.java 类，代码如下：

```java
public class Run3 {
    public static void main(String[] args) {
        try {
            MyThread thread = new MyThread();
            thread.start();
            Thread.sleep(1000);
            thread.interrupt();
            System.out.println("是否停止1? ="+thread.isInterrupted());
            System.out.println("是否停止2? ="+thread.isInterrupted());
        } catch (InterruptedException e) {
            System.out.println("main catch");
            e.printStackTrace();
        }
        System.out.println("end!");
    }
}
```

程序运行结果如图 1-43 所示。

从结果中可以看到，isInterrupted() 方法并未清除状态标志，不具有此功能，所以输出两个 true。

图 1-43 已经是停止状态

综上所述，这两个方法的区别如下。

1）this.interrupted()：测试当前线程是否已经是中断状态，执行后具有清除状态标志值为 false 的功能。

2）this.isInterrupted()：测试线程 Thread 对象是否已经是中断状态，不清除状态标志。

1.11.3 能停止的线程——异常法

根据前面所学知识，只需要通过线程的 for 语句来判断一下线程是否处于停止状态即可判断后面的代码是否可运行，如果线程处于停止状态，则后面的代码不再运行。

创建实验用的项目 t13，类 MyThread.java 代码如下：

```java
public class MyThread extends Thread {
    @Override
    public void run() {
        super.run();
        for (int i = 0; i < 500000; i++) {
            if (this.interrupted()) {
                System.out.println("已经是停止状态了!我要退出了!");
                break;
            }
            System.out.println("i=" + (i + 1));
        }
    }
}
```

        }
    }

类 Run.java 代码如下：

```java
public class Run {

    public static void main(String[] args) {
        try {
            MyThread thread = new MyThread();
            thread.start();
            Thread.sleep(2000);
            thread.interrupt();
        } catch (InterruptedException e) {
            System.out.println("main catch");
            e.printStackTrace();
        }
        System.out.println("end!");
    }

}
```

程序运行结果如图 1-44 所示。

图 1-44　线程可以退出了

上面示例虽然停止了线程，但如果 for 语句下面还有语句，那么程序还是会继续运行。创建测试项目 t13forprint，类 MyThread.java 代码如下：

```java
package exthread;

public class MyThread extends Thread {
    @Override
    public void run() {
```

```java
        super.run();
        for (int i = 0; i < 500000; i++) {
            if (this.interrupted()) {
                System.out.println("已经是停止状态了!我要退出了!");
                break;
            }
            System.out.println("i=" + (i + 1));
        }
        System.out.println("我被输出,如果此代码是for又继续运行,线程并未停止!");
    }
}
```

文件 Run.java 代码如下:

```java
package test;

import exthread.MyThread;
import exthread.MyThread;

public class Run {

    public static void main(String[] args) {
        try {
            MyThread thread = new MyThread();
            thread.start();
            Thread.sleep(2000);
            thread.interrupt();
        } catch (InterruptedException e) {
            System.out.println("main catch");
            e.printStackTrace();
        }
        System.out.println("end!");
    }

}
```

程序运行结果如图 1-45 所示。

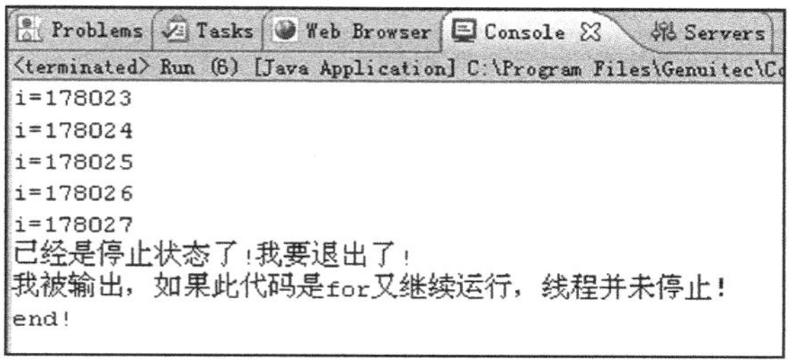

图 1-45 for 后面的语句继续运行

如何解决语句继续运行的问题呢？看一下更新后的代码。

创建 t13_1 项目,类 MyThread.java 代码如下:

```java
package exthread;

public class MyThread extends Thread {
@Override
    public void run() {
        super.run();
        try {
            for (int i = 0; i < 500000; i++) {
                if (this.interrupted()) {
                    System.out.println("已经是停止状态了!我要退出了!");
                    throw new InterruptedException();
                }
                System.out.println("i=" + (i + 1));
            }
            System.out.println("我在for下面");
        } catch (InterruptedException e) {
            System.out.println("进MyThread.java类run方法中的catch了!");
            e.printStackTrace();
        }
    }
}
```

类 Run.java 代码如下:

```java
package test;

import exthread.MyThread;

public class Run {

    public static void main(String[] args) {
        try {
            MyThread thread = new MyThread();
            thread.start();
            Thread.sleep(2000);
            thread.interrupt();
        } catch (InterruptedException e) {
            System.out.println("main catch");
            e.printStackTrace();
        }
        System.out.println("end!");
    }

}
```

程序运行结果如图 1-46 所示。

由程序运行结果可以看出,线程终于被正确停止了。这种方式就是 1.11 节介绍的第三种停止线程的方法——使用 interrupt() 方法中断线程。

图 1-46　线程正确停止

1.11.4　在 sleep 状态下停止线程

如果线程在 sleep 状态下，则停止线程会是什么效果呢？

新建项目 t14，类 MyThread.java 代码如下：

```java
public class MyThread extends Thread {
    @Override
    public void run() {
        super.run();
        try {
            System.out.println("run begin");
            Thread.sleep(200000);
            System.out.println("run end");
        } catch (InterruptedException e) {
            System.out.println("在沉睡中被停止!进入catch!"+this.isInterrupted());
            e.printStackTrace();
        }
    }
}
```

文件 Run.java 代码如下：

```java
public class Run {

    public static void main(String[] args) {
        try {
            MyThread thread = new MyThread();
            thread.start();
            Thread.sleep(200);
            thread.interrupt();
        } catch (InterruptedException e) {
            System.out.println("main catch");
            e.printStackTrace();
        }
        System.out.println("end!");
```

}
}

程序运行效果如图 1-47 所示。

```
run begin
end!
在沉睡中被停止!进入catch!false
java.lang.InterruptedException: sleep interrupted
        at java.lang.Thread.sleep(Native Method)
        at exthread.MyThread.run(MyThread.java:9)
```

图 1-47　程序运行结果

从运行结果来看，如果线程在 sleep 状态下停止，则该线程会进入 catch 语句，并且清除停止状态值，变成 false。

标例是先调用 sleep() 方法，再调用 interrupt() 方法停止线程，还有一个反操作在学习线程时也要注意，即先调用 interrupt() 方法，再调用 sleep() 方法，这种情况下也会出现异常。

新建项目 t15，类 MyThread.java 代码如下：

```java
public class MyThread extends Thread {
    @Override
    public void run() {
        super.run();
        try {
            for(int i=0;i<100000;i++){
                System.out.println("i="+(i+1));
            }
            System.out.println("run begin");
            Thread.sleep(200000);
            System.out.println("run end");
        } catch (InterruptedException e) {
            System.out.println("先停止,再遇到了sleep!进入catch!");
            e.printStackTrace();
        }
    }
}
```

类 Run.java 代码如下：

```java
public class Run {
    public static void main(String[] args) {
        MyThread thread = new MyThread();
        thread.start();
```

```
        thread.interrupt();
        System.out.println("end!");
    }
}
```

程序运行结果如图 1-48 所示。

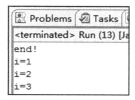

图 1-48　输出"end!"说明 interrupt() 方法先执行

控制台最下面的输出如图 1-49 所示。

图 1-49　执行 interrupt() 方法停止后遇到 sleep() 方法报异常

不管其调用的顺序，只要 interrupt() 和 sleep() 方法碰到一起就会出现异常：
1）在 sleep 状态执行 interrupt() 方法会出现异常；
2）调用 interrupt() 方法给线程打了中断的标记，再执行 sleep() 方法也会出现异常。

1.11.5　用 stop() 方法暴力停止线程

使用 stop() 方法可以强行停止线程，即暴力停止线程。
新建项目 useStopMethodThreadTest，文件 MyThread.java 代码如下：

```
package testpackage;
```

```
public class MyThread extends Thread {
    private int i = 0;

    @Override
    public void run() {
        try {
            while (true) {
                i++;
                System.out.println("i=" + i);
                Thread.sleep(1000);
            }
        } catch (InterruptedException e) {
            // TODO Auto-generated catch block
            e.printStackTrace();
        }
    }

}
```

文件 Run.java 代码如下：

```
package test.run;

import testpackage.MyThread;

public class Run {

    public static void main(String[] args) {
        try {
            MyThread thread = new MyThread();
            thread.start();
            Thread.sleep(8000);
            thread.stop();
        } catch (InterruptedException e) {
            // TODO Auto-generated catch block
            e.printStackTrace();
        }
    }

}
```

程序运行结果如图 1-50 所示。

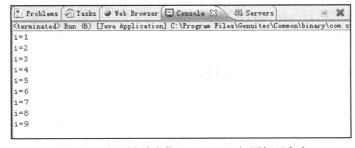

图 1-50　线程被暴力停止，stop 运行图标呈灰色

由运行结果可以看出,线程被暴力停止了,这种方式就是 1.11 节介绍的第二种停止线程的方法——使用 stop() 方法强行终止线程。

stop() 方法呈删除线程状态,是不再被采用的方法,原因是 stop() 方法容易造成业务处理的不确定性。例如,A 线程执行如下业务:

增加数据1
增加数据2
增加数据3
增加数据4
增加数据5
增加数据6
增加数据7

这时在任意时机对 A 线程调用 stop() 方法,A 线程并不能确定在哪里被停止了,造成数据增加得不完整。

1.11.6　stop() 方法与 java.lang.ThreadDeath 异常

调用 stop() 方法时会抛出 java.lang.ThreadDeath 异常,但在通常情况下,此异常不需要显式地捕捉。

创建测试用的项目 runMethodUseStopMethod,文件 MyThread.java 代码如下:

```java
package testpackage;

public class MyThread extends Thread {
    @Override
    public void run() {
        try {
            this.stop();
        } catch (ThreadDeath e) {
            System.out.println("进入了catch()方法!");
            e.printStackTrace();
        }
    }
}
```

文件 Run.java 代码如下:

```java
package test.run;

import testpackage.MyThread;

public class Run {
    public static void main(String[] args) {
        MyThread thread = new MyThread();
        thread.start();
    }
}
```

程序运行结果如图 1-51 所示。

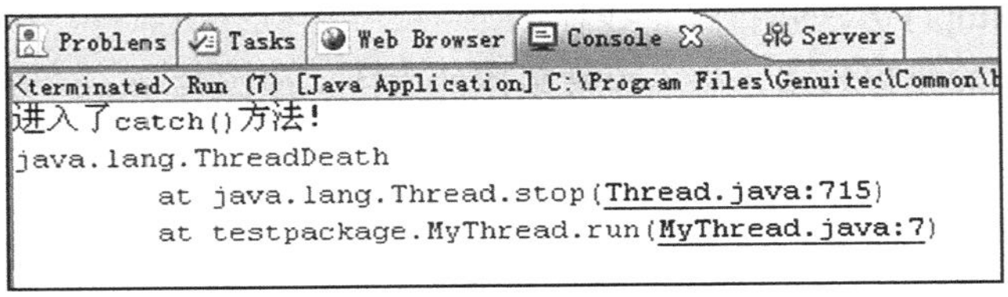

图 1-51　进入 catch 异常

stop() 方法已经是作废的方法，因为如果暴力性地强制让线程停止，则一些清理性的工作可能得不到完成，或者数据添加不完整。

1.11.7　使用 stop() 释放锁给数据造成不一致的结果

对锁定的对象进行"解锁"，会导致数据得不到同步的处理，进而出现数据不一致的问题。本节将会讲解使用 stop() 释放锁给数据造成不一致性的结果，如果出现这样的情况，则程序处理的数据完全有可能遭到破坏，最终导致程序执行的流程是错误的，在此一定要注意。下面来看一个示例。

创建项目 stopThrowLock，文件 MyService.java 代码如下：

```java
package testpackage;

public class MyService {

    private String username = "a";
    private String password = "aa";

    synchronized public String getUsername() {
        return username;
    }

    synchronized public String getPassword() {
        return password;
    }

    synchronized public void printString(String username, String password) {
        try {
            this.username = username;
            Thread.sleep(100000000);
            this.password = password;
        } catch (InterruptedException e) {
            e.printStackTrace();
        }
    }
}
```

调用业务方法 printString() 的线程代码如下:

```java
package testpackage;

public class MyThreadA extends Thread {

    private MyService object;

    public MyThreadA(MyService object) {
        super();
        this.object = object;
    }

    @Override
    public void run() {
        object.printString("b", "bb");
    }
}
```

输出 username 和 password 的线程代码如下:

```java
package testpackage;

public class MyThreadB extends Thread {

    private MyService object;

    public MyThreadB(MyService object) {
        super();
        this.object = object;
    }

    @Override
    public void run() {
        System.out.println("username=" + object.getUsername());
        System.out.println("password=" + object.getPassword());
    }
}
```

文件 Run.java 代码如下:

```java
package test.run;

import testpackage.MyService;
import testpackage.MyThreadA;
import testpackage.MyThreadB;

public class Run {
    public static void main(String[] args) {
        try {
            MyService object = new MyService();
            MyThreadA threadA = new MyThreadA(object);
```

```
            threadA.start();
            Thread.sleep(100);
            MyThreadB threadB = new MyThreadB(object);
            threadB.start();
            Thread.sleep(3000);
            threadA.stop();
            System.out.println("stop()执行后，在下方开始打印username和password。");
        } catch (InterruptedException e) {
            e.printStackTrace();
        }
    }
}
```

程序运行结果如图 1-52 所示。

```
stop()执行后，在下方开始打印username和password。
username=b
password=aa
```

图 1-52　强制停止线程造成数据不一致

由于 stop() 方法已经在 JDK 中被标明是"作废/过期"的方法，显然它在功能上具有缺陷，所以不建议在程序中使用 stop() 方法停止线程。

1.11.8　使用 "return;" 语句停止线程的缺点与解决方案

将 interrupt() 方法与 "return;" 语句结合使用也能实现停止线程的效果。

创建测试用的项目 useReturnInterrupt，线程类 MyThread.java 代码如下：

```
package extthread;

public class MyThread extends Thread {

    @Override
    public void run() {
        while (true) {
            if (this.isInterrupted()) {
                System.out.println("停止了!");
                return;
            }
            System.out.println("timer=" + System.currentTimeMillis());
        }
    }
}
```

运行类 Run.java 代码如下：

```
package test.run;

import extthread.MyThread;
```

```java
public class Run {

    public static void main(String[] args) throws InterruptedException {
        MyThread t=new MyThread();
        t.start();
        Thread.sleep(2000);
        t.interrupt();
    }

}
```

程序运行结果如图 1-53 所示。

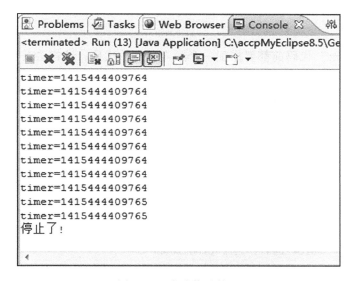

图 1-53　成功停止线程

虽然使用"return;"较"抛异常"法在代码结构上可以更加方便地实现线程的停止，不过还是建议使用"抛异常"法，因为在 catch 块中可以对异常的信息进行统一的处理，例如，使用"return;"来设计代码：

```java
public class MyThread extends Thread {
    @Override
    public void run() {
        // insert操作
        if (this.interrupted()) {
            System.out.println("写入log info");
            return;
        }
        // update操作
        if (this.interrupted()) {
            System.out.println("写入log info");
            return;
```

```
        }
        // delete操作
        if (this.interrupted()) {
            System.out.println("写入log info");
            return;
        }
        // select操作
        if (this.interrupted()) {
            System.out.println("写入log info");
            return;
        }
        System.out.println("for for for for for");
    }
}
```

在每个"return;"代码前都要搭配一个写入日志的代码,这样会使代码出现冗余,不利于代码的阅读与扩展,这时可以使用"抛异常"法来简化这段代码:

```
public class MyThread2 extends Thread {
    @Override
    public void run() {
        try {
            // insert操作
            if (this.interrupted()) {
                throw new InterruptedException();
            }
            // update操作
            if (this.interrupted()) {
                throw new InterruptedException();
            }
            // delete操作
            if (this.interrupted()) {
                throw new InterruptedException();
            }
            // select操作
            if (this.interrupted()) {
                throw new InterruptedException();
            }
            System.out.println("for for for for for");
        } catch (InterruptedException e) {
            System.out.println("写入log info");
            e.printStackTrace();
        }
    }
}
```

写入日志的功能在 catch 块中被统一处理了,代码风格更加标准。

1.12 暂停线程

暂停线程意味着此线程还可以恢复运行,在 Java 多线程中,可以使用 suspend() 方法暂停线程,使用 resume() 方法来恢复线程的执行。

1.12.1 suspend() 方法与 resume() 方法的使用

本节将讲述 suspend() 方法与 resume() 方法的使用。

创建测试用的项目 suspend_resume_test,文件 MyThread.java 代码如下:

```java
package mythread;

public class MyThread extends Thread {

    private long i = 0;

    public long getI() {
        return i;
    }

    public void setI(long i) {
        this.i = i;
    }

    @Override
    public void run() {
        while (true) {
            i++;
        }
    }

}
```

文件 Run.java 代码如下:

```java
package test.run;

import mythread.MyThread;

public class Run {

    public static void main(String[] args) {
        try {
            MyThread thread = new MyThread();
            thread.start();
            Thread.sleep(5000);
            // A段
            thread.suspend();
```

```java
                System.out.println("A= " + System.currentTimeMillis() + " i="
                        + thread.getI());
                Thread.sleep(5000);
                System.out.println("A= " + System.currentTimeMillis() + " i="
                        + thread.getI());
                // B段
                thread.resume();
                Thread.sleep(5000);

                // C段
                thread.suspend();
                System.out.println("B= " + System.currentTimeMillis() + " i="
                        + thread.getI());
                Thread.sleep(5000);
                System.out.println("B= " + System.currentTimeMillis() + " i="
                        + thread.getI());
            } catch (InterruptedException e) {
                e.printStackTrace();
            }
        }
    }
}
```

程序运行结果如图 1-54 所示。

stop() 方法用于销毁线程对象，如果想继续运行线程，则必须使用 start() 方法重新启动线程，而 suspend() 方法用于让线程不再执行任务，线程对象并不销毁，在当前所执行的代码处暂停，未来还可以恢复运行。

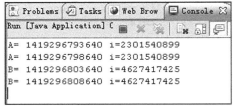

图 1-54 暂停与恢复的测试

从控制台输出的时间上来看，线程的确被暂停了，而且可以恢复成运行状态。

1.12.2 suspend() 方法与 resume() 方法的缺点——独占

如果 suspend() 方法与 resume() 方法使用不当，极易造成公共同步对象被独占，其他线程无法访问公共同步对象的结果。

创建 suspend_resume_deal_lock 项目，文件 SynchronizedObject.java 代码如下：

```java
package testpackage;

public class SynchronizedObject {

    synchronized public void printString() {
        System.out.println("begin");
        if (Thread.currentThread().getName().equals("a")) {
            System.out.println("a线程永远 suspend了！");
            Thread.currentThread().suspend();
        }
```

```
            System.out.println("end");
        }
    }
```

文件 Run.java 代码如下：

```
package test.run;

import testpackage.SynchronizedObject;

public class Run {
    public static void main(String[] args) {
        try {
            final SynchronizedObject object = new SynchronizedObject();

            Thread thread1 = new Thread() {
                @Override
                public void run() {
                    object.printString();
                }
            };

            thread1.setName("a");
            thread1.start();

            Thread.sleep(1000);

            Thread thread2 = new Thread() {
                @Override
                public void run() {
                    System.out.println("thread2启动了，但进入不了printString()方法！"
                            + "只打印1个begin");
                    System.out.println("因为printString()方法被a线程锁定并且永远"
                            + "suspend暂停了！");
                    object.printString();
                }
            };
            thread2.start();
        } catch (InterruptedException e) {
            // TODO Auto-generated catch block
            e.printStackTrace();
        }
    }
}
```

程序运行结果如图 1-55 所示。

另外一种独占锁的情况也需要格外注意，稍有不注意，就会掉进"坑"里。创建测试用的项目 suspend_resume_LockStop，类 MyThread.java 代码如下：

图1-55 独占并锁死 printString() 方法

```java
package mythread;

public class MyThread extends Thread {
    private long i = 0;

    @Override
    public void run() {
        while (true) {
            i++;
        }
    }
}
```

类 Run.java 代码如下：

```java
package test.run;

import mythread.MyThread;

public class Run {

    public static void main(String[] args) {

        try {
            MyThread thread = new MyThread();
            thread.start();
            Thread.sleep(1000);
            thread.suspend();
            System.out.println("main end!");
        } catch (InterruptedException e) {
            e.printStackTrace();
        }
    }
}
```

程序运行结果如图 1-56 所示。

进程状态在控制台中呈红色按钮显示，说明进程并未销毁。虽然 main 线程销毁了，但是 MyThread 呈暂停状态，所以进程不会销毁。

但如果将线程类 MyThread.java 更改如下：

图 1-56 控制台输出 main end 信息

```
package mythread;

public class MyThread extends Thread {
    private long i = 0;

    @Override
    public void run() {
        while (true) {
            i++;
            System.out.println(i);
        }
    }
}
```

再次运行程序,控制台不输出 main end,如图 1-57 所示。

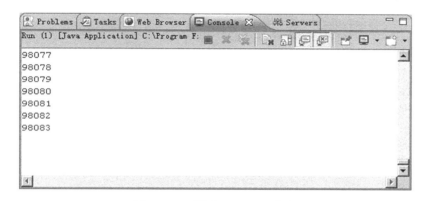

图 1-57 不输出 main end 信息

出现这种情况的原因是当程序运行到 System.out.println(i) 方法内部停止时,同步锁是不释放的,println() 方法源代码如图 1-58 所示。

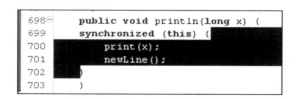

图 1-58 锁不释放

当前 PrintStream 对象的 println() 方法一直呈 "暂停" 状态,并且 "锁未释放",而 main() 方法中的代码 " System.out.println("main end!"); " 也需要这把锁,main 线程并未销毁,造成迟迟不能输出 main end。

虽然 suspend() 方法是过期作废的方法,但研究其过期作废的原因是很有必要的。

1.12.3 suspend() 方法与 resume() 方法的缺点——数据不完整

在使用 suspend() 方法与 resume() 方法时也容易出现线程暂停，进而导致数据不完整的情况。
创建项目 suspend_resume_nosameValue，文件 MyObject.java 代码如下：

```java
package myobject;

public class MyObject {

    private String username = "1";
    private String password = "11";

    public void setValue(String u, String p) {
        this.username = u;
        if (Thread.currentThread().getName().equals("a")) {
            System.out.println("停止a线程！");
            Thread.currentThread().suspend();
        }
        this.password = p;
    }

    public void printUsernamePassword() {
        System.out.println(username + " " + password);
    }
}
```

文件 Run.java 代码如下：

```java
package test;

import myobject.MyObject;

public class Run {

    public static void main(String[] args) throws InterruptedException {

        final MyObject myobject = new MyObject();

        Thread thread1 = new Thread() {
            public void run() {
                myobject.setValue("a", "aa");
            };
        };
        thread1.setName("a");
        thread1.start();

        Thread.sleep(500);

        Thread thread2 = new Thread() {
            public void run() {
                myobject.printUsernamePassword();
```

```
            };
        };
        thread2.start();
    }
}
```

程序运行结果如图 1-59 所示。

程序运行结果出现值不完整的情况，所以在程序中使用 suspend() 方法要格外注意。

这两个方法被标识为作废过期的，想要实现对线程进行暂停与恢复的处理，可使用 wait()、notify() 或 notifyAll() 方法。

图 1-59　程序运行结果

1.13　yield() 方法

yield() 方法的作用是放弃当前的 CPU 资源，让其他任务去占用 CPU 执行时间，放弃的时间不确定，有可能刚刚放弃，马上又获得 CPU 时间片。在本示例中可以取得运行的时间为结果，以测试 yield() 方法的使用效果。

创建 t17 项目，MyThread.java 文件代码如下：

```
package extthread;

public class MyThread extends Thread {

    @Override
    public void run() {
        long beginTime = System.currentTimeMillis();
        int count = 0;
        for (int i = 0; i < 50000000; i++) {
            // Thread.yield();
            count = count + (i + 1);
        }
        long endTime = System.currentTimeMillis();
        System.out.println("用时: " + (endTime - beginTime) + "毫秒!");
    }

}
```

文件 Run.java 代码如下：

```
package test;

import extthread.MyThread;

import extthread.MyThread;

public class Run {
```

```
public static void main(String[] args) {
    MyThread thread = new MyThread();
    thread.start();
}
}
```

程序运行结果如图 1-60 所示。

将代码：

```
// Thread.yield();
```

去掉注释，再次运行，结果如图 1-61 所示。

图 1-60　CPU 独占时间片，　　　　图 1-61　将 CPU 资源让给其他
　　　　　速度很快　　　　　　　　　　　　　　资源，导致速度变慢

1.14 线程的优先级

在操作系统中，线程可以划分优先级，优先级较高的线程得到 CPU 资源较多，也就是 CPU 优先执行优先级较高的线程对象中的任务，其实就是让高优先级的线程获得更多的 CPU 时间片。

设置线程优先级有助于"线程规划器"确定在下一次选择哪一个线程来优先执行。

设置线程的优先级使用 setPriority() 方法，此方法在 JDK 中的源代码如下：

```
public final void setPriority(int newPriority) {
    ThreadGroup g;
    checkAccess();
    if (newPriority > MAX_PRIORITY || newPriority < MIN_PRIORITY) {
        throw new IllegalArgumentException();
    }
    if ((g = getThreadGroup()) != null) {
        if (newPriority > g.getMaxPriority()) {
            newPriority = g.getMaxPriority();
        }
        setPriority0(priority = newPriority);
    }
}
```

在 Java 中，线程的优先级分为 1~10 共 10 个等级，如果优先级小于 1 或大于 10，则 JDK 抛出异常 throw new IllegalArgumentException()。

JDK 使用 3 个常量来预置定义优先级的值，代码如下：

```
public final static int MIN_PRIORITY = 1;
```

```
public final static int NORM_PRIORITY = 5;
public final static int MAX_PRIORITY = 10;
```

1.14.1 线程优先级的继承特性

在 Java 中，线程的优先级具有继承性，例如，A 线程启动 B 线程，则 B 线程的优先级与 A 线程是一样的。

创建 t18 项目，创建 MyThread1.java 文件，代码如下：

```java
package extthread;

public class MyThread1 extends Thread {
    @Override
    public void run() {
        System.out.println("MyThread1 run priority=" + this.getPriority());
        MyThread2 thread2 = new MyThread2();
        thread2.start();
    }
}
```

创建 MyThread2.java 文件，代码如下：

```java
package extthread;

public class MyThread2 extends Thread {
    @Override
    public void run() {
        System.out.println("MyThread2 run priority=" + this.getPriority());
    }
}
```

文件 Run.java 的代码如下：

```java
package test;

import extthread.MyThread1;

public class Run {
    public static void main(String[] args) {
        System.out.println("main thread begin priority="
                + Thread.currentThread().getPriority());
        // Thread.currentThread().setPriority(6);
        System.out.println("main thread end   priority="
                + Thread.currentThread().getPriority());
        MyThread1 thread1 = new MyThread1();
        thread1.start();
    }
}
```

程序运行结果如图 1-62 所示。

将以下注释去掉：

```
// Thread.currentThread().setPriority(6);
```

再次运行 Run.java 文件，程序运行结果如图 1-63 所示。

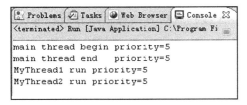

图 1-62　优先级被继承

图 1-63　优先级被更改再继续继承

1.14.2　优先级的规律性

虽然使用 setPriority() 方法可以设置线程的优先级，但还没有看到设置优先级所带来的效果。

创建名称为 t19 的项目，文件 MyThread1.java 代码如下：

```java
package extthread;

import java.util.Random;

public class MyThread1 extends Thread {
    @Override
    public void run() {
        long beginTime = System.currentTimeMillis();
        long addResult = 0;
        for (int j = 0; j < 10; j++) {
            for (int i = 0; i < 50000; i++) {
                Random random = new Random();
                random.nextInt();
                addResult = addResult + i;
            }
        }
        long endTime = System.currentTimeMillis();
        System.out.println("★★★★thread 1 use time=" + (endTime - beginTime));
    }
}
```

文件 MyThread2.java 代码如下：

```java
package extthread;

import java.util.Random;

public class MyThread2 extends Thread {
    @Override
    public void run() {
```

```java
            long beginTime = System.currentTimeMillis();
            long addResult = 0;
            for (int j = 0; j < 10; j++) {
                for (int i = 0; i < 50000; i++) {
                    Random random = new Random();
                    random.nextInt();
                    addResult = addResult + i;
                }
            }
            long endTime = System.currentTimeMillis();
            System.out.println("☆☆☆☆☆thread 2 use time=" + (endTime - beginTime));
        }
    }
```

文件 Run.java 代码如下：

```java
package test;

import extthread.MyThread1;
import extthread.MyThread2;

public class Run {
    public static void main(String[] args) {
        for (int i = 0; i < 5; i++) {
            MyThread1 thread1 = new MyThread1();
            thread1.setPriority(10);
            thread1.start();

            MyThread2 thread2 = new MyThread2();
            thread2.setPriority(1);
            thread2.start();
        }
    }
}
```

文件 Run.java 运行 3 次后的输出结果如图 1-64 所示。

```
★★★★★thread 1 use time=486    ★★★★★thread 1 use time=468    ★★★★★thread 1 use time=477
★★★★★thread 1 use time=515    ★★★★★thread 1 use time=500    ★★★★★thread 1 use time=500
★★★★★thread 1 use time=511    ★★★★★thread 1 use time=511    ★★★★★thread 1 use time=513
☆☆☆☆☆thread 2 use time=528    ★★★★★thread 1 use time=514    ☆☆☆☆☆thread 2 use time=516
★★★★★thread 1 use time=533    ★★★★★thread 1 use time=516    ★★★★★thread 1 use time=533
★★★★★thread 1 use time=550    ☆☆☆☆☆thread 2 use time=565    ★★★★★thread 1 use time=548
☆☆☆☆☆thread 2 use time=562    ☆☆☆☆☆thread 2 use time=571    ☆☆☆☆☆thread 2 use time=569
☆☆☆☆☆thread 2 use time=562    ☆☆☆☆☆thread 2 use time=584    ☆☆☆☆☆thread 2 use time=583
☆☆☆☆☆thread 2 use time=571    ☆☆☆☆☆thread 2 use time=575    ☆☆☆☆☆thread 2 use time=587
☆☆☆☆☆thread 2 use time=573    ☆☆☆☆☆thread 2 use time=584    ☆☆☆☆☆thread 2 use time=590
```

图 1-64 高优先级的线程总是先执行完

从图 1-64 中可以发现，高优先级的线程总是大部分先执行完，但不代表高优先级的线程全部先执行完。另外，并不是 MyThread1 线程先被 main 线程调用就先执行完，出现这种结果是因为 MyThread1 线程的优先级是最高级别 10。当线程优先级的等级差距很大时，谁先执行完和代码的调用顺序无关，为了验证这个结论，修改 Run.java 代码，如下所示：

```java
public class Run {
    public static void main(String[] args) {
        for (int i = 0; i < 5; i++) {
            MyThread1 thread1 = new MyThread1();
            thread1.setPriority(1);
            thread1.start();

            MyThread2 thread2 = new MyThread2();
            thread2.setPriority(10);
            thread2.start();
        }
    }
}
```

文件 Run.java 运行 3 次后的输出结果如图 1-65 所示。

```
☆☆☆☆☆thread 2 use time=483    ☆☆☆☆☆thread 2 use time=448    ☆☆☆☆☆thread 2 use time=472
☆☆☆☆☆thread 2 use time=490    ☆☆☆☆☆thread 2 use time=503    ☆☆☆☆☆thread 2 use time=485
☆☆☆☆☆thread 2 use time=525    ☆☆☆☆☆thread 2 use time=505    ☆☆☆☆☆thread 2 use time=508
★★★★★thread 1 use time=527    ☆☆☆☆☆thread 2 use time=499    ☆☆☆☆☆thread 2 use time=479
★★★★★thread 1 use time=535    ☆☆☆☆☆thread 2 use time=532    ☆☆☆☆☆thread 2 use time=515
☆☆☆☆☆thread 2 use time=523    ★★★★★thread 1 use time=528    ★★★★★thread 1 use time=552
☆☆☆☆☆thread 2 use time=561    ★★★★★thread 1 use time=550    ★★★★★thread 1 use time=551
★★★★★thread 1 use time=564    ★★★★★thread 1 use time=572    ★★★★★thread 1 use time=571
★★★★★thread 1 use time=576    ★★★★★thread 1 use time=580    ★★★★★thread 1 use time=574
★★★★★thread 1 use time=579    ★★★★★thread 1 use time=580    ★★★★★thread 1 use time=585
```

图 1-65 大部分 thread2 先执行完

从图 1-65 中可以发现，大部分 thread2 先执行完，这就验证线程的优先级与代码执行顺序无关，出现这种结果是因为 MyThread2 的优先级是最高的，说明线程的优先级具有一定的规律性，即 CPU 尽量将执行资源让给优先级比较高的线程。

1.14.3 优先级的随机性

前面案例介绍了线程具有优先级，优先级高的线程往往优先执行完，但这个结果不是绝对的，因为线程的优先级还具有 "随机性"，即优先级较高的线程不一定每一次都先执行完。

创建名称为 t20 的项目，文件 MyThread1.java 代码如下：

```java
package extthread;

import java.util.Random;
public class MyThread1 extends Thread {
```

```java
    @Override
    public void run() {
        long beginTime = System.currentTimeMillis();
        for (int i = 0; i < 1000; i++) {
            Random random = new Random();
            random.nextInt();
        }
        long endTime = System.currentTimeMillis();
        System.out.println("★★★★thread 1 use time=" + (endTime - beginTime));
    }
}
```

文件 MyThread2.java 代码如下：

```java
package extthread;

import java.util.Random;

public class MyThread2 extends Thread {
    @Override
    public void run() {
        long beginTime = System.currentTimeMillis();
        for (int i = 0; i < 1000; i++) {
            Random random = new Random();
            random.nextInt();
        }
        long endTime = System.currentTimeMillis();
        System.out.println("☆☆☆☆thread 2 use time=" + (endTime - beginTime));
    }
}
```

文件 Run.java 代码如下：

```java
package test;

import extthread.MyThread1;
import extthread.MyThread2;

public class Run {
    public static void main(String[] args) {
        for (int i = 0; i < 5; i++) {
            MyThread1 thread1 = new MyThread1();
            thread1.setPriority(5);
            thread1.start();

            MyThread2 thread2 = new MyThread2();
            thread2.setPriority(6);
            thread2.start();
        }
    }
}
```

为了让结果体现"随机性"，将两个线程的优先级分别设置为 5、6，让优先级接近一些。文件 Run.java 运行 6 次后的输出结果如图 1-66 所示。

```
*****thread 1 use time=3        *****thread 1 use time=3        *****thread 1 use time=3
☆☆☆☆☆thread 2 use time=1        *****thread 1 use time=1        ☆☆☆☆☆thread 2 use time=3
*****thread 1 use time=4        ☆☆☆☆☆thread 2 use time=1        *****thread 1 use time=2
*****thread 1 use time=2        *****thread 1 use time=1        *****thread 1 use time=0
*****thread 1 use time=2        ☆☆☆☆☆thread 2 use time=2        *****thread 1 use time=2
☆☆☆☆☆thread 2 use time=3        ☆☆☆☆☆thread 2 use time=0        ☆☆☆☆☆thread 2 use time=3
☆☆☆☆☆thread 2 use time=3        ☆☆☆☆☆thread 2 use time=0        ☆☆☆☆☆thread 2 use time=1
☆☆☆☆☆thread 2 use time=3        ☆☆☆☆☆thread 2 use time=2        *****thread 1 use time=0
*****thread 1 use time=1        *****thread 1 use time=3        *****thread 1 use time=0
☆☆☆☆☆thread 2 use time=0        ☆☆☆☆☆thread 2 use time=1        ☆☆☆☆☆thread 2 use time=2

*****thread 1 use time=2        *****thread 1 use time=2        *****thread 1 use time=2
☆☆☆☆☆thread 2 use time=0        *****thread 1 use time=0        *****thread 1 use time=2
☆☆☆☆☆thread 2 use time=2        *****thread 1 use time=3        *****thread 1 use time=0
*****thread 1 use time=2        ☆☆☆☆☆thread 2 use time=0        ☆☆☆☆☆thread 2 use time=0
*****thread 1 use time=0        ☆☆☆☆☆thread 2 use time=1        ☆☆☆☆☆thread 2 use time=3
☆☆☆☆☆thread 2 use time=0        ☆☆☆☆☆thread 2 use time=2        *****thread 1 use time=1
*****thread 1 use time=0        ☆☆☆☆☆thread 2 use time=2        *****thread 1 use time=0
*****thread 1 use time=3        *****thread 1 use time=1        ☆☆☆☆☆thread 2 use time=1
☆☆☆☆☆thread 2 use time=1        ☆☆☆☆☆thread 2 use time=0        ☆☆☆☆☆thread 2 use time=1
☆☆☆☆☆thread 2 use time=1        ☆☆☆☆☆thread 2 use time=0        ☆☆☆☆☆thread 2 use time=1
```

图 1-66 交错执行完

根据此案例可以得出一个结论，优先级较高的线程并不一定每一次都先执行完 run() 方法中的任务，也就是线程优先级与输出顺序无关，这两者并没有依赖关系，它们具有不确定性、随机性。

1.14.4 优先级对线程运行速度的影响

创建测试用的项目 countPriority，创建两个线程类，代码如图 1-67 所示。

```java
package extthread;

public class ThreadA extends Thread {

    private int count = 0;

    public int getCount() {
        return count;
    }

    @Override
    public void run() {
        while (true) {
            count++;
        }
    }
}
```

```java
package extthread;

public class ThreadB extends Thread {

    private int count = 0;

    public int getCount() {
        return count;
    }

    @Override
    public void run() {
        while (true) {
            count++;
        }
    }
}
```

图 1-67 两个线程类代码

创建类 Run.java，代码如下：

```java
package test;

import extthread.ThreadA;
import extthread.ThreadB;

public class Run {

    public static void main(String[] args) {
        try {
            ThreadA a = new ThreadA();
            a.setPriority(Thread.NORM_PRIORITY - 3);
            a.start();

            ThreadB b = new ThreadB();
            b.setPriority(Thread.NORM_PRIORITY + 3);
            b.start();

            Thread.sleep(20000);
            a.stop();
            b.stop();

            System.out.println("a=" + a.getCount());
            System.out.println("b=" + b.getCount());
        } catch (InterruptedException e) {
            e.printStackTrace();
        }
    }
}
```

程序运行结果如图 1-68 所示。

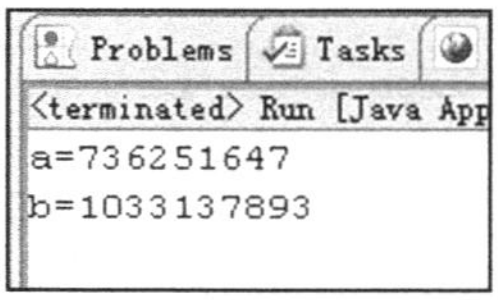

图 1-68　优先级高的运行得快

1.15　守护线程

Java 中有两种线程：一种是用户线程，也称非守护线程；另一种是守护线程。

什么是守护线程？守护线程是一种特殊的线程，当进程中不存在非守护线程了，则守护线程自动销毁。典型的守护线程是垃圾回收线程，当进程中没有非守护线程了，则垃圾回收线程也就没有存在的必要了，自动销毁。用一个比较通俗的比喻来解释一下"守护线程"，任何一个守护线程都可以看作整个 JVM 中所有非守护线程的"保姆"，只要当前 JVM 实例中存在任何一个非守护线程没有结束（好比幼儿园中有小朋友），那么守护线程（也就是"保姆"）就要工作，只有当最后一个非守护线程结束时（好比幼儿园中没有小朋友了），则守护线程（也就是"保姆"）随着 JVM 一同结束工作。守护 Daemon 线程的作用是为其他线程的运行提供便利服务,最典型的应用就是 GC(垃圾回收器)。综上所述,当最后一个用户线程销毁了,守护线程退出,进程也随即结束了。

主线程 main 在本章节中属于用户线程,凡是调用 setDaemon(true) 代码并且传入 true 值的线程才是守护线程。

创建项目 daemonThread,文件 MyThread.java 代码如下:

```java
package testpackage;

public class MyThread extends Thread {
    private int i = 0;

    @Override
    public void run() {
        try {
            while (true) {
                i++;
                System.out.println("i=" + (i));
                Thread.sleep(1000);
            }
        } catch (InterruptedException e) {
            // TODO Auto-generated catch block
            e.printStackTrace();
        }
    }
}
```

文件 Run.java 代码如下:

```java
package test.run;

import testpackage.MyThread;

public class Run {
    public static void main(String[] args) {
        try {
            MyThread thread = new MyThread();
            thread.setDaemon(true);
            thread.start();
            Thread.sleep(5000);
            System.out.println("我离开thread对象也不再打印了,也就是停止了!");
        } catch (InterruptedException e) {
            // TODO Auto-generated catch block
            e.printStackTrace();
        }
    }
}
```

程序运行结果如图 1-69 所示。

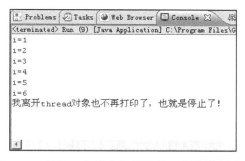

图 1-69 最后一个用户线程销毁，守护线程退出

要在执行 start() 方法之前执行方法，不然会出现异常，示例代码如下：

```
package test.run;

import testpackage.MyThread;

public class Run2 {
    public static void main(String[] args) {
        MyThread thread = new MyThread();
        thread.start();
        thread.setDaemon(true);
    }
}
```

程序运行出现异常如下：

```
Exception in thread "main" java.lang.IllegalThreadStateException
    at java.lang.Thread.setDaemon(Thread.java:1275)
    at test.run.Run2.main(Run2.java:9)
```

1.16 本章小结

本章介绍了 Thread 类的 API，在使用这些 API 的过程中，会出现一些意想不到的情况，其实这也是多线程不可预知性的一个体现，学习并掌握了这些常用情况，也就掌握了多线程开发的命脉与特点，为进一步学习多线程相关知识打下基础。

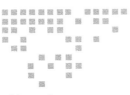

第 2 章 对象及变量的并发访问

本章主要介绍 Java 多线程中的同步，也就是如何在 Java 语言中写出线程安全的程序，解决非线程安全的相关问题。在多线程中解决同步问题，是学习多线程的重中之重。

本章中，读者应该着重掌握如下技术点：
- synchronized 对象监视器为 Object 时的使用方法；
- synchronized 对象监视器为 Class 时的使用方法；
- 非线程安全问题是如何出现的；
- 关键字 volatile 的主要作用；
- 关键字 volatile 与 synchronized 的区别及使用情况。

2.1 synchronized 同步方法

关键字 synchronized 可用来保障原子性、可见性和有序性。

在第 1 章中已经接触线程安全与非线程安全相关的技术点，它们是学习多线程技术时一定会遇到的常见问题。非线程安全问题会在多个线程对同一个对象中的实例变量进行并发访问时发生，产生的后果就是"脏读"，也就是读取到的数据其实是被更改过的。而线程安全是指获得实例变量的值是经过同步处理的，不会出现脏读的现象。此知识点在第 1 章中介绍过，但本章将细化线程并发访问的内容，在细节上更多地涉及在并发时变量值的处理方法。

2.1.1 方法内的变量为线程安全

非线程安全问题存在于实例变量中，对于方法内部的私有变量，则不存在非线程安全问题，结果是"线程安全"的。

下面的示例用于演示在方法内部声明一个变量时，是不存在非线程安全问题的。

创建 t1 项目，HasSelfPrivateNum.java 文件代码如下：

```java
package service;

public class HasSelfPrivateNum {

    public void addI(String username) {
        try {
            int num = 0;
            if (username.equals("a")) {
                num = 100;
                System.out.println("a set over!");
                Thread.sleep(2000);
            } else {
                num = 200;
                System.out.println("b set over!");
            }
            System.out.println(username + " num=" + num);
        } catch (InterruptedException e) {
            // TODO Auto-generated catch block
            e.printStackTrace();
        }
    }

}
```

文件 ThreadA.java 代码如下：

```java
package extthread;

import service.HasSelfPrivateNum;

public class ThreadA extends Thread {

    private HasSelfPrivateNum numRef;

    public ThreadA(HasSelfPrivateNum numRef) {
        super();
        this.numRef = numRef;
    }

    @Override
    public void run() {
        super.run();
        numRef.addI("a");
    }

}
```

文件 ThreadB.java 代码如下：

```java
package extthread;

import service.HasSelfPrivateNum;

public class ThreadB extends Thread {

    private HasSelfPrivateNum numRef;

    public ThreadB(HasSelfPrivateNum numRef) {
        super();
        this.numRef = numRef;
    }

    @Override
    public void run() {
        super.run();
        numRef.addI("b");
    }

}
```

文件 Run.java 代码如下：

```java
package test;

import service.HasSelfPrivateNum;
import extthread.ThreadA;
import extthread.ThreadB;

public class Run {

    public static void main(String[] args) {

        HasSelfPrivateNum numRef = new HasSelfPrivateNum();

        ThreadA athread = new ThreadA(numRef);
        athread.start();

        ThreadB bthread = new ThreadB(numRef);
        bthread.start();

    }

}
```

程序运行结果如图 2-1 所示。

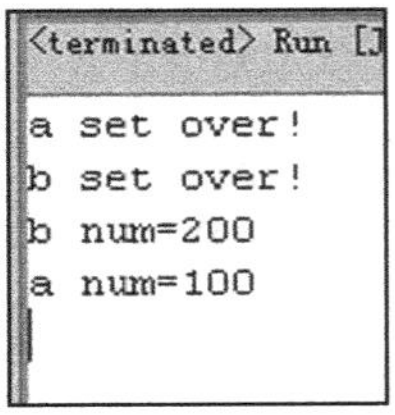

```
<terminated> Run [
a set over!
b set over!
b num=200
a num=100
```

图 2-1　方法中的变量呈线程安全状态

方法中的变量不存在非线程安全问题，永远都是线程安全的，这是因为方法内部的变量具有私有特性。

2.1.2 实例变量非线程安全问题与解决方案

如果多个线程共同访问一个对象中的实例变量，则有可能出现非线程安全问题。

用线程访问的对象中如果有多个实例变量，则运行的结果有可能出现交叉的情况。此情况已经在第 1 章的非线程安全的案例中演示过。

如果对象仅有一个实例变量，则有可能出现覆盖的情况。创建 t2 项目进行测试，HasSelfPrivateNum.java 文件代码如下：

```java
package service;

public class HasSelfPrivateNum {

    private int num = 0;

    public void addI(String username) {
        try {
            if (username.equals("a")) {
                num = 100;
                System.out.println("a set over!");
                Thread.sleep(2000);
            } else {
                num = 200;
                System.out.println("b set over!");
            }
            System.out.println(username + " num=" + num);
        } catch (InterruptedException e) {
            // TODO Auto-generated catch block
            e.printStackTrace();
        }
    }

}
```

文件 ThreadA.java 代码如下：

```java
package extthread;

import service.HasSelfPrivateNum;

public class ThreadA extends Thread {

    private HasSelfPrivateNum numRef;

    public ThreadA(HasSelfPrivateNum numRef) {
        super();
        this.numRef = numRef;
```

```java
    }

    @Override
    public void run() {
        super.run();
        numRef.addI("a");
    }

}
```

文件 ThreadB.java 代码如下：

```java
package extthread;

import service.HasSelfPrivateNum;

public class ThreadB extends Thread {

    private HasSelfPrivateNum numRef;

    public ThreadB(HasSelfPrivateNum numRef) {
        super();
        this.numRef = numRef;
    }

    @Override
    public void run() {
        super.run();
        numRef.addI("b");
    }

}
```

文件 Run.java 代码如下：

```java
package test;

import service.HasSelfPrivateNum;
import extthread.ThreadA;
import extthread.ThreadB;

public class Run {

    public static void main(String[] args) {

        HasSelfPrivateNum numRef = new HasSelfPrivateNum();

        ThreadA athread = new ThreadA(numRef);
        athread.start();

        ThreadB bthread = new ThreadB(numRef);
```

```
            bthread.start();

        }

}
```

程序运行结果如图 2-2 所示。

上面的实验是两个线程同时访问同一个业务对象中的一个没有同步的方法，如果两个线程同时操作业务对象中的实例变量，则有可能出现非线程安全问题，此示例的知识点在前面已经介绍过，只需要在 public void addI(String username) 方法前加关键字 synchronized 即可，更改后的代码如下：

图 2-2　单例模式中的实例变量呈非线程安全状态

```
package service;

public class HasSelfPrivateNum {

    private int num = 0;

    synchronized public void addI(String username) {
        try {
            if (username.equals("a")) {
                num = 100;
                System.out.println("a set over!");
                Thread.sleep(2000);
            } else {
                num = 200;
                System.out.println("b set over!");
            }
            System.out.println(username + " num=" + num);
        } catch (InterruptedException e) {
            // TODO Auto-generated catch block
            e.printStackTrace();
        }
    }

}
```

再次运行程序，结果如图 2-3 所示。

结论：两个线程同时访问同一个对象中的同步方法时一定是线程安全的。在本实验中，由于线程是同步访问的，并且 a 线程先执行，所以先输出 a，然后输出 b，但是完全有可能 b 线程先运行，那么就先输出 b，后输出 a，不管哪个线程先运行，这个线程进入用 synchronized 声明的方法时就上锁，方法执行完成后自动解锁，之后下一个线程才会进入用 synchronized 声明的方法里，不解锁其他线程执行不了用 synchronized 声明的方法。

图 2-3　同步了，线程安全了

2.1.3 同步 synchronized 在字节码指令中的原理

在方法中使用 synchronized 关键字实现同步的原因是使用了 flag 标记 ACC_SYN-CHRONIZED，当调用方法时，调用指令会检查方法的 ACC_SYNCHRONIZED 访问标志是否设置，如果设置了，执行线程先持有同步锁，然后执行方法，最后在方法完成时释放锁。

测试代码如下：

```java
public class Test {
    synchronized public static void testMethod() {
    }
    public static void main(String[] args) throws InterruptedException {
        testMethod();
    }
}
```

在 cmd 中使用 javap 命令将 class 文件转成字节码指令，参数 -v 用于输出附加信息，参数 -c 用于对代码进行反汇编。

使用 javap.exe 命令如下：

```
javap -c -v Test.class
```

生成这个 class 文件对应的字节码指令，指令的核心代码如下：

```
public synchronized void myMethod();
    descriptor: ()V
    flags: ACC_PUBLIC, ACC_SYNCHRONIZED
    Code:
        stack=1, locals=2, args_size=1
            0: bipush        100
            2: istore_1
            3: return
        LineNumberTable:
            line 5: 0
            line 6: 3
        LocalVariableTable:
            Start  Length  Slot  Name  Signature
                0       4     0  this  Ltest56/Test;
                3       1     1  age   I
```

在反编译的字节码指令中，对 public synchronized void myMethod() 方法使用了 flag 标记 ACC_SYNCHRONIZED，说明此方法是同步的。

如果使用 synchronized 代码块，则使用 monitorenter 和 monitorexit 指令进行同步处理，测试代码如下：

```java
public class Test2 {
    public void myMethod() {
        synchronized (this) {
            int age = 100;
        }
    }
}
```

```
    }
    public static void main(String[] args) throws InterruptedException {
        Test2 test = new Test2();
        test.myMethod();
    }
}
```

在 cmd 中使用命令：

```
javap -c -v Test2.class
```

生成这个 class 文件对应的字节码指令，指令核心代码如下：

```
public void myMethod();
    descriptor: ()V
     ags: ACC_PUBLIC
    Code:
        stack=2, locals=3, args_size=1
            0: aload_0
            1: dup
            2: astore_1
            3: monitorenter
            4: bipush        100
            6: istore_2
            7: aload_1
            8: monitorexit
            9: goto          15
           12: aload_1
           13: monitorexit
           14: athrow
           15: return
```

字节码中使用 monitorenter 和 monitorexit 指令进行同步处理。

经过前面若干章节的测试，我们可以明白同步与异步的区别。

同步：按顺序执行 A 和 B 这两个业务，就是同步。

异步：执行 A 业务的时候，B 业务也在同时执行，就是异步。

2.1.4　多个对象多个锁

下面来看一个实验。创建的项目名称为 twoObjectTwoLock，创建 HasSelfPrivateNum.java 类，代码如下：

```
package service;

public class HasSelfPrivateNum {

    private int num = 0;

    synchronized public void addI(String username) {
```

```
        try {
            if (username.equals("a")) {
                num = 100;
                System.out.println("a set over!");
                Thread.sleep(2000);
            } else {
                num = 200;
                System.out.println("b set over!");
            }
            System.out.println(username + " num=" + num);
        } catch (InterruptedException e) {
            // TODO Auto-generated catch block
            e.printStackTrace();
        }
    }
}
```

上面的代码中有同步方法 addI()，说明此方法应该被顺序调用。

创建线程 ThreadA.java 和 ThreadB.java 代码，如图 2-4 所示。

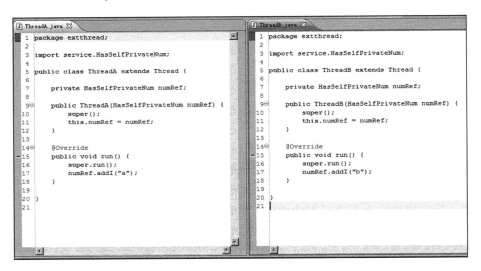

图 2-4　两个线程类代码

类 Run.java 代码如下：

```
package test;

import service.HasSelfPrivateNum;
import extthread.ThreadA;
import extthread.ThreadB;

public class Run {
```

```
    public static void main(String[] args) {

        HasSelfPrivateNum numRef1 = new HasSelfPrivateNum();
        HasSelfPrivateNum numRef2 = new HasSelfPrivateNum();

        ThreadA athread = new ThreadA(numRef1);
        athread.start();

        ThreadB bthread = new ThreadB(numRef2);
        bthread.start();

    }

}
```

这里创建了两个 HasSelfPrivateNum.java 类的对象,程序运行结果如图 2-5 所示。

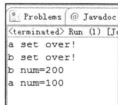

图 2-5　无同步各有各锁

上面的示例演示了两个线程分别访问同一个类的两个不同实例的相同名称的同步方法,输出"a set over!"后继续输出"b set over!",却不是前面测试中输出的结果:

```
a set over!
a num=100
b set over!
b num=200
```

如果先输出:

```
a set over!
a num=100
```

然后输出:

```
b set over!
b num=200
```

就是同步的。

如果输出"a set over!"后继续输出"b set over!",就是以异步的方式运行的。

本示例创建了两个业务对象,在系统中产生两个锁,线程和业务对象属于一对一的关系,每个线程执行自己所属业务对象中的同步方法,不存在争抢关系,所以运行结果是异步的,另外,在这种情况下,synchronized 可以不需要,因为不会出现非线程安全问题。

只有多个线程执行相同的业务对象中的同步方法时,线程和业务对象属于多对一的关系,为了避免出现非线程安全问题,所以使用了 synchronized。

从上面的程序运行结果来看,虽然在 HasSelfPrivateNum.java 中使用了 synchronized 关键字,但输出顺序不是同步的,而是交叉的,为什么是这样的结果呢?关键字 synchronized 取得的锁都是对象锁,而不是把一段代码或方法当作锁,所以在上面的示例中,哪个线程先执行带 synchronized 关键字的方法，哪个线程就持有该方法所属对象的锁 Lock，那么其他线程只能处于等待状态，前提是多个线程访问的是同一个对象。但如果多个线程访问多个对象，也就是每个线程访问自己所属的业务对象（上面的示例就是这种情况），则 JVM 会创建多个锁，不存在锁争抢的情况。另外，更具体地讲，由于本示例创建了两个业务对象，所以产生两份实例变量，每个线程访问自己的实例变量，所以加不加 synchronized 关键字都是线程安全的。

2.1.5 将 synchronized 方法与对象作为锁

为了证明可将对象作为锁，创建测试用的项目 synchronizedMethodLockObject，类 MyObject.java 文件代码如下：

```java
package extobject;

public class MyObject {

    public void methodA() {
        try {
            System.out.println("begin methodA threadName="
                    + Thread.currentThread().getName());
            Thread.sleep(5000);
            System.out.println("end");
        } catch (InterruptedException e) {
            e.printStackTrace();
        }
    }

}
```

自定义线程类 ThreadA.java 代码如下：

```java
package extthread;

import extobject.MyObject;

public class ThreadA extends Thread {

    private MyObject object;

    public ThreadA(MyObject object) {
        super();
        this.object = object;
    }

    @Override
    public void run() {
        super.run();
        object.methcdA();
```

自定义线程类 ThreadB.java 代码如下:

```java
package extthread;

import extobject.MyObject;

public class ThreadB extends Thread {

    private MyObject object;

    public ThreadB(MyObject object) {
        super();
        this.object = object;
    }

    @Override
    public void run() {
        super.run();
        object.methodA();
    }
}
```

运行类 Run.java 代码如下:

```java
package test.run;

import extobject.MyObject;
import extthread.ThreadA;
import extthread.ThreadB;

public class Run {

    public static void main(String[] args) {
        MyObject object = new MyObject();
        ThreadA a = new ThreadA(object);
        a.setName("A");
        ThreadB b = new ThreadB(object);
        b.setName("B");

        a.start();
        b.start();
    }
}
```

程序运行结果如图 2-6 所示。

两个线程可一同进入 methodA() 方法,因为该方法并

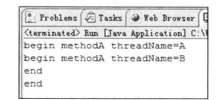

图 2-6　两个线程可一同进入 methodA() 方法

没有同步化。

更改 MyObject.java 代码如下：

```java
package extobject;

public class MyObject {

    synchronized public void methodA() {
        try {
            System.out.println("begin methodA threadName="
                    + Thread.currentThread().getName());
            Thread.sleep(5000);
            System.out.println("end");
        } catch (InterruptedException e) {
            e.printStackTrace();
        }
    }

}
```

如上面的代码所示，在 methodA() 方法前加入了关键字 synchronized 进行同步处理。程序再次运行的结果如图 2-7 所示。

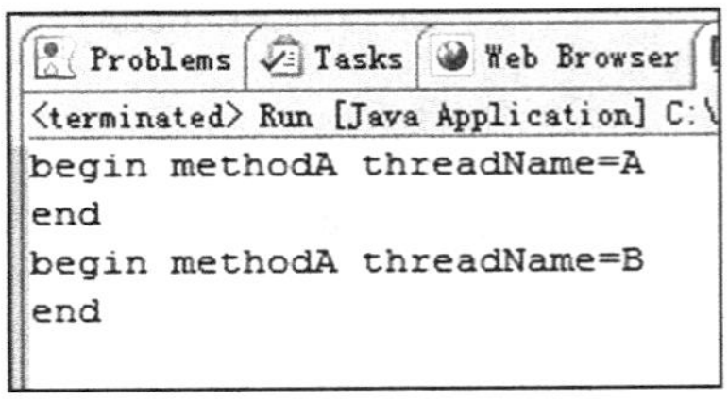

通过上面的示例得到结论，调用用关键字 synchronized 声明的方法一定是排队进行运行的。另外，需要牢牢记住"共享"这两个字，只有共享资源的读写访问才需要同步化，如果不是共享资源，那么就没有同步的必要。

那其他方法在被调用时会是什么效果呢？如何查看将对象作为 Lock 锁的效果呢？继续新建实验用的项目 synchronizedMethodLockObject2，类文件 MyObject.java 代码如下：

图 2-7　排队进入方法

```java
package extobject;

public class MyObject {

    synchronized public void methodA() {
        try {
            System.out.println("begin methodA threadName="
                    + Thread.currentThread().getName());
            Thread.sleep(5000);
            System.out.println("end endTime=" + System.currentTimeMillis());
        } catch (InterruptedException e) {
            e.printStackTrace();
        }
    }

    public void methodB() {
        try {
```

```
            System.out.println("begin methodB threadName="
                    + Thread.currentThread().getName() + " begin time="
                    + System.currentTimeMillis());
            Thread.sleep(5000);
            System.out.println("end");
        } catch (InterruptedException e) {
            e.printStackTrace();
        }
    }
}
```

两个自定义线程类分别调用不同的方法,代码如图 2-8 所示。

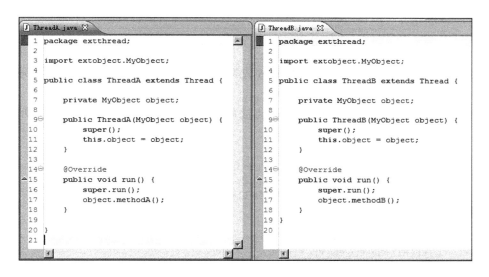

图 2-8　调用不同方法的线程类

文件 Run.java 代码如下:

```
package test.run;

import extobject.MyObject;
import extthread.ThreadA;
import extthread.ThreadB;

public class Run {

    public static void main(String[] args) {
        MyObject object = new MyObject();
        ThreadA a = new ThreadA(object);
        a.setName("A");
        ThreadB b = new ThreadB(object);
        b.setName("B");
```

```
        a.start();
        b.start();
    }
}
```

程序运行结果如图 2-9 所示。

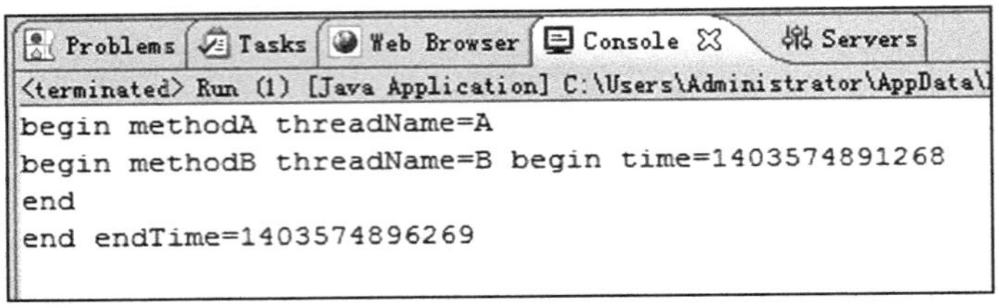

图 2-9　线程 B 异步调用非同步方法

通过上面的示例可以得知，虽然线程 A 先持有了 object 对象的锁，但线程 B 完全可以异步调用非 synchronized 类型的方法。

在 MyObject.java 文件中的 methodB() 方法前加入 synchronized 关键字，代码如下：

```
synchronized public void methodB() {
    try {
        System.out.println("begin methodB threadName="
                + Thread.currentThread().getName() + " begin time="
                + System.currentTimeMillis());
        Thread.sleep(5000);
        System.out.println("end");
    } catch (InterruptedException e) {
        e.printStackTrace();
    }
}
```

本示例演示了两个线程访问同一个对象的两个同步的方法，运行结果如图 2-10 所示。

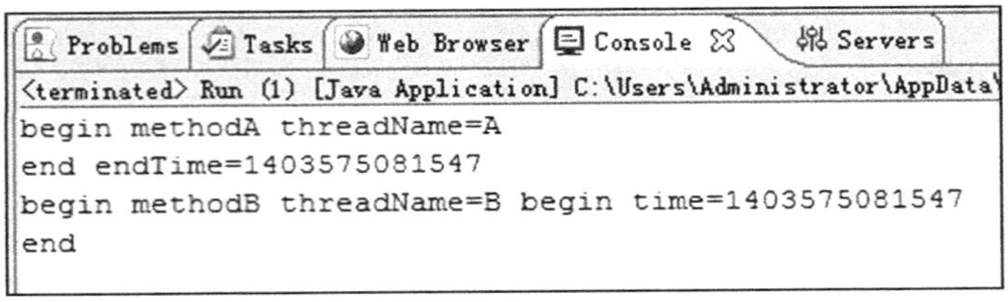

图 2-10　同步运行

结论如下：

1）A 线程先持有 object 对象的 Lock 锁，B 线程可以以异步的方式调用 object 对象中的非 synchronized 类型的方法。

2）A 线程先持有 object 对象的 Lock 锁，B 线程如果在这时调用 object 对象中的 synchronized

类型的方法，则需要等待，也就是同步。

3）在方法声明处添加 synchronized 并不是锁方法，而是锁当前类的对象。

4）在 Java 中只有"将对象作为锁"这种说法，并没有"锁方法"这种说法。

5）在 Java 语言中，"锁"就是"对象"，"对象"可以映射成"锁"，哪个线程拿到这把锁，哪个线程就可以执行这个对象中的 synchronized 同步方法。

6）如果在 X 对象中使用了 synchronized 关键字声明非静态方法，则 X 对象就被当成锁。

2.1.6 脏读

2.1.4 节示例中已经演示了在多个线程调用同一个方法时为了避免数据出现交叉的情况，使用 synchronized 关键字来进行同步。

虽然在赋值时进行了同步，但在取值时有可能出现一些意想不到的情况，这种情况就是脏读（dirty read），发生脏读的原因是在读取实例变量时，此值已经被其他线程更改过了。

创建 t3 项目，PublicVar.java 文件代码如下：

```java
package entity;

public class PublicVar {

    public String username = "A";
    public String password = "AA";

    synchronized public void setValue(String username, String password) {
        try {
            this.username = username;
            Thread.sleep(5000);
            this.password = password;

            System.out.println("setValue method thread name="
                    + Thread.currentThread().getName() + " username="
                    + username + " password=" + password);
        } catch (InterruptedException e) {
            e.printStackTrace();
        }
    }

    public void getValue() {
        System.out.println("getValue method thread name="
                + Thread.currentThread().getName() + " username=" + username + "  
                    password=" + password);
    }
}
```

同步方法 setValue() 的锁属于类 PublicVar 的实例。

创建线程类 ThreadA.java 的代码如下：

```java
package extthread;

import entity.PublicVar;

public class ThreadA extends Thread {

    private PublicVar publicVar;

    public ThreadA(PublicVar publicVar) {
        super();
        this.publicVar = publicVar;
    }

    @Override
    public void run() {
        super.run();
        publicVar.setValue("B", "BB");
    }
}
```

文件 Test.java 代码如下：

```java
package test;

import entity.PublicVar;
import extthread.ThreadA;

public class Test {

    public static void main(String[] args) {
        try {
            PublicVar publicVarRef = new PublicVar();
            ThreadA thread = new ThreadA(publicVarRef);
            thread.start();

            Thread.sleep(200);// 输出结果受此值大小影响

            publicVarRef.getValue();
        } catch (InterruptedException e) {
            // TODO Auto-generated catch block
            e.printStackTrace();
        }

    }
}
```

程序运行结果如图 2-11 所示。

出现脏读是因为 public void getValue() 方法并不是同步的，所以可以在任意时刻进行调用，解决办法是加上同步 synchronized 关键字，代码如下：

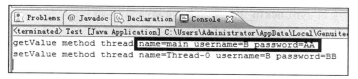

图 2-11　出现脏读情况

```
synchronized public void getValue() {
    System.out.println("getValue method thread name="
        + Thread.currentThread().getName() + " username=" + username + " "
        password=" + password);
}
```

程序运行结果如图 2-12 所示。

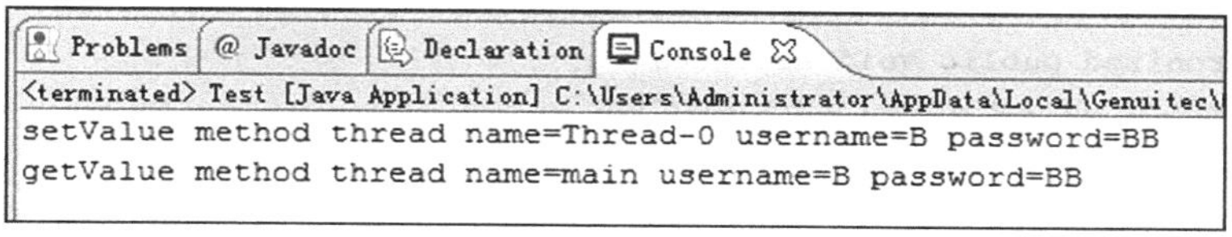

图 2-12　不出现脏读了

方法 setValue() 和 getValue() 被依次执行，通过这个示例可以知道脏读是通过 synchronized 关键字解决的。

当 A 线程调用 anyObject 对象加入 synchronized 关键字的 X 方法时，A 线程就获得了 X 方法锁，更准确地讲，是获得了对象的锁，所以其他线程必须等 A 线程执行完毕后才可以调用 X 方法，但 B 线程可以随意调用其他非 synchronized 同步方法。

当 A 线程调用 anyObject 对象加入 synchronized 关键字的 X 方法时，A 线程就获得了 X 方法所在对象的锁，所以其他线程必须等 A 线程执行完毕后才可以调用 X 方法，而 B 线程如果调用声明了 synchronized 关键字的非 X 方法，必须等 A 线程将 X 方法执行完，也就是将对象锁进行释放后才可以调用，这时 A 线程已经执行了一个完整的任务，也就是说 username 和 password 这两个实例变量已经同时被赋值，不存在脏读的基本环境。

多个线程执行同一个业务对象中的不同同步方法时，是按顺序同步的方式调用的。

脏读前一定会出现不同线程一起去写实例变量的情况，这就是不同线程"争抢"实例变量的结果。

2.1.7　synchronized 锁重入

关键字 synchronized 拥有重入锁的功能，即在使用 synchronized 时，当一个线程得到一个对象锁后，再次请求此对象锁时是可以得到该对象锁的，这也证明在一个 synchronized 方法 / 块的内部调用本类的其他 synchronized 方法 / 块时，是永远可以得到锁的。

创建实验用的项目 synLockIn_1，类 Service.java 代码如下：

```java
package myservice;

public class Service {

    synchronized public void service1() {
        System.out.println("service1");
        service2();
    }

    synchronized public void service2() {
        System.out.println("service2");
        service3();
    }

    synchronized public void service3() {
        System.out.println("service3");
    }

}
```

线程类 MyThread.java 代码如下：

```java
package extthread;

import myservice.Service;

public class MyThread extends Thread {
    @Override
    public void run() {
        Service service = new Service();
        service.service1();
    }

}
```

运行类 Run.java 代码如下：

```java
package test;

import extthread.MyThread;

public class Run {
    public static void main(String[] args) {
        MyThread t = new MyThread();
        t.start();
    }
}
```

程序运行结果如图 2-13 所示。

"可重入锁"是指自己可以再次获取自己的内部锁。例如,一个线程获得了某个对象锁,此时这个对象锁还没有释放,当其再次想要获取这个对象锁时还是可以获取的,如果不可重入锁,则方法 service2() 不会被调用,方法 service3() 更不会被调用。

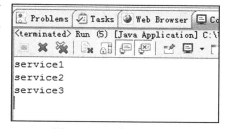

图 2-13 程序运行结果

2.1.8 锁重入支持继承的环境

锁重入也支持父子类继承的环境。

创建实验用的项目 synLockIn_2,类 Main.java 代码如下:

```java
package myservice;

public class Main {

    public int i = 10;

    synchronized public void operateIMainMethod() {
        try {
            i--;
            System.out.println("main print i=" + i);
            Thread.sleep(100);
        } catch (InterruptedException e) {
            // TODO Auto-generated catch block
            e.printStackTrace();
        }
    }

}
```

子类 Sub.java 代码如下:

```java
package myservice;

public class Sub extends Main {

    synchronized public void operateISubMethod() {
        try {
            while (i > 0) {
                i--;
                System.out.println("sub print i=" + i);
                Thread.sleep(100);
                super.operateIMainMethod();
            }
        } catch (InterruptedException e) {
            e.printStackTrace();
        }
    }

}
```

自定义线程类 MyThread.java 代码如下:

```
package extthread;

import myservice.Main;
import myservice.Sub;

public class MyThread extends Thread {
    @Override
    public void run() {
        Sub sub = new Sub();
        sub.operateISubMethod();
    }

}
```

运行类 Run.java 代码如下：

```
package test;

import extthread.MyThread;

public class Run {
    public static void main(String[] args) {
        MyThread t = new MyThread();
        t.start();
    }
}
```

程序运行结果如图 2-14 所示。

此示例说明，当存在父子类继承关系时，子类是完全可以通过锁重入调用父类的同步方法的。

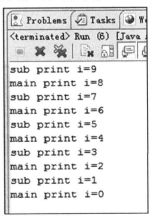

图 2-14　重入到父类中的锁

2.1.9　出现异常，锁自动释放

当一个线程执行的代码出现异常时，其所持有的锁会自动释放。

创建实验用的项目 throwExceptionNoLock，类 Service.java 代码如下：

```
package service;
```

```java
public class Service {
    synchronized public void testMethod() {
        if (Thread.currentThread().getName().equals("a")) {
            System.out.println("ThreadName=" + Thread.currentThread().getName()
                    + " run beginTime=" + System.currentTimeMillis());
            int i = 1;
            while (i == 1) {
                if (("" + Math.random()).substring(0, 8).equals("0.123456")) {
                    System.out.println("ThreadName="
                            + Thread.currentThread().getName()
                            + " run    exceptionTime="
                            + System.currentTimeMillis());
                    Integer.parseInt("a");
                }
            }
        } else {
            System.out.println("Thread B run Time="
                    + System.currentTimeMillis());
        }
    }
}
```

两个自定义线程类代码如图 2-15 所示。

```
1  package extthread;
2
3  import service.Service;
4
5  public class ThreadA extends Thread {
6
7      private Service service;
8
9      public ThreadA(Service service) {
10         super();
11         this.service = service;
12     }
13
14     @Override
15     public void run() {
16         service.testMethod();
17     }
18
19 }
20
```

```
1  package extthread;
2
3  import service.Service;
4
5  public class ThreadB extends Thread {
6      private Service service;
7
8      public ThreadB(Service service) {
9          super();
10         this.service = service;
11     }
12
13     @Override
14     public void run() {
15         service.testMethod();
16     }
17
18 }
19
```

图 2-15　两个自定义线程类代码

运行类 Run.java 代码如下：

```java
package controller;

import service.Service;
import extthread.ThreadA;
import extthread.ThreadB;

public class Test {
    public static void main(String[] args) {
        try {
```

```java
            Service service = new Service();

            ThreadA a = new ThreadA(service);
            a.setName("a");
            a.start();
            Thread.sleep(500);

            ThreadB b = new ThreadB(service);
            b.setName("b");
            b.start();
        } catch (InterruptedException e) {
            e.printStackTrace();
        }
    }

}
```

程序运行结果如图 2-16 所示。

```
ThreadName=a run beginTime=1405322590103
ThreadName=a run   exceptionTime=1405322593190
Exception in thread "a" Thread B run Time=1405322593190
java.lang.NumberFormatException: For input string: "a"
        at java.lang.NumberFormatException.forInputString(NumberFormatException.java:48)
        at java.lang.Integer.parseInt(Integer.java:447)
        at java.lang.Integer.parseInt(Integer.java:497)
        at service.Service.testMethod(Service.java:15)
        at extthread.ThreadA.run(ThreadA.java:16)
```

图 2-16 程序运行结果

线程 a 出现异常并释放锁，线程 b 进入方法正常输出。本示例说明，当出现异常时，锁可以自动释放。

要注意，类 Thread.java 中的 suspend() 方法和 sleep(millis) 方法被调用后并不释放锁。

2.1.10　重写方法不使用 synchronized

重写方法如果不使用 synchronized 关键字，即是非同步方法，使用后变成同步方法。

创建测试用的项目 synNotExtends，类 Main.java 代码如下：

```java
package service;

public class Main {

    synchronized public void serviceMethod() {
        try {
            System.out.println("int main 下一步sleep begin threadName="
                    + Thread.currentThread().getName() + " time="
                    + System.currentTimeMillis());
            Thread.sleep(5000);
            System.out.println("int main 下一步sleep   end threadName="
                    + Thread.currentThread().getName() + " time="
```

```
                + System.currentTimeMillis());
        } catch (InterruptedException e) {
            e.printStackTrace();
        }
    }
}
```

类 Sub.java 代码如下：

```
package service;

public class Sub extends Main {

    @Override
    public void serviceMethod() {
        try {
            System.out.println("int sub 下一步sleep begin threadName="
                    + Thread.currentThread().getName() + " time="
                    + System.currentTimeMillis());
            Thread.sleep(5000);
            System.out.println("int sub 下一步sleep  end threadName="
                    + Thread.currentThread().getName() + " time="
                    + System.currentTimeMillis());
            super.serviceMethod();
        } catch (InterruptedException e) {
            // TODO Auto-generated catch block
            e.printStackTrace();
        }
    }

}
```

类 MyThreadA.java 和 MyThreadB.java 代码如图 2-17 所示。

```
package extthread;

import service.Sub;

public class MyThreadA extends Thread {

    private Sub sub;

    public MyThreadA(Sub sub) {
        super();
        this.sub = sub;
    }

    @Override
    public void run() {
        sub.serviceMethod();
    }
}
```

```
package extthread;

import service.Sub;

public class MyThreadB extends Thread {

    private Sub sub;

    public MyThreadB(Sub sub) {
        super();
        this.sub = sub;
    }

    @Override
    public void run() {
        sub.serviceMethod();
    }
}
```

图 2-17　两个线程代码

类 Test.java 代码如下：

```java
package controller;

import service.Sub;
import extthread.MyThreadA;
import extthread.MyThreadB;

public class Test {

    public static void main(String[] args) {
        Sub subRef = new Sub();

        MyThreadA a = new MyThreadA(subRef);
        a.setName("A");
        a.start();

        MyThreadB b = new MyThreadB(subRef);
        b.setName("B");
        b.start();
    }

}
```

程序运行结果如图 2-18 所示。

```
int sub 下一步sleep begin threadName=A time=1405324477242   ← 非同步调用
int sub 下一步sleep begin threadName=B time=1405324477242
int sub 下一步sleep   end threadName=A time=1405324482242
int sub 下一步sleep   end threadName=B time=1405324482242
int main 下一步sleep begin threadName=A time=1405324482242
int main 下一步sleep   end threadName=A time=1405324487242
int main 下一步sleep begin threadName=B time=1405324487242
int main 下一步sleep   end threadName=B time=1405324492243
```

图 2-18 程序运行结果

从输出结果可以看到，线程以异步的方式进行输出，所以需要在子类的重写方法中添加 synchronized 关键字。程序运行结果如图 2-19 所示。

```
int sub 下一步sleep begin threadName=A time=1405324605424   ← 同步了
int sub 下一步sleep   end threadName=A time=1405324610424
int main 下一步sleep begin threadName=A time=1405324610424
int main 下一步sleep   end threadName=A time=1405324615425
int sub 下一步sleep begin threadName=B time=1405324615425
int sub 下一步sleep   end threadName=B time=1405324620425
int main 下一步sleep begin threadName=B time=1405324620425
int main 下一步sleep   end threadName=B time=1405324625425
```

图 2-19 同步了

2.1.11 public static boolean holdsLock(Object obj) 方法的使用

public static native boolean holdsLock(Object obj) 方法的作用是当 currentThread 在指定的对象上保持锁定时，才返回 true。

创建测试用的代码如下：

```java
package test9;

public class Test1 {
    public static void main(String[] args) {
        System.out.println("A " + Thread.currentThread().holdsLock(Test1.class));
        synchronized (Test1.class) {
            System.out.println("B " + Thread.currentThread().holdsLock(Test1.class));
        }
        System.out.println("C " + Thread.currentThread().holdsLock(Test1.class));
    }
}
```

程序运行结果如下：

```
A false
B true
C false
```

2.2 synchronized 同步语句块

用关键字 synchronized 声明方法在某些情况下是有弊端的，例如，A 线程调用同步方法执行一个长时间的任务，那么 B 线程等待的时间就比较长，这种情况可以使用 synchronized 同步语句块来解决，以提高运行效率。

synchronized 方法是将当前对象作为锁，而 synchronized 代码块是将任意对象作为锁。可以将锁看成一个标识，哪个线程持有这个标识，就可以执行同步方法。

2.2.1 synchronized 方法的弊端

为了证明用 synchronized 关键字声明方法时是有弊端的，创建 t5 项目进行测试。

文件 Task.java 代码如下：

```java
package mytask;

import commonutils.CommonUtils;

public class Task {

    private String getData1;
    private String getData2;

    public synchronized void doLongTimeTask() {
```

```
            try {
                System.out.println("begin task");
                Thread.sleep(3000);
                getData1 = "长时间处理任务后从远程返回的值1 threadName="
                        + Thread.currentThread().getName();
                getData2 = "长时间处理任务后从远程返回的值2 threadName="
                        + Thread.currentThread().getName();
                System.out.println(getData1);
                System.out.println(getData2);
                System.out.println("end task");
            } catch (InterruptedException e) {
                // TODO Auto-generated catch block
                e.printStackTrace();
            }
        }
    }
```

用 synchronized 声明方法时，将其放在 public 之前或之后没有区别。

文件 CommonUtils.java 代码如下：

```
package commonutils;

public class CommonUtils {

    public static long beginTime1;
    public static long endTime1;

    public static long beginTime2;
    public static long endTime2;
}
```

文件 MyThread1.java 代码如下：

```
package mythread;

import commonutils.CommonUtils;

import mytask.Task;

public class MyThread1 extends Thread {

    private Task task;

    public MyThread1(Task task) {
        super();
        this.task = task;
    }

    @Override
    public void run() {
        super.run();
```

```java
        CommonUtils.beginTime1 = System.currentTimeMillis();
        task.doLongTimeTask();
        CommonUtils.endTime1 = System.currentTimeMillis();
    }

}
```

文件 MyThread2.java 代码如下:

```java
package mythread;

import commonutils.CommonUtils;

import mytask.Task;

public class MyThread2 extends Thread {

    private Task task;

    public MyThread2(Task task) {
        super();
        this.task = task;
    }

    @Override
    public void run() {
        super.run();
        CommonUtils.beginTime2 = System.currentTimeMillis();
        task.doLongTimeTask();
        CommonUtils.endTime2 = System.currentTimeMillis();
    }

}
```

文件 Run.java 代码如下:

```java
package test;

import mytask.Task;
import mythread.MyThread1;
import mythread.MyThread2;

import commonutils.CommonUtils;

public class Run {

    public static void main(String[] args) {
        Task task = new Task();

        MyThread1 thread1 = new MyThread1(task);
        thread1.start();
```

```
        MyThread2 thread2 = new MyThread2(task);
        thread2.start();

        try {
            Thread.sleep(10000);
        } catch (InterruptedException e) {
            e.printStackTrace();
        }

        long beginTime = CommonUtils.beginTime1;
        if (CommonUtils.beginTime2 < CommonUtils.beginTime1) {
            beginTime = CommonUtils.beginTime2;
        }

        long endTime = CommonUtils.endTime1;
        if (CommonUtils.endTime2 > CommonUtils.endTime1) {
            endTime = CommonUtils.endTime2;
        }

        System.out.println("耗时: " + ((endTime - beginTime) / 1000));
    }
}
```

程序运行结果如图 2-20 所示。

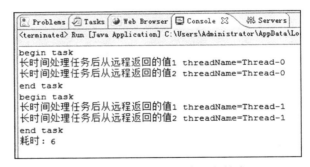

图 2-20　大约 6s 后程序运行结束

使用 synchronized 关键字来声明方法 public synchronized void doLongTimeTask()，从运行时间上来看，弊端突现，解决这样的问题可以使用 synchronized 同步块。

2.2.2　synchronized 同步代码块的使用

当两个并发线程访问同一个对象 object 中的 synchronized(this) 同步代码块时，一段时间内只能有一个线程得到执行，另一个线程必须等待当前线程执行完这个代码块以后才能执行该代码块。

创建测试用的项目 synchronizedOneThreadIn，类文件 ObjectService.java 代码如下：

```
package service;
```

```
public class ObjectService {

    public void serviceMethod() {
        try {
            synchronized (this) {
                System.out.println("begin time=" + System.currentTimeMillis());
                Thread.sleep(2000);
                System.out.println("end     end=" + System.currentTimeMillis());
            }
        } catch (InterruptedException e) {
            e.printStackTrace();
        }
    }
}
```

自定义线程 ThreadA.java 代码如下：

```
package extthread;

import service.ObjectService;

public class ThreadA extends Thread {

    private ObjectService service;

    public ThreadA(ObjectService service) {
        super();
        this.service = service;
    }

    @Override
    public void run() {
        super.run();
        service.serviceMethod();
    }
}
```

自定义线程 ThreadB.java 代码如下：

```
package extthread;

import service.ObjectService;

public class ThreadB extends Thread {
    private ObjectService service;

    public ThreadB(ObjectService service) {
        super();
        this.service = service;
    }
```

```
    @Override
    public void run() {
        super.run();
        service.serviceMethod();
    }
}
```

运行类 Run.java 代码如下：

```
package test.run;

import service.ObjectService;
import extthread.ThreadA;
import extthread.ThreadB;

public class Run {

    public static void main(String[] args) {
        ObjectService service = new ObjectService();

        ThreadA a = new ThreadA(service);
        a.setName("a");
        a.start();

        ThreadB b = new ThreadB(service);
        b.setName("b");
        b.start();
    }
}
```

程序运行结果如图 2-21 所示。

上面的示例，虽然使用了 synchronized 同步代码块，但执行的效率还是没有提高，仍是同步运行。

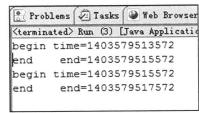

图 2-21 同步调用的结果

2.2.3 用同步代码块解决同步方法的弊端

如何用 synchronized 同步代码块解决程序执行效率慢的问题呢？创建 t6 项目，将 t5 项目的所有文件复制到 t6 项目中，并更改文件 Task.java 代码如下：

```
package mytask;

public class Task {

    private String getData1;
    private String getData2;

    public void doLongTimeTask() {
```

```java
        try {
            System.out.println("begin task");
            Thread.sleep(3000);

            String privateGetData1 = "长时间处理任务后从远程返回的值1 threadName=" +
                Thread.currentThread().getName();
            String privateGetData2 = "长时间处理任务后从远程返回的值2 threadName=" +
                Thread.currentThread().getName();

            synchronized (this) {
                getData1 = privateGetData1;
                getData2 = privateGetData2;
                System.out.println(getData1);
                System.out.println(getData2);
                System.out.println("end task");
            }

        } catch (InterruptedException e) {
            e.printStackTrace();
        }
    }

}
```

程序运行结果如图 2-22 所示。

```
begin task
begin task
长时间处理任务后从远程返回的值1 threadName=Thread-1
长时间处理任务后从远程返回的值2 threadName=Thread-1
end task
长时间处理任务后从远程返回的值1 threadName=Thread-0
长时间处理任务后从远程返回的值2 threadName=Thread-0
end task
耗时：3
```

图 2-22　运行速度很快

通过上面的示例可以知道，当一个线程访问 object 的一个 synchronized 同步代码块时，另一个线程仍然可以访问该 object 对象中的非 synchronized(this) 同步代码块。

在本示例中，运行时间虽然缩短，加快了运行效率，但同步 synchronized(this) 代码块真的是同步的吗？真的持有当前调用对象的锁吗？答案为是，但必须用代码的方式来进行验证。

2.2.4　一半异步，一半同步

下面通过示例说明不在 synchronized 块中就是异步执行，在 synchronized 块中就是同步执行。

创建 t7 项目，文件 Task.java 代码如下：

```java
package mytask;

public class Task {

    public void doLongTimeTask() {
        for (int i = 0; i < 100; i++) {
            System.out.println("nosynchronized threadName="
                    + Thread.currentThread().getName() + " i=" + (i + 1));
        }
        System.out.println("");
        synchronized (this) {
            for (int i = 0; i < 100; i++) {
                System.out.println("synchronized threadName="
                        + Thread.currentThread().getName() + " i=" + (i + 1));
            }
        }
    }
}
```

两个线程类代码如图 2-23 所示。

```java
package mythread;

import mytask.Task;

public class MyThread1 extends Thread {

    private Task task;

    public MyThread1(Task task) {
        super();
        this.task = task;
    }

    @Override
    public void run() {
        super.run();
        task.doLongTimeTask();
    }
}
```

```java
package mythread;

import mytask.Task;

public class MyThread2 extends Thread {

    private Task task;

    public MyThread2(Task task) {
        super();
        this.task = task;
    }

    @Override
    public void run() {
        super.run();
        task.doLongTimeTask();
    }
}
```

图 2-23　两个线程类代码

文件 Run.java 代码如下：

```java
package test;

import mytask.Task;
import mythread.MyThread1;
import mythread.MyThread2;
```

```java
public class Run {

    public static void main(String[] args) {
        Task task = new Task();

        MyThread1 thread1 = new MyThread1(task);
        thread1.start();

        MyThread2 thread2 = new MyThread2(task);
        thread2.start();
    }
}
```

程序运行结果如图 2-24 所示。

进入 synchronized 代码块后排队执行，程序运行结果如图 2-25 所示。

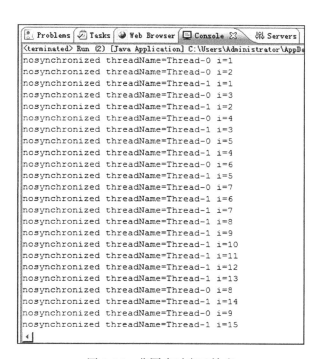

图 2-24　非同步时交叉输出　　　　　　图 2-25　排队执行

2.2.5 synchronized 代码块间的同步性

在使用同步 synchronized(this) 代码块时需要注意，当一个线程访问 object 的一个 synchronized(this) 同步代码块时，其他线程对同一个 object 中所有其他 synchronized(this) 同步代码块的访问将被阻塞，这说明 synchronized 使用的对象监视器是同一个，即使用的锁是同一个。

创建验证用的项目 doubleSynBlockOneTwo，类文件 ObjectService.java 代码如下：

```java
package service;

public class ObjectService {

    public void serviceMethodA() {
        try {
            synchronized (this) {
                System.out.println("A begin time=" + System.currentTimeMillis());
                Thread.sleep(2000);
                System.out.println("A end   end=" + System.currentTimeMillis());
            }
        } catch (InterruptedException e) {
            e.printStackTrace();
        }
    }

    public void serviceMethodB() {
        synchronized (this) {
            System.out.println("B begin time=" + System.currentTimeMillis());
            System.out.println("B end   end=" + System.currentTimeMillis());
        }
    }
}
```

自定义线程类 ThreadA.java 代码如下：

```java
package extthread;

import service.ObjectService;

public class ThreadA extends Thread {

    private ObjectService service;

    public ThreadA(ObjectService service) {
        super();
        this.service = service;
    }

    @Override
    public void run() {
```

```
            super.run();
            service.serviceMethodA();
        }

    }
```

自定义线程类 ThreadB.java 代码如下:

```
package extthread;

import service.ObjectService;

public class ThreadB extends Thread {
    private ObjectService service;

    public ThreadB(ObjectService service) {
        super();
        this.service = service;
    }

    @Override
    public void run() {
        super.run();
        service.serviceMethodB();
    }
}
```

运行类 Run.java 代码如下:

```
package test.run;

import service.ObjectService;
import extthread.ThreadA;
import extthread.ThreadB;

public class Run {

    public static void main(String[] args) {
        ObjectService service = new ObjectService();

        ThreadA a = new ThreadA(service);
        a.setName("a");
        a.start();

        ThreadB b = new ThreadB(service);
        b.setName("b");
        b.start();
    }

}
```

程序运行结果如图 2-26 所示。

```
 Problems  Tasks  Web Browser
<terminated> Run (4) [Java Application]
A begin time=1403580267856
A end      end=1403580269856
B begin time=1403580269856
B end      end=1403580269856
```

图 2-26 两个同步代码块按顺序执行

2.2.6 println() 方法也是同步的

JDK 的源代码中也有 synchronized (this) 使用的体现，PrintStream.java 类中的 println() 重载方法代码如下：

```java
public void println(String x) {
    synchronized (this) {
        print(x);
        newLine();
    }
}

public void println(Object x) {
    String s = String.valueOf(x);
    synchronized (this) {
        print(s);
        newLine();
    }
}
```

以上两个方法都用到了 synchronized (this) 同步代码块，说明 public void println(String x) 方法和 public void println(Object x) 方法在 synchronized (this) 同步代码块中执行的方式是按顺序同步执行的，这样输出的数据是完整的，不会出现信息交叉混乱的情况。

2.2.7 验证同步 synchronized(this) 代码块是锁定当前对象的

和 synchronized 方法一样，synchronized(this) 代码块也是锁定当前对象的。

创建 t8 项目，文件 Task.java 代码如下：

```java
package mytask;

public class Task {

    public void otherMethod() {
        System.out.println("------------------------run--otherMethod");
    }

    public void doLongTimeTask() {
```

```java
        synchronized (this) {
            for (int i = 0; i < 10000; i++) {
                System.out.println("synchronized threadName="
                        + Thread.currentThread().getName() + " i=" + (i + 1));
            }
        }
    }
}
```

文件 MyThread1.java 代码如下：

```java
package mythread;

import mytask.Task;

public class MyThread1 extends Thread {

    private Task task;

    public MyThread1(Task task) {
        super();
        this.task = task;
    }

    @Override
    public void run() {
        super.run();
        task.doLongTimeTask();
    }

}
```

文件 MyThread2.java 代码如下：

```java
package mythread;

import mytask.Task;

public class MyThread2 extends Thread {

    private Task task;

    public MyThread2(Task task) {
        super();
        this.task = task;
    }

    @Override
    public void run() {
        super.run();
        task.otherMethod();
```

        }
}

文件 Run.java 代码如下：

```
package test;

import mytask.Task;
import mythread.MyThread1;
import mythread.MyThread2;

public class Run {

    public static void main(String[] args)
        throws InterruptedException {
        Task task = new Task();

        MyThread1 thread1 = new MyThread1(task);
        thread1.start();

        Thread.sleep(100);

        MyThread2 thread2 = new MyThread2(task);
        thread2.start();
    }
}
```

程序运行结果如图 2-27 所示。

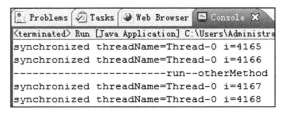

图 2-27　异步输出

更改 Task.java 代码如下：

```
package mytask;

public class Task {

    synchronized public void otherMethod() {
        System.out.println("-----------------------run--otherMethod");
    }

    public void doLongTimeTask() {
        synchronized (this) {
```

```
            for (int i = 0; i < 10000; i++) {
                System.out.println("synchronized threadName="
                    + Thread.currentThread().getName() + " i=" + (i + 1));
            }
        }
    }
}
```

程序运行结果如图 2-28 所示。

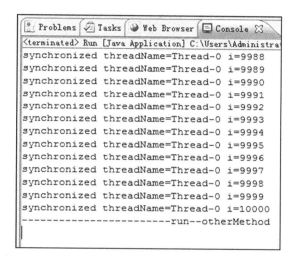

图 2-28　同步输出

同步输出的原因是使用 synchronized (this) 同步代码块将当前类的对象作为锁，使用 synchronized public void otherMethod() 同步方法将当前方法所在类的对象作为锁，都是一把锁，所以运行结果呈同步的效果。

2.2.8　将任意对象作为锁

多个线程调用同一个对象中的不同名称的 synchronized 同步方法或 synchronized(this) 同步代码块时，调用的效果是按顺序执行，即同步。

synchronized 同步方法或 synchronized(this) 同步代码块分别有两种作用。

synchronized 同步方法的作用：

1）对其他 synchronized 同步方法或 synchronized(this) 同步代码块调用呈同步效果。

2）同一时间只有一个线程可以执行 synchronized 同步方法中的代码。

synchronized(this) 同步代码块的作用：

1）对其他 synchronized 同步方法或 synchronized(this) 同步代码块调用呈同步效果。

2）同一时间只有一个线程可以执行 synchronized(this) 同步代码块中的代码。

除了使用 synchronized(this) 格式来创建同步代码块，其实 Java 还支持将"任意对象"作为锁来实现同步的功能，这个"任意对象"大多数是实例变量及方法的参数。使用格式为 synchronized（非 this 对象）。

synchronized（非 this 对象 x）同步代码块的作用：当多个线程争抢相同的"非 this 对象 x"的锁时，同一时间只有一个线程可以执行 synchronized（非 this 对象 x）同步代码块中的代码。

synchronized（非 this 对象 x）同步代码块的作用可以通过下面的示例验证。

创建测试用的项目 synBlockString，类文件 Service.java 代码如下：

```java
package service;

public class Service {

    private String usernameParam;
    private String passwordParam;

    private String anyString = new String();

    public void setUsernamePassword(String username, String password) {
        try {

            synchronized (anyString) {
                System.out.println("线程名称为: " + Thread.currentThread().getName()
                    + "在"
                    + System.currentTimeMillis() + "进入同步块");
                usernameParam = username;
                Thread.sleep(3000);//模拟处理数据需要的耗时
                passwordParam = password;
                System.out.println("线程名称为: " + Thread.currentThread().
                    getName() + "在"
                    + System.currentTimeMillis() + "离开同步块");
            }
        } catch (InterruptedException e) {
            // TODO Auto-generated catch block
            e.printStackTrace();
        }
    }

}
```

自定义线程 ThreadA.java 代码如下：

```java
package extthread;

import service.Service;

public class ThreadA extends Thread {
    private Service service;
```

```java
    public ThreadA(Service service) {
        super();
        this.service = service;
    }

    @Override
    public void run() {
        service.setUsernamePassword("a", "aa");
    }

}
```

自定义线程 ThreadB.java 代码如下：

```java
package extthread;

import service.Service;

public class ThreadB extends Thread {

    private Service service;

    public ThreadB(Service service) {
        super();
        this.service = service;
    }

    @Override
    public void run() {
        service.setUsernamePassword("b", "bb");
    }

}
```

运行类 Run.java 代码如下：

```java
package test;

import service.Service;
import extthread.ThreadA;
import extthread.ThreadB;

public class Run {

    public static void main(String[] args) {
        Service service = new Service();

        ThreadA a = new ThreadA(service);
        a.setName("A");
        a.start();
```

```
        ThreadB b = new ThreadB(service);
        b.setName("B");
        b.start();
    }

}
```

锁非 this 对象具有一定的优点：如果一个类中有很多个 synchronized 方法，则这时虽然能实现同步，但影响运行效率，如果使用同步代码块锁非 this 对象，则 synchronized（非 this）代码块中的程序与同步方法是异步的，因为有两把锁，不与其他锁 this 同步方法争抢 this 锁，可大大提高运行效率。

程序运行结果如图 2-29 所示。

图 2-29　同步效果

2.2.9　多个锁就是异步执行

更改 Service.java 文件代码如下：

```
package service;

public class Service {

    private String usernameParam;
    private String passwordParam;

    public void setUsernamePassword(String username, String password) {
        try {
            String anyString = new String();
            synchronized (anyString) {
                System.out.println("线程名称为：" + Thread.currentThread().
                        getName() + "在"
                        + System.currentTimeMillis() + "进入同步块");
                usernameParam = username;
                Thread.sleep(3000);
                passwordParam = password;
                System.out.println("线程名称为：" + Thread.currentThread().
                        getName() + "在"
                        + System.currentTimeMillis() + "离开同步块");
            }
        } catch (InterruptedException e) {
            // TODO Auto-generated catch block
            e.printStackTrace();
        }
    }

}
```

程序运行结果如图 2-30 所示。

可见，想使用"synchronized（非 this 对象 x）同步代码块"格式进行同步操作时，锁必须是同一个，如果不是同一个锁，则运行结果就是异步调用，交叉运行。

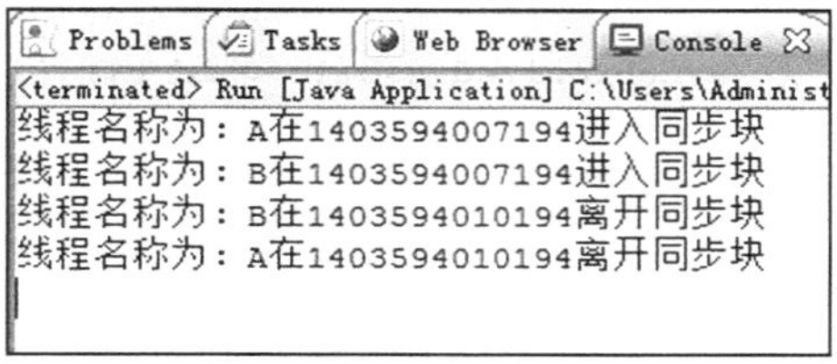

下面再用另一个示例来验证一下使用"synchronized（非 this 对象 x）同步代码块"格式时持有不同的锁是异步的效果。

创建新的项目，名称为 synBlockString2，验证 synchronized（非 this 对象）与同步 synchronized 方法是异步调用的效果。

图 2-30　不是同步的而是异步，因为不是同一个锁

两个自定义线程类代码如图 2-31 所示。

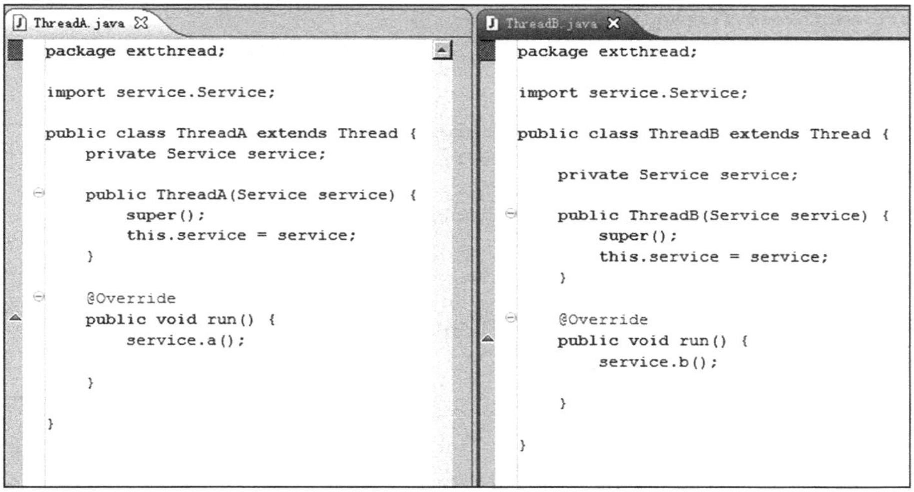

图 2-31　两个自定义线程类代码

类 Service.java 代码如下：

```
package service;

public class Service {

    private String anyString = new String();

    public void a() {
        try {
            synchronized (anyString) {
                System.out.println("a begin");
                Thread.sleep(3000);
                System.out.println("a   end");
            }
```

```
            } catch (InterruptedException e) {
                e.printStackTrace();
            }
        }
    synchronized public void b() {
        System.out.println("b begin");
        System.out.println("b   end");
    }
}
```

类 Run.java 代码如下:

```
package test;

import service.Service;
import extthread.ThreadA;
import extthread.ThreadB;

public class Run {

    public static void main(String[] args) {
        Service service = new Service();

        ThreadA a = new ThreadA(service);
        a.setName("A");
        a.start();

        ThreadB b = new ThreadB(service);
        b.setName("B");
        b.start();

    }
}
```

程序运行结果如图 2-32 所示。

由于锁不同,所以运行结果是异步的。

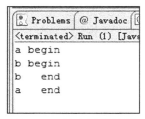

图 2-32　异步运行效果

2.2.10　验证方法被调用是随机的

同步代码块放在非同步 synchronized 方法中进行声明,并不能保证调用方法的线程的执行同步(顺序性),也就是线程调用方法的顺序是无序的,虽然在同步块中执行的顺序是同步的。下面通过示例验证多个线程调用同一个方法是随机的。

创建测试用的项目,项目名称为 syn_Out_asyn,类 MyList.java 代码如下:

```
package mylist;

import java.util.ArrayList;
```

```
import java.util.List;

public class MyList {

    private List list = new ArrayList();

    synchronized public void add(String username) {
        System.out.println("ThreadName=" + Thread.currentThread().getName()
                + "执行了add方法!");
        list.add(username);
        System.out.println("ThreadName=" + Thread.currentThread().getName()
                + "退出了add方法!");
    }

    synchronized public int getSize() {
        System.out.println("ThreadName=" + Thread.currentThread().getName()
                + "执行了getSize方法!");
        int sizeValue = list.size();
        System.out.println("ThreadName=" + Thread.currentThread().getName()
                + "退出了getSize方法!");
        return sizeValue;
    }

}
```

两个线程对象代码如图 2-33 所示。

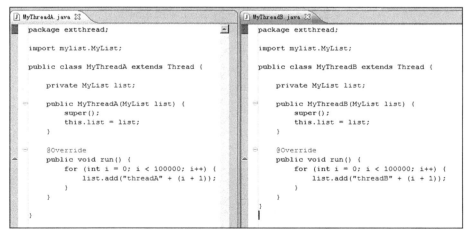

图 2-33　两个线程对象代码

类 Test.java 代码如下：

```
package test;

import mylist.MyList;
import extthread.MyThreadA;
```

```
import extthread.MyThreadB;

public class Test {

    public static void main(String[] args) {
        MyList mylist = new MyList();

        MyThreadA a = new MyThreadA(mylist);
        a.setName("A");
        a.start();

        MyThreadB b = new MyThreadB(mylist);
        b.setName("B");
        b.start();
    }

}
```

程序运行结果如图 2-34 所示。

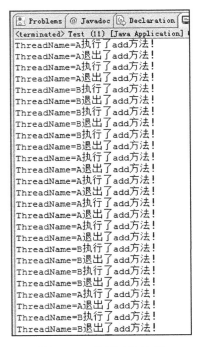

图 2-34　程序运行结果

从运行结果来看，同步方法中的代码是同步输出的，所以线程的"进入"与"退出"是成对出现的，但方法被调用的时机是随机的，即线程 A 和线程 B 的执行是异步的。

2.2.11 不同步导致的逻辑错误及其解决方法

如果方法不被同步化,则会出现逻辑上的错误。

创建 t9 项目,创建一个只能放一个元素的自定义集合工具类 MyOneList.java,代码如下:

```java
package mylist;

import java.util.ArrayList;
import java.util.List;

public class MyOneList {

    private List list = new ArrayList();

    synchronized public void add(String data) {
        list.add(data);
    };

    synchronized public int getSize() {
        return list.size();
    };

}
```

创建业务类 MyService.java,代码如下:

```java
package service;

import mylist.MyOneList;

public class MyService {

    public MyOneList addServiceMethod(MyOneList list, String data) {
        try {
            if (list.getSize() < 1) {
                Thread.sleep(2000);//模拟从远程端花费2s取回数据
                list.add(data);
            }
        } catch (InterruptedException e) {
            e.printStackTrace();
        }
        return list;
    }
}
```

创建线程类 MyThread1.java,代码如下:

```java
package mythread;

import mylist.MyOneList;
```

```java
import service.MyService;

public class MyThread1 extends Thread {

    private MyOneList list;

    public MyThread1(MyOneList list) {
        super();
        this.list = list;
    }

    @Override
    public void run() {
        MyService msRef = new MyService();
        msRef.addServiceMethod(list, "A");
    }

}
```

创建线程类 MyThread2.java，代码如下：

```java
package mythread;

import service.MyService;
import mylist.MyOneList;

public class MyThread2 extends Thread {

    private MyOneList list;

    public MyThread2(MyOneList list) {
        super();
        this.list = list;
    }

    @Override
    public void run() {
        MyService msRef = new MyService();
        msRef.addServiceMethod(list, "B");
    }

}
```

创建 Run.java，代码如下：

```java
package test;

import mylist.MyOneList;
import mythread.MyThread1;
import mythread.MyThread2;

public class Run {
```

```java
    public static void main(String[] args) throws InterruptedException {
        MyOneList list = new MyOneList();

        MyThread1 thread1 = new MyThread1(list);
        thread1.setName("A");
        thread1.start();

        MyThread2 thread2 = new MyThread2(list);
        thread2.setName("B");
        thread2.start();

        Thread.sleep(6000);

        System.out.println("listSize=" + list.getSize());
    }

}
```

程序运行结果如图 2-35 所示。

出现错误的原因是两个线程以异步的方式返回 list 参数的 size() 大小，解决办法就是"同步化"。

更改 MyService.java 类文件代码如下：

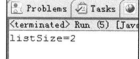

图 2-35　无序性带来的错误结果

```java
package service;

import mylist.MyOneList;

public class MyService {

    public MyOneList addServiceMethod(MyOneList list, String data) {
        try {
            synchronized (list) {
                if (list.getSize() < 1) {
                    Thread.sleep(2000);
                    list.add(data);
                }
            }
        } catch (InterruptedException e) {
            e.printStackTrace();
        }
        return list;
    }

}
```

由于 list 参数对象在项目中是一份实例，是单例的，而且也正需要对 list 参数的 getSize() 方法做同步的调用，所以就对 list 参数进行同步处理。

程序运行结果如图 2-36 所示。

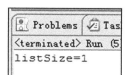

图 2-36　正确的运行结果

2.2.12 细化验证 3 个结论

synchronized（非 this 对象 x）格式的写法是将 x 对象本身作为"对象监视器"，这样就可以分析出 3 个结论：

- 当多个线程同时执行 synchronized(x){} 同步代码块时呈同步效果。
- 当其他线程执行 x 对象中 synchronized 同步方法时呈同步效果。
- 当其他线程执行 x 对象方法里面的 synchronized(this) 代码块时呈现同步效果。

需要注意，如果其他线程调用不加 synchronized 关键字的方法，则还是异步调用。

为了验证上面的 3 个结论，创建实验用的 synchronizedBlockLockAll 项目。

先来验证第 1 个结论：当多个线程同时执行 synchronized(x){} 同步代码块时呈同步效果。

创建名称为 test1 的包。

类 MyObject.java 代码如下：

```java
package test1.extobject;

public class MyObject {
}
```

类 Service.java 代码如下：

```java
package test1.service;

import test1.extobject.MyObject;

public class Service {

    public void testMethod1(MyObject object) {
        synchronized (object) {
            try {
                System.out.println("testMethod1 ____getLock time="
                        + System.currentTimeMillis() + " run ThreadName="
                        + Thread.currentThread().getName());
                Thread.sleep(2000);
                System.out.println("testMethod1 releaseLock time="
                        + System.currentTimeMillis() + " run ThreadName="
                        + Thread.currentThread().getName());
            } catch (InterruptedException e) {
                e.printStackTrace();
            }
        }
    }
}
```

2 个自定义线程代码如图 2-37 所示。

类文件 Run1_1.java 代码如下：

图 2-37 自定义线程代码

```java
package test1.run;

import test1.extobject.MyObject;
import test1.extthread.ThreadA;
import test1.extthread.ThreadB;
import test1.service.Service;

public class Run1_1 {

    public static void main(String[] args) {
        Service service = new Service();
        MyObject object = new MyObject();

        ThreadA a = new ThreadA(service, object);
        a.setName("a");
        a.start();

        ThreadB b = new ThreadB(service, object);
        b.setName("b");
        b.start();
    }

}
```

程序运行结果如图 2-38 所示。

图 2-38 同步效果

同步的原因是使用同锁，如果使用不同的锁，会出现什么效果呢？
创建类文件 Run1_2.java，代码如下：

```java
package test1.run;

import test1.extobject.MyObject;
import test1.extthread.ThreadA;
import test1.extthread.ThreadB;
import test1.service.Service;

public class Run1_2 {

    public static void main(String[] args) {
        Service service = new Service();
        MyObject object1 = new MyObject();
        MyObject object2 = new MyObject();

        ThreadA a = new ThreadA(service, object1);
        a.setName("a");
        a.start();

        ThreadB b = new ThreadB(service, object2);
        b.setName("b");
        b.start();
    }

}
```

程序运行结果如图 2-39 所示。

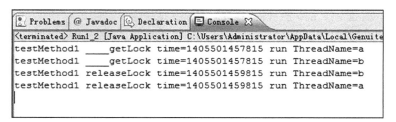

图 2-39　异步调用因为是不同的锁

继续验证第 2 个结论：当其他线程执行 x 对象中 synchronized 同步方法时呈同步效果。
创建名称为 test2 的包。
类 MyObject.java 代码如下：

```java
package test2.extobject;

public class MyObject {
    synchronized public void speedPrintString() {
        System.out.println("speedPrintString ____getLock time="
                + System.currentTimeMillis() + " run ThreadName="
```

```java
                + Thread.currentThread().getName());
        System.out.println("-----------------");
        System.out.println("speedPrintString releaseLock time="
                + System.currentTimeMillis() + " run ThreadName="
                + Thread.currentThread().getName());
    }
}
```

类 Service.java 代码如下:

```java
package test2.service;

import test2.extobject.MyObject;

public class Service {

    public void testMethod1(MyObject object) {
        synchronized (object) {
            try {
                System.out.println("testMethod1 ____getLock time="
                        + System.currentTimeMillis() + " run ThreadName="
                        + Thread.currentThread().getName());
                Thread.sleep(5000);
                System.out.println("testMethod1 releaseLock time="
                        + System.currentTimeMillis() + " run ThreadName="
                        + Thread.currentThread().getName());
            } catch (InterruptedException e) {
                e.printStackTrace();
            }
        }
    }

}
```

2 个自定义线程代码如图 2-40 所示。

```java
package test2.extthread;

import test2.extobject.MyObject;

public class ThreadA extends Thread {

    private Service service;
    private MyObject object;

    public ThreadA(Service service, MyObject object) {
        super();
        this.service = service;
        this.object = object;
    }

    @Override
    public void run() {
        super.run();
        service.testMethod1(object);
    }

}
```

```java
package test2.extthread;

import test2.extobject.MyObject;

public class ThreadB extends Thread {
    private MyObject object;

    public ThreadB(MyObject object) {
        super();
        this.object = object;
    }

    @Override
    public void run() {
        super.run();
        object.speedPrintString();
    }
}
```

图 2-40 自定义线程代码

类 Run.java 代码如下：

```java
package test2.run;

import test2.extobject.MyObject;
import test2.extthread.ThreadA;
import test2.extthread.ThreadB;
import test2.service.Service;

public class Run {

    public static void main(String[] args) throws InterruptedException {
        Service service = new Service();
        MyObject object = new MyObject();

        ThreadA a = new ThreadA(service, object);
        a.setName("a");
        a.start();

        Thread.sleep(100);

        ThreadB b = new ThreadB(object);
        b.setName("b");
        b.start();
    }

}
```

程序运行结果如图 2-41 所示。

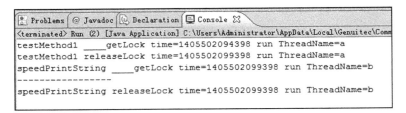

图 2-41　同步效果

继续验证第 3 个结论：当其他线程执行 x 对象方法里面的 synchronized(this) 代码块时呈现同步效果。

创建名称为 test3 的包。

创建名称为 MyObject.java 的类，代码如下：

```java
package test3.extobject;

public class MyObject {
    public void speedPrintString() {
        synchronized (this) {
```

```
            System.out.println("speedPrintString ____getLock time="
                    + System.currentTimeMillis() + " run ThreadName="
                    + Thread.currentThread().getName());
            System.out.println("----------------");
            System.out.println("speedPrintString releaseLock time="
                    + System.currentTimeMillis() + " run ThreadName="
                    + Thread.currentThread().getName());
        }
    }
}
```

其他代码与 test2 包中 Java 类的代码一样，程序运行结果如图 2-42 所示。

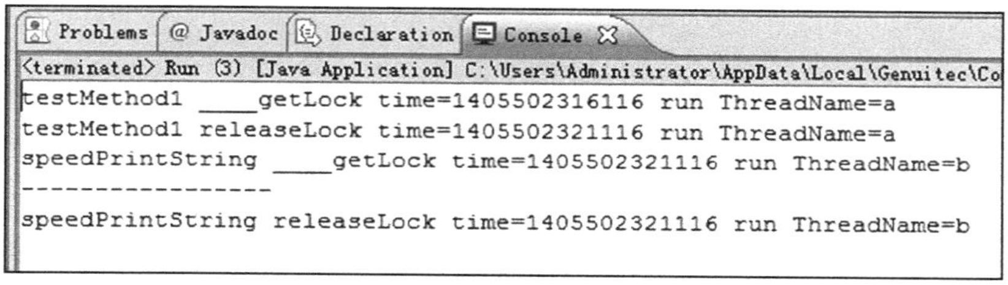

```
testMethod1 ____getLock time=1405502316116 run ThreadName=a
testMethod1 releaseLock time=1405502321116 run ThreadName=a
speedPrintString ____getLock time=1405502321116 run ThreadName=b
----------------
speedPrintString releaseLock time=1405502321116 run ThreadName=b
```

图 2-42　同样也是同步效果

2.2.13　类 Class 的单例性

每一个 *.java 文件对应 Class 类的实例都是一个，在内存中是单例的，测试代码如下：

```
class MyTest {
}

public class Test {
    public static void main(String[] args) throws InterruptedException {
        MyTest test1 = new MyTest();
        MyTest test2 = new MyTest();
        MyTest test3 = new MyTest();
        MyTest test4 = new MyTest();

        System.out.println(test1.getClass() == test1.getClass());
        System.out.println(test1.getClass() == test2.getClass());
        System.out.println(test1.getClass() == test3.getClass());
        System.out.println(test1.getClass() == test4.getClass());

        //
        System.out.println();
        //

        SimpleDateFormat format1 = new SimpleDateFormat("");
        SimpleDateFormat format2 = new SimpleDateFormat("");
        SimpleDateFormat format3 = new SimpleDateFormat("");
        SimpleDateFormat format4 = new SimpleDateFormat("");
```

```
            System.out.println(format1.getClass() == format1.getClass());
            System.out.println(format1.getClass() == format2.getClass());
            System.out.println(format1.getClass() == format3.getClass());
            System.out.println(format1.getClass() == format4.getClass());
    }
}
```

程序运行结果如下：

```
true
true
true
true

true
true
true
true
```

Class 类用于描述类的基本信息，包括有多少个字段，有多少个构造方法，有多少个普通方法等，为了减少对内存的高占用率，在内存中只需要存在一份 Class 类对象就可以了，所以被设计成是单例的。

2.2.14 静态同步 synchronized 方法与 synchronized(class) 代码块

关键字 synchronized 还可以应用在 static 静态方法上，如果这样写，那是对当前的 *.java 文件对应的 Class 类对象进行持锁，Class 类的对象是单例的，更具体地说，在静态 static 方法上使用 synchronized 关键字声明同步方法时，使用当前静态方法所在类对应 Class 类的单例对象作为锁。

测试项目在 synStaticMethod 中，类文件 Service.java 代码如下：

```java
package service;

public class Service {

    synchronized public static void printA() {
        try {
            System.out.println("线程名称为: " + Thread.currentThread().getName()
                    + "在" + System.currentTimeMillis() + "进入printA");
            Thread.sleep(3000);
            System.out.println("线程名称为: " + Thread.currentThread().getName()
                    + "在" + System.currentTimeMillis() + "离开printA");
        } catch (InterruptedException e) {
            e.printStackTrace();
        }
    }

    synchronized public static void printB() {
        System.out.println("线程名称为: " + Thread.currentThread().getName()
```

```
                    + "在" + System.currentTimeMillis() + "进入printB");
            System.out.println("线程名称为: " + Thread.currentThread().getName()
                    + "在" + System.currentTimeMillis() + "离开printB");
    }

}
```

自定义线程类 ThreadA.java 代码如下:

```
package extthread;

import service.Service;

public class ThreadA extends Thread {
    @Override
    public void run() {
        Service.printA();
    }

}
```

自定义线程类 ThreadB.java 代码如下:

```
package extthread;

import service.Service;

public class ThreadB extends Thread {
    @Override
    public void run() {
        Service.printB();
    }
}
```

运行类 Run.java 代码如下:

```
package test;

import service.Service;
import extthread.ThreadA;
import extthread.ThreadB;

public class Run {

    public static void main(String[] args) {

        ThreadA a = new ThreadA();
        a.setName("A");
        a.start();

        ThreadB b = new ThreadB();
        b.setName("B");
```

```
        b.start();
    }
}
```

程序运行结果如图 2-43 所示。

虽然该程序运行结果和将 synchronized 关键字加到非 static 方法上的效果是一样的——同步,但两者还是有本质上的不同,synchronized 关键字加到 static 静态方法上的方式是将 Class 类对象作为锁,而 synchronized 关键字加到非 static 静态方法上的方式是将方法所在类的对象作为锁。

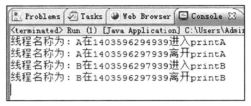

图 2-43 运行结果是同步效果

为了验证不是同一个锁,创建新的项目 synTwoLock,文件 Service.java 代码如下:

```
package service;

public class Service {

    synchronized public static void printA() {
        try {
            System.out.println("线程名称为: " + Thread.currentThread().getName()
                    + "在" + System.currentTimeMillis() + "进入printA");
            Thread.sleep(3000);
            System.out.println("线程名称为: " + Thread.currentThread().getName()
                    + "在" + System.currentTimeMillis() + "离开printA");
        } catch (InterruptedException e) {
            e.printStackTrace();
        }
    }

    synchronized public static void printB() {
        System.out.println("线程名称为: " + Thread.currentThread().getName() + "在"
                + System.currentTimeMillis() + "进入printB");
        System.out.println("线程名称为: " + Thread.currentThread().getName() + "在"
                + System.currentTimeMillis() + "离开printB");
    }

    synchronized public void printC() {
        System.out.println("线程名称为: " + Thread.currentThread().getName() + "在"
                + System.currentTimeMillis() + "进入printC");
        System.out.println("线程名称为: " + Thread.currentThread().getName() + "在"
                + System.currentTimeMillis() + "离开printC");
    }

}
```

自定义线程类 ThreadA.java 代码如下：

```java
package extthread;

import service.Service;

public class ThreadA extends Thread {
    private Service service;

    public ThreadA(Service service) {
        super();
        this.service = service;
    }

    @Override
    public void run() {
        service.printA();
    }
}
```

自定义线程类 ThreadB.java 代码如下：

```java
package extthread;

import service.Service;

public class ThreadB extends Thread {
    private Service service;

    public ThreadB(Service service) {
        super();
        this.service = service;
    }

    @Override
    public void run() {
        service.printB();
    }
}
```

自定义线程类 ThreadC.java 代码如下：

```java
package extthread;

import service.Service;

public class ThreadC extends Thread {

    private Service service;

    public ThreadC(Service service) {
```

```java
        super();
        this.service = service;
    }

    @Override
    public void run() {
        service.printC();
    }
}
```

运行类 Run.java 代码如下：

```java
package test;

import service.Service;
import extthread.ThreadA;
import extthread.ThreadB;
import extthread.ThreadC;

public class Run {

    public static void main(String[] args) {

        Service service = new Service();

        ThreadA a = new ThreadA(service);
        a.setName("A");
        a.start();

        ThreadB b = new ThreadB(service);
        b.setName("B");
        b.start();

        ThreadC c = new ThreadC(service);
        c.setName("C");
        c.start();

    }

}
```

程序运行结果如图 2-44 所示。

异步运行的原因是持有不同的锁，一个是将类 Service 的对象作为锁，另一个是将 Service 类对应 Class 类的对象作为锁，A、B 线程和 C 线程是异步的关系，而 A 线程和 B 线程是同步的关系。

图 2-44 方法 printC() 为异步运行

2.2.15 同步 syn static 方法可以对类的所有对象实例起作用

Class 锁可以对类的所有对象实例起作用，用项目 synMoreObjectStaticOneLock 来验证，类文件 Service.java 代码如下：

```java
package service;

public class Service {

    synchronized public static void printA() {
        try {
            System.out.println("线程名称为: " + Thread.currentThread().getName()
                    + "在" + System.currentTimeMillis() + "进入printA");
            Thread.sleep(3000);
            System.out.println("线程名称为: " + Thread.currentThread().getName()
                    + "在" + System.currentTimeMillis() + "离开printA");
        } catch (InterruptedException e) {
            e.printStackTrace();
        }
    }

    synchronized public static void printB() {
        System.out.println("线程名称为: " + Thread.currentThread().getName() + "在"
                + System.currentTimeMillis() + "进入printB");
        System.out.println("线程名称为: " + Thread.currentThread().getName() + "在"
                + System.currentTimeMillis() + "离开printB");
    }

}
```

自定义线程 ThreadA.java 代码如下：

```java
package extthread;

import service.Service;

public class ThreadA extends Thread {
    private Service service;

    public ThreadA(Service service) {
        super();
        this.service = service;
    }

    @Override
    public void run() {
        service.printA();
    }
}
```

自定义线程 ThreadB.java 代码如下：

```
package extthread;

import service.Service;

public class ThreadB extends Thread {
    private Service service;

    public ThreadB(Service service) {
        super();
        this.service = service;
    }

    @Override
    public void run() {
        service.printB();
    }
}
```

运行类 Run.java 代码如下：

```
package test;

import service.Service;
import extthread.ThreadA;
import extthread.ThreadB;

public class Run {

    public static void main(String[] args) {

        Service service1 = new Service();
        Service service2 = new Service();

        ThreadA a = new ThreadA(service1);
        a.setName("A");
        a.start();

        ThreadB b = new ThreadB(service2);
        b.setName("B");
        b.start();

    }

}
```

程序运行结果如图 2-45 所示。

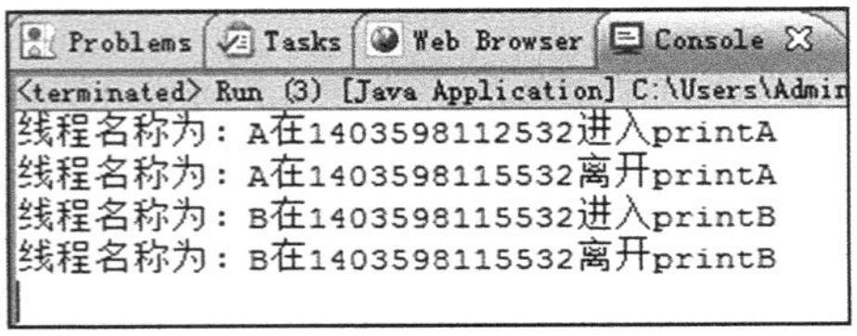
Problems
Tasks
Web Browser
Console
<terminated> Run (3) [Java Application] C:\Users\Admin
线程名称为：A在1403598112532进入printA
线程名称为：A在1403598115532离开printA
线程名称为：B在1403598115532进入printB
线程名称为：B在1403598115532离开printB

图 2-45　虽然是不同对象，但静态的同步方法还是同步运行

2.2.16　同步 syn(class) 代码块可以对类的所有对象实例起作用

同步 synchronized(class) 代码块的作用其实和 synchronized static 方法的作用一样。创建测试用的项目 synBlockMoreObjectOneLock，类文件 Service.java 代码如下：

```java
package service;

public class Service {

    public void printA() {
        synchronized (Service.class) {
            try {
                System.out.println("线程名称为：" + Thread.currentThread().getName()
                        + "在" + System.currentTimeMillis() + "进入printA");
                Thread.sleep(3000);
                System.out.println("线程名称为：" + Thread.currentThread().getName()
                        + "在" + System.currentTimeMillis() + "离开printA");
            } catch (InterruptedException e) {
                e.printStackTrace();
            }
        }

    }

    public void printB() {
        synchronized (Service.class) {
            System.out.println("线程名称为：" + Thread.currentThread().getName()
                    + "在" + System.currentTimeMillis() + "进入printB");
            System.out.println("线程名称为：" + Thread.currentThread().getName()
                    + "在" + System.currentTimeMillis() + "离开printB");
        }
    }
}
```

两个自定义线程代码如图 2-46 所示。

```java
package extthread;

import service.Service;

public class ThreadA extends Thread {
    private Service service;

    public ThreadA(Service service) {
        super();
        this.service = service;
    }

    @Override
    public void run() {
        service.printA();
    }
}
```

```java
package extthread;

import service.Service;

public class ThreadB extends Thread {
    private Service service;

    public ThreadB(Service service) {
        super();
        this.service = service;
    }

    @Override
    public void run() {
        service.printB();
    }
}
```

图 2-46　两个自定义线程代码

运行类 Run.java 代码如下：

```java
package test;

import service.Service;
import extthread.ThreadA;
import extthread.ThreadB;

public class Run {

    public static void main(String[] args) {

        Service service1 = new Service();
        Service service2 = new Service();

        ThreadA a = new ThreadA(service1);
        a.setName("A");
        a.start();

        ThreadB b = new ThreadB(service2);
        b.setName("B");
        b.start();

    }

}
```

程序运行结果如图 2-47 所示。

图 2-47　同步运行

2.2.17　String 常量池特性与同步相关的问题与解决方案

JVM 具有 String 常量池的功能，所以如图 2-48 所示的结果为 true。

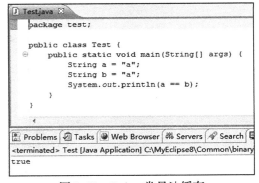

图 2-48　String 常量池缓存

当将 synchronized(string) 同步块与 String 联合使用时，要注意常量池会带来一些意外。新建名称为 StringAndSyn 的项目，类文件 Service.java 代码如下：

```java
package service;

public class Service {
    public static void print(String stringParam) {
        try {
            synchronized (stringParam) {
                while (true) {
                    System.out.println(Thread.currentThread().getName());
                    Thread.sleep(1000);
                }
            }
        } catch (InterruptedException e) {
            e.printStackTrace();
        }
    }
}
```

两个自定义线程代码如图 2-49 所示。

```java
package extthread;

import service.Service;

public class ThreadA extends Thread {
    private Service service;
    public ThreadA(Service service) {
        super();
        this.service = service;
    }

    @Override
    public void run() {
        service.print("AA");
    }
}
```

```java
package extthread;

import service.Service;

public class ThreadB extends Thread {
    private Service service;
    public ThreadB(Service service) {
        super();
        this.service = service;
    }

    @Override
    public void run() {
        service.print("AA");
    }
}
```

图 2-49 两个自定义线程代码

运行类 Run.java 代码如下：

```java
package test;

import service.Service;
import extthread.ThreadA;
import extthread.ThreadB;

public class Run {

    public static void main(String[] args) {

        Service service = new Service();

        ThreadA a = new ThreadA(service);
        a.setName("A");
        a.start();
```

```
        ThreadB b = new ThreadB(service);
        b.setName("B");
        b.start();
    }
}
```

程序运行结果如图 2-50 所示。

出现这种情况是因为 String 的两个值都是 "AA"，两个线程持有相同的锁，造成 B 线程不能执行。这就是 String 常量池所带来的问题，所以大多数情况下，同步 synchronized 代码块不使用 String 作为锁对象，而改用其他，例如，new Object() 实例化一个新的 Object 对象，它并不放入缓存池中，或者执行 new String() 创建不同的字符串对象，形成不同的锁。

图 2-50　死循环

继续创建名称为 StringAndSyn2 的项目，类文件 Service.java 代码如下：

```
package service;

public class Service {
    public static void print(Object object) {
        try {
            synchronized (object) {
                while (true) {
                    System.out.println(Thread.currentThread()
                            .getName());
                    Thread.sleep(1000);
                }
            }
        } catch (InterruptedException e) {
            e.printStackTrace();
        }
    }
}
```

两个自定义线程代码如图 2-51 所示。

```
ThreadA.java
 1 package extthread;
 2
 3 import service.Service;
 4
 5 public class ThreadA extends Thread {
 6     private Service service;
 7
 8     public ThreadA(Service service) {
 9         super();
10         this.service = service;
11     }
12
13     @Override
14     public void run() {
15         service.print(new Object());
16     }
17 }
18
```

```
ThreadB.java
 1 package extthread;
 2
 3 import service.Service;
 4
 5 public class ThreadB extends Thread {
 6     private Service service;
 7
 8     public ThreadB(Service service) {
 9         super();
10         this.service = service;
11     }
12
13     @Override
14     public void run() {
15         service.print(new Object());
16     }
17 }
18
```

图 2-51　自定义线程代码

运行类 Run.java 代码如下：

```java
package test;

import service.Service;
import extthread.ThreadA;
import extthread.ThreadB;

public class Run {

    public static void main(String[] args) {

        Service service = new Service();

        ThreadA a = new ThreadA(service);
        a.setName("A");
        a.start();

        ThreadB b = new ThreadB(service);
        b.setName("B");
        b.start();

    }
}
```

程序运行结果如图 2-52 所示。

交替输出的原因是持有的锁不是一个。

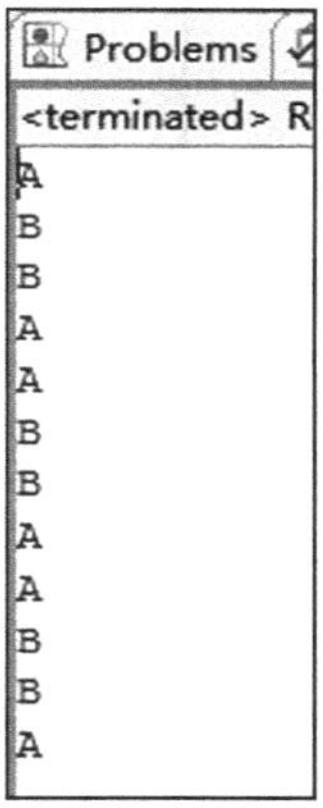

图 2-52　交替输出

2.2.18　同步 synchronized 方法无限等待问题与解决方案

使用同步方法会导致锁资源被长期占用，得不到运行的机会，创建示例项目 twoStop，类 Service.java 代码如下：

```java
package service;

public class Service {
    synchronized public void methodA() {
        System.out.println("methodA begin");
        boolean isContinueRun = true;
        while (isContinueRun) {
        }
        System.out.println("methodA end");
    }

    synchronized public void methodB() {
        System.out.println("methodB begin");
        System.out.println("methodB end");
    }
}
```

自定义线程类代码如图 2-53 所示。

```java
package extthread;

import service.Service;

public class ThreadA extends Thread {
    private Service service;

    public ThreadA(Service service) {
        super();
        this.service = service;
    }

    @Override
    public void run() {
        service.methodA();
    }
}
```

```java
package extthread;

import service.Service;

public class ThreadB extends Thread {
    private Service service;

    public ThreadB(Service service) {
        super();
        this.service = service;
    }

    @Override
    public void run() {
        service.methodB();
    }
}
```

图 2-53 自定义线程类代码

运行类 Run.java 代码如下：

```java
package test.run;

import service.Service;
import extthread.ThreadA;
import extthread.ThreadB;

public class Run {

    public static void main(String[] args) {
        Service service = new Service();

        ThreadA athread = new ThreadA(service);
        athread.start();

        ThreadB bthread = new ThreadB(service);
        bthread.start();
    }

}
```

程序运行结果如图 2-54 所示。

B 线程永远得不到运行的机会，这时就可以使用同步块来解决这样的问题，更改后的 Service.java 文件代码如下：

```java
package service;

public class Service {
    Object object1 = new Object();
```

图 2-54 运行结果是死循环

```
    public void methodA() {
        synchronized (object1) {
            System.out.println("methodA begin");
            boolean isContinueRun = true;
            while (isContinueRun) {
            }
            System.out.println("methodA end");
        }
    }

    Object object2 = new Object();

    public void methodB() {
        synchronized (object2) {
            System.out.println("methodB begin");
            System.out.println("methodB end");
        }
    }
}
```

程序运行结果如图 2-55 所示。

本示例代码在项目 twoNoStop 中。

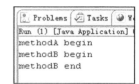

图 2-55　不再出现同步等待的情况

2.2.19　多线程的死锁

Java 线程死锁是一个经典的多线程问题，因为不同的线程都在等待根本不可能被释放的锁，从而导致所有的任务都无法继续完成。在多线程技术中，"死锁"是必须避免的，因为这会造成线程"假死"。

创建名称为 deadLockTest 的项目，DealThread.java 类代码如下：

```
package test;

public class DealThread implements Runnable {

    public String username;
    public Object lock1 = new Object();
    public Object lock2 = new Object();

    public void setFlag(String username) {
        this.username = username;
    }

    @Override
    public void run() {
        if (username.equals("a")) {
            synchronized (lock1) {
                try {
                    System.out.println("username = " + username);
                    Thread.sleep(3000);
```

```java
                } catch (InterruptedException e) {
                    // TODO Auto-generated catch block
                    e.printStackTrace();
                }
                synchronized (lock2) {
                    System.out.println("按lock1->lock2代码顺序执行了");
                }
            }
        }
        if (username.equals("b")) {
            synchronized (lock2) {
                try {
                    System.out.println("username = " + username);
                    Thread.sleep(3000);
                } catch (InterruptedException e) {
                    // TODO Auto-generated catch block
                    e.printStackTrace();
                }
                synchronized (lock1) {
                    System.out.println("按lock2->lock1代码顺序执行了");
                }
            }
        }
    }
}
```

运行类 Run.java 代码如下：

```java
package test;

public class Run {
    public static void main(String[] args) {
        try {
            DealThread t1 = new DealThread();
            t1.setFlag("a");

            Thread thread1 = new Thread(t1);
            thread1.start();

            Thread.sleep(100);

            t1.setFlag("b");
            Thread thread2 = new Thread(t1);
            thread2.start();
        } catch (InterruptedException e) {
            // TODO Auto-generated catch block
            e.printStackTrace();
        }
    }
}
```

程序运行结果如图 2-56 所示。

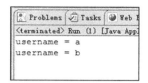

图 2-56　死锁了

可以使用 JDK 自带的工具来监测是否有死锁的现象，首先进入 CMD 工具，再进入 JDK 安装文件夹中的 bin 目录，执行 jps 命令，如图 2-57 所示。

图 2-57　执行 jps 命令

得到运行线程 Run 的 id 值是 3244，再执行 jstack 命令，结果如图 2-58 所示。

图 2-58　执行 jstack 命令

监测出有死锁现象，如图 2-59 所示。

图 2-59　出现死锁

死锁是程序设计的 Bug，在设计程序时要避免双方互相持有对方的锁，只要互相等待对方释放锁，就有可能出现死锁。

2.2.20 内置类与静态内置类

关键字 synchronized 的知识点还涉及内置类的使用，先来看一下简单内置类的测试方法，创建 innerClass 项目，类 PublicClass.java 代码如下：

```java
package test;

public class PublicClass {

    private String username;
    private String password;

    class PrivateClass {
        private String age;
        private String address;

        public String getAge() {
            return age;
        }

        public void setAge(String age) {
            this.age = age;
        }

        public String getAddress() {
            return address;
        }

        public void setAddress(String address) {
            this.address = address;
        }

        public void printPublicProperty() {
            System.out.println(username + " " + password);
        }
    }

    public String getUsername() {
        return username;
    }

    public void setUsername(String username) {
        this.username = username;
    }

    public String getPassword() {
        return password;
    }
```

```java
    public void setPassword(String password) {
        this.password = password;
    }

}
```

创建运行类 Run.java，代码如下：

```java
package test;

import test.PublicClass.PrivateClass;

public class Run {

    public static void main(String[] args) {
        PublicClass publicClass = new PublicClass();
        publicClass.setUsername("usernameValue");
        publicClass.setPassword("passwordValue");

        System.out.println(publicClass.getUsername() + " "
                + publicClass.getPassword());

        PrivateClass privateClass = publicClass.new PrivateClass();
        privateClass.setAge("ageValue");
        privateClass.setAddress("addressValue");

        System.out.println(privateClass.getAge() + " "
                + privateClass.getAddress());

    }

}
```

如果 PublicClass.java 类和 Run.java 类不在同一个包中，则需要将 PrivateClass 内置类声明成 public 的。

程序运行结果如图 2-60 所示。

想要实例化内置类必须使用如下代码：

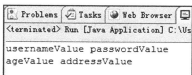

图 2-60　程序运行结果（内置类）

```java
PrivateClass privateClass = publicClass.new PrivateClass();
```

下面来了解静态内置类的使用方法。创建实验用的项目 innerStaticClass，类 PublicClass.java 代码如下：

```java
package test;

public class PublicClass {

    static private String username;
    static private String password;
```

```java
    static class PrivateClass {
        private String age;
        private String address;

        public String getAge() {
            return age;
        }

        public void setAge(String age) {
            this.age = age;
        }

        public String getAddress() {
            return address;
        }

        public void setAddress(String address) {
            this.address = address;
        }

        public void printPublicProperty() {
            System.out.println(username + " " + password);
        }
    }

    public String getUsername() {
        return username;
    }

    public void setUsername(String username) {
        this.username = username;
    }

    public String getPassword() {
        return password;
    }

    public void setPassword(String password) {
        this.password = password;
    }
}
```

运行类 Run.java,代码如下:

```java
package test;

import test.PublicClass.PrivateClass;

public class Run {

    public static void main(String[] args) {
```

```
        PublicClass publicClass = new PublicClass();
        publicClass.setUsername("usernameValue");
        publicClass.setPassword("passwordValue");

        System.out.println(publicClass.getUsername() + " "
                + publicClass.getPassword());

        PrivateClass privateClass = new PrivateClass();
        privateClass.setAge("ageValue");
        privateClass.setAddress("addressValue");

        System.out.println(privateClass.getAge() + " "
                + privateClass.getAddress());
    }

}
```

程序运行结果如图 2-61 所示。

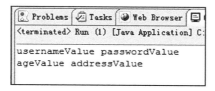

图 2-61　程序运行结果（静态内置类）

2.2.21　内置类与同步：实验 1

本实验测试的案例是在内置类中有两个同步方法，但使用的是不同的锁，输出结果也是异步的。

创建实验用的项目 innerTest1，类 OutClass.java 代码如下：

```
package test;

public class OutClass {

    static class Inner {
        public void method1() {
            synchronized ("其他的锁") {
                for (int i = 1; i <= 10; i++) {
                    System.out.println(Thread.currentThread().getName() + "  i="
                            + i);
                    try {
                        Thread.sleep(100);
                    } catch (InterruptedException e) {
```

```
                    }
                }
            }
        }

        public synchronized void method2() {
            for (int i = 11; i <= 20; i++) {
                System.out.println(Thread.currentThread().
                       getName() + " i=" + i);
                try {
                    Thread.sleep(100);
                } catch (InterruptedException e) {
                }
            }
        }
    }
}
```

运行类 Run.java 代码如下：

```
package test;

import test.OutClass.Inner;

public class Run {
    public static void main(String[] args) {

        final Inner inner = new Inner();

        Thread t1 = new Thread(new Runnable() {
            public void run() {
                inner.method1();
            }
        }, "A");

        Thread t2 = new Thread(new Runnable() {
            public void run() {
                inner.method2();
            }
        }, "B");

        t1.start();
        t2.start();

    }
}
```

程序运行结果如图 2-62 所示。

由于持有不同的锁，所以输出结果是乱序的。

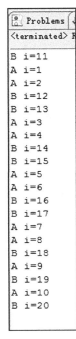

图 2-62　乱序输出

2.2.22 内置类与同步：实验 2

本实验测试同步代码块 synchronized (lock) 对 lock 上锁后，其他线程只能以同步的方式调用 lock 中的同步方法。

创建测试用的项目 innerTest2，类 OutClass.java 代码如下：

```java
package test;

public class OutClass {
    static class InnerClass1 {
        public void method1(InnerClass2 class2) {
            String threadName = Thread.currentThread().getName();
            synchronized (class2) {
                System.out.println(threadName + " 进入InnerClass1类中的method1方法");
                for (int i = 0; i < 10; i++) {
                    System.out.println("i=" + i);
                    try {
                        Thread.sleep(100);
                    } catch (InterruptedException e) {

                    }
                }
                System.out.println(threadName + " 离开InnerClass1类中的method1方法");
            }
        }

        public synchronized void method2() {
            String threadName = Thread.currentThread().getName();
            System.out.println(threadName + " 进入InnerClass1类中的method2方法");
            for (int j = 0; j < 10; j++) {
                System.out.println("j=" + j);
                try {
                    Thread.sleep(100);
                } catch (InterruptedException e) {

                }
            }
            System.out.println(threadName + " 离开InnerClass1类中的method2方法");
        }
    }

    static class InnerClass2 {
        public synchronized void method1() {
            String threadName = Thread.currentThread().getName();
            System.out.println(threadName + " 进入InnerClass2类中的method1方法");
            for (int k = 0; k < 10; k++) {
                System.out.println("k=" + k);
                try {
```

```
                    Thread.sleep(100);
                } catch (InterruptedException e) {

                }
            }
            System.out.println(threadName + " 离开InnerClass2类中的method1方法");
        }
    }
}
```

运行类 Run.java 代码如下:

```
package test;

import test.OutClass.InnerClass1;
import test.OutClass.InnerClass2;

public class Run {

    public static void main(String[] args) {
        final InnerClass1 in1 = new InnerClass1();
        final InnerClass2 in2 = new InnerClass2();
        Thread t1 = new Thread(new Runnable() {
            public void run() {
                in1.method1(in2);
            }
        }, "T1");
        Thread t2 = new Thread(new Runnable() {
            public void run() {
                in1.method2();
            }
        }, "T2");
        // //
        // //
        Thread t3 = new Thread(new Runnable() {
            public void run() {
                in2.method1();
            }
        }, "T3");
        t1.start();
        t2.start();
        t3.start();
    }
}
```

程序运行结果如图 2-63 所示。

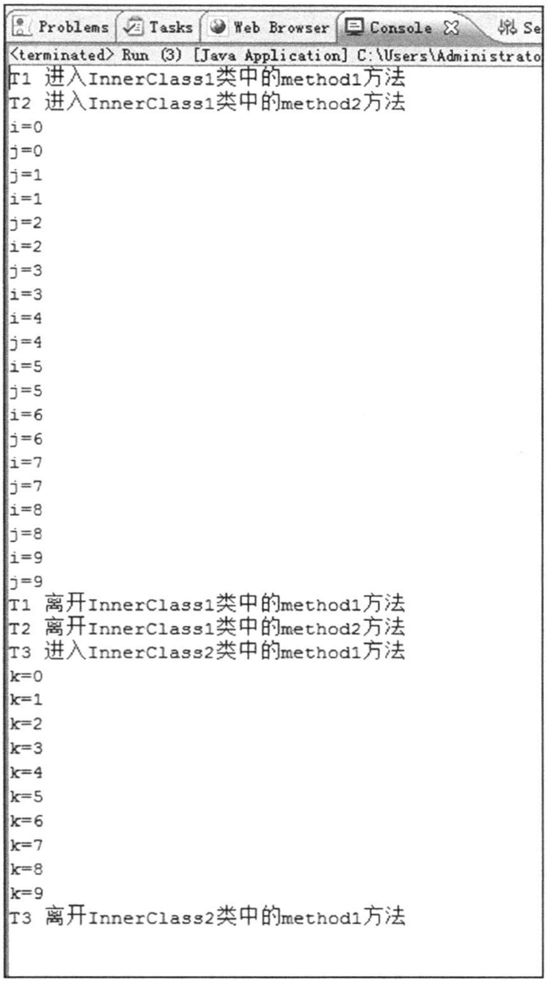

图 2-63　程序运行结果

2.2.23　锁对象改变导致异步执行

在将任何数据类型作为同步锁时，需要注意是否有多个线程同时争抢锁对象。如果多个线程同时争抢相同的锁对象，则这些线程之间就是同步的；如果多个线程分别获得自己的锁，则这些线程之间就是异步的。

通常情况下，一旦持有锁后就不再对锁对象进行更改，因为一旦更改就有可能出现一些错误。

创建测试用的项目 setNewStringTwoLock，MyService.java 类代码如下：

```
package myservice;
```

```java
public class MyService {
    private String lock = "123";

    public void testMethod() {
        try {
            synchronized (lock) {
                System.out.println(Thread.currentThread().getName() + " begin "
                        + System.currentTimeMillis());
                lock = "456";
                Thread.sleep(2000);
                System.out.println(Thread.currentThread().getName() + "   end "
                        + System.currentTimeMillis());
            }
        } catch (InterruptedException e) {
            e.printStackTrace();
        }
    }
}
```

两个自定义线程类代码如图 2-64 所示。

```
package extthread;

import myservice.MyService;

public class ThreadA extends Thread {

    private MyService service;

    public ThreadA(MyService service) {
        super();
        this.service = service;
    }

    @Override
    public void run() {
        service.testMethod();
    }
}
```

```
package extthread;

import myservice.MyService;

public class ThreadB extends Thread {

    private MyService service;

    public ThreadB(MyService service) {
        super();
        this.service = service;
    }

    @Override
    public void run() {
        service.testMethod();
    }
}
```

图 2-64　两个自定义线程类代码

运行类 Run1.java 代码如下：

```java
package test.run;

import myservice.MyService;
import extthread.ThreadA;
import extthread.ThreadB;

public class Run1 {

    public static void main(String[] args) throws InterruptedException {
```

```
            MyService service = new MyService();

            ThreadA a = new ThreadA(service);
            a.setName("A");

            ThreadB b = new ThreadB(service);
            b.setName("B");

            a.start();
            Thread.sleep(50);// 存在50ms
            b.start();
    }
}
```

程序运行结果如图 2-65 所示。

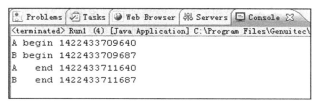

图 2-65　异步输出

50ms 后，B 线程取得的锁是 "456"。

继续实验，创建运行类 Run2.java 代码如下：

```
package test.run;

import myservice.MyService;
import extthread.ThreadA;
import extthread.ThreadB;

public class Run2 {

    public static void main(String[] args) throws InterruptedException {

        MyService service = new MyService();

        ThreadA a = new ThreadA(service);
        a.setName("A");

        ThreadB b = new ThreadB(service);
        b.setName("B");

        a.start();
        b.start();
    }
}
```

去掉代码 Thread.sleep(50)。

程序运行结果如图 2-66 所示。

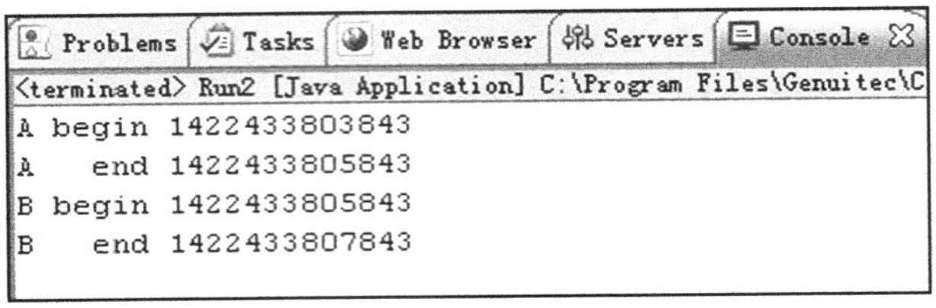

图 2-66 同步输出

需要注意的是，字符串 String 类型是不可变的，用于创建新的内存空间来存储新的字符。

控制台输出的信息说明 A 线程和 B 线程检测到锁对象的值 "123"，"123" 存储到 A 线程内存空间中，虽然将锁改成了 "456"，"456" 存储到 B 线程内存空间中，但结果还是同步的，因为 A 线程和 B 线程共同争抢的锁是 A 线程内存空间中的 "123"，不是 B 线程内存空间中的 "456"。

但是，还是会有很小的机会会出现一起输出两个 begin 的情况，那是因为 A 线程将锁值改变之后，B 线程才启动去执行 run() 方法，不存在 A 线程和 B 线程争抢锁的情况，所以导致 B 线程获得更改值后的锁值是 "456"，并连续输出两个 begin。

2.2.24 锁对象不改变依然同步执行

只要对象不变，运行的结果即为同步，此结论的实验代码在 setNewPropertiesLockOne 项目中进行演示，创建类 Userinfo.java，其结构如图 2-67 所示。

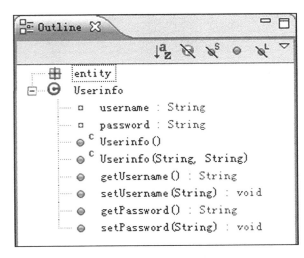

图 2-67 类结构

类 Service.java 代码如下：

```java
package service;

import entity.Userinfo;

public class Service {

    public void serviceMethodA(Userinfo userinfo) {
        synchronized (userinfo) {
            try {
                System.out.println(Thread.currentThread().getName());
                userinfo.setUsername("abcabcabc");
                Thread.sleep(3000);
                System.out.println("end! time=" + System.currentTimeMillis());
            } catch (InterruptedException e) {
                // TODO Auto-generated catch block
                e.printStackTrace();
            }
        }
    }
}
```

两个线程类代码如图 2-68 所示。

```java
package extthread;

import service.Service;

public class ThreadA extends Thread {

    private Service service;
    private Userinfo userinfo;

    public ThreadA(Service service,
            Userinfo userinfo) {
        super();
        this.service = service;
        this.userinfo = userinfo;
    }

    @Override
    public void run() {
        service.serviceMethodA(userinfo);
    }
}
```

```java
package extthread;

import service.Service;

public class ThreadB extends Thread {

    private Service service;
    private Userinfo userinfo;

    public ThreadB(Service service,
            Userinfo userinfo) {
        super();
        this.service = service;
        this.userinfo = userinfo;
    }

    @Override
    public void run() {
        service.serviceMethodA(userinfo);
    }
}
```

图 2-68　两个线程类代码

运行类 Run.java 代码如下：

```java
package test.run;

import service.Service;
import entity.Userinfo;
import extthread.ThreadA;
```

```
import extthread.ThreadB;

public class Run {

    public static void main(String[] args) {

        try {
            Service service = new Service();
            Userinfo userinfo = new Userinfo();

            ThreadA a = new ThreadA(service, userinfo);
            a.setName("a");
            a.start();
            Thread.sleep(50);
            ThreadB b = new ThreadB(service, userinfo);
            b.setName("b");
            b.start();

        } catch (InterruptedException e) {
            e.printStackTrace();
        }

    }
}
```

程序运行结果如图 2-69 所示。

只要对象不变就是同步效果,因为 A 线程和 B 线程持有的锁对象永远为同一个,仅仅对象的属性改变了,但对象未发生改变。

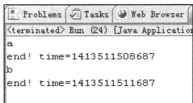

图 2-69　只要对象不变就是同步效果

2.2.25　同步写法案例比较

使用关键字 synchronized 的写法比较多,常用的有如下几种,代码如下:

```
public class MyService {
    synchronized public static void testMethod1() {
    }
    public void testMethod2() {
        synchronized (MyService.class) {
        }
    }
    synchronized public void testMethod3() {
    }
    public void testMethod4() {
        synchronized (this) {
        }
    }
    public void testMethod5() {
        synchronized ("abc") {
```

            }
        }
    }
}

上面的代码中出现了 3 种类型的锁对象：

（A）testMethod1() 和 testMethod2() 持有的锁是同一个，即 MyService.java 对应 Class 类的对象。

（B）testMethod3() 和 testMethod4() 持有的锁是同一个，即 MyService.java 类的对象。

（C）testMethod5() 持有的锁是字符串 abc。

说明 testMethod1() 和 testMethod2() 是同步关系，testMethod3() 和 testMethod4() 是同步关系。A 和 C 之间是异步关系，B 和 C 之间是异步关系，A 和 B 之间是异步关系。

2.3 volatile 关键字

在 Java 中，关键字 volatile 就像一个神话，几乎在各种博客、微信订阅号、聊天群中反复被提起，可见 volatile 在多线程领域是多么的重要。volatile 在使用上具有以下特性：

1）可见性：B 线程能马上看到 A 线程更改的数据。

2）原子性：在 32 位系统中，针对未使用 volatile 声明的 long 或 double 数据类型没有实现写原子性，如果想实现，则声明变量时添加 volatile，而在 64 位系统中，原子性取决于具体的实现，在 X86 架构 64 位 JDK 版本中，写 double 或 long 是原子的。另外，针对用 volatile 声明的 int i 变量进行 i++ 操作时是非原子的。这些知识点都在后面的章节有代码进行验证。

3）禁止代码重排序。

下面就让我们依次对这 3 个知识点进行论证与学习。

2.3.1 可见性的测试

关键字 volatile 具有可见性，可见性是指 A 线程更改变量的值后，B 线程马上就能看到更改后的变量的值，提高了软件的灵敏度。

1. 单线程出现死循环

创建测试用的项目 t99，创建 PrintString.java 类，代码如下：

```java
public class PrintString {

    private boolean isContinuePrint = true;

    public boolean isContinuePrint() {
        return isContinuePrint;
```

```java
    }

    public void setContinuePrint(boolean isContinuePrint) {
        this.isContinuePrint = isContinuePrint;
    }

    public void printStringMethod() {
        try {
            while (isContinuePrint == true) {
                System.out.println("run printStringMethod threadName="
                        + Thread.currentThread().getName());
                Thread.sleep(1000);
            }
        } catch (InterruptedException e) {
            // TODO Auto-generated catch block
            e.printStackTrace();
        }
    }
}
```

运行类 Run.java 代码如下：

```java
public class Run {

    public static void main(String[] args) {
        PrintString printStringService = new PrintString();
        printStringService.printStringMethod();
        System.out.println("我要停止它！stopThread="
                + Thread.currentThread().getName());
        printStringService.setContinuePrint(false);
    }

}
```

程序运行后根本停不下来，结果如图 2-70 所示。

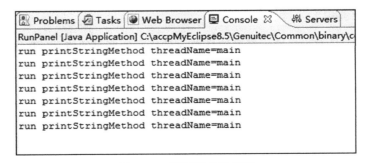

图 2-70 停不下来的程序

程序停不下来的原因主要是 main 线程一直在处理 while() 循环，导致程序不能继续执行后面的代码。可以使用多线程技术解决这种问题。

2. 使用多线程解决死循环

继续创建新的项目 t10，更改 PrintString.java 类，代码如下：

```java
public class PrintString implements Runnable {

    private boolean isContinuePrint = true;

    public boolean isContinuePrint() {
        return isContinuePrint;
    }

    public void setContinuePrint(boolean isContinuePrint) {
        this.isContinuePrint = isContinuePrint;
    }

    public void printStringMethod() {
        try {
            while (isContinuePrint == true) {
                System.out.println("run printStringMethod threadName="
                        + Thread.currentThread().getName());
                Thread.sleep(1000);
            }
        } catch (InterruptedException e) {
            // TODO Auto-generated catch block
            e.printStackTrace();
        }
    }

    @Override
    public void run() {
        printStringMethod();
    }
}
```

运行 Run.java 类代码如下：

```java
public class Run {

    public static void main(String[] args) {
        PrintString printStringService = new PrintString();
        new Thread(printStringService).start();

        System.out.println("我要停止它! stopThread="
                + Thread.currentThread().getName());
        printStringService.setContinuePrint(false);
    }

}
```

程序运行结果如图 2-71 所示。

图 2-71 程序停了下来

3. 使用多线程有可能出现死循环

下面通过一个示例演示使用多线程有可能出现死循环。创建 t16 项目，RunThread.java 类代码如下：

```java
package extthread;

public class RunThread extends Thread {

    private boolean isRunning = true;

    public boolean isRunning() {
        return isRunning;
    }

    public void setRunning(boolean isRunning) {
        this.isRunning = isRunning;
    }

    @Override
    public void run() {
        System.out.println("进入run了");
        while (isRunning == true) {
        }
        System.out.println("线程被停止了！");
    }

}
```

需要注意的是，在 while 语句中一定要执行空循环，里面不要有代码，即使 System.out.println() 输出语句也不要出现。

类 Run.java 代码如下：

```java
package test;

import extthread.RunThread;

public class Run {
    public static void main(String[] args) {
        try {
            RunThread thread = new RunThread();
```

```
            thread.start();
            Thread.sleep(1000);
            thread.setRunning(false);
            System.out.println("已经赋值为false");
        } catch (InterruptedException e) {
            // TODO Auto-generated catch block
            e.printStackTrace();
        }
    }
}
```

在 jdk1.8_161 版本环境中，使用 Eclipse 开发环境运行后的结果有可能不出现死循环，也有可能出现死循环，笔者测试的结果是出现死循环。

如果不出现死循环，则可以在 JVM 中设置当前运行模式为服务器模式，配置的步骤是在 Eclipse 中对 JVM 添加运行参数 -server，即更改运行模式为服务器模式，设置界面如图 2-72 所示。

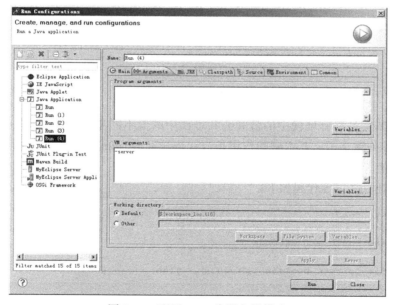

图 2-72　配置 JVM 为服务器模式

单击 Run 按钮后即出现了死循环的效果，如图 2-73 所示。

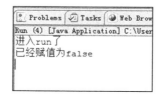

图 2-73　出现了死循环线程且未销毁

代码"System.out.println("线程被停止了！")"永远不会被执行。

4. 使用 volatile 关键字解决多线程出现的死循环

是什么原因造成出现死循环呢？在启动 RunThread.java 线程时，变量" private boolean isRunning = true;"存在于公共堆栈及线程的私有堆栈中，线程运行后一直在线程的私有堆栈中取得的 isRunning 的值是 true，而代码" thread.setRunning(false);"虽然被执行，更新的却是公共堆栈中的 isRunning 变量，改为了 false，操作的是两块内存地址中的数据，所以线程一直处于死循环的状态，内存结构如图 2-74 所示。

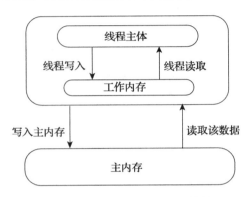

图 2-74　线程具有私有内存结构

这个问题其实是私有堆栈中的值和公共堆栈中的值不同步造成的，解决这样的问题就要使用 volatile 关键字了，其主要的作用是当线程访问 isRunning 这个变量时，强制从公共堆栈中进行取值。

更改 RunThread.java 代码如下：

```java
package extthread;

public class RunThread extends Thread {

    volatile private boolean isRunning = true;

    public boolean isRunning() {
        return isRunning;
    }

    public void setRunning(boolean isRunning) {
        this.isRunning = isRunning;
    }

    @Override
    public void run() {
        System.out.println("进入run了");
        while (isRunning == true) {
```

```
            }
            System.out.println("线程被停止了!");
        }

    }
```

程序运行结果如图 2-75 所示。

由程序运行结果可知,线程被正确停止了。这种方式就是前面章节介绍过的第一种停止线程的方法:**使用退出标志使线程正常退出**。

通过使用 volatile 关键字,可以强制从公共内存中读取变量的值,内存结构如图 2-76 所示。

图 2-75 运行后不出现死循环

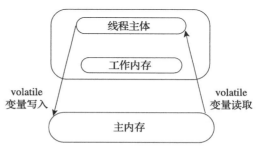

图 2-76 读取公共内存结构

使用 volatile 关键字,可以增加实例变量在多个线程之间的可见性。

5. synchronized 代码块具有增加可见性的作用

关键字 synchronized 可以使多个线程访问同一个资源具有同步性,而且具有使线程工作内存中的私有变量与公共内存中的变量同步的特性,即可见性,下面通过示例对此进行验证。

创建测试用的项目 synchronizedUpdateNewValue,类 Service.java 代码如下:

```
package service;

public class Service {

    private boolean isContinueRun = true;

    public void runMethod() {
        while (isContinueRun == true) {
        }
        System.out.println("停下来了!");
    }

    public void stopMethod() {
        isContinueRun = false;
    }
```

线程类 ThreadA.java 代码如下：

```
package extthread;

import service.Service;

public class ThreadA extends Thread {
    private Service service;

    public ThreadA(Service service) {
        super();
        this.service = service;
    }

    @Override
    public void run() {
        service.runMethod();
    }
}
```

线程类 ThreadB.java 代码如下：

```
package extthread;

import service.Service;

public class ThreadB extends Thread {
    private Service service;

    public ThreadB(Service service) {
        super();
        this.service = service;
    }

    @Override
    public void run() {
        service.stopMethod();
    }
}
```

运行类 Run.java 代码如下：

```
package test;

import service.Service;
import extthread.ThreadA;
import extthread.ThreadB;

public class Run {
```

```java
    public static void main(String[] args) {
        try {
            Service service = new Service();

            ThreadA a = new ThreadA(service);
            a.start();

            Thread.sleep(1000);

            ThreadB b = new ThreadB(service);
            b.start();

            System.out.println("已经发起停止的命令了！");
        } catch (InterruptedException e) {
            // TODO Auto-generated catch block
            e.printStackTrace();
        }
    }
}
```

程序运行后出现死循环，如图 2-77 所示。

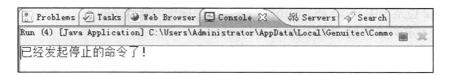

图 2-77　永不变灰的按钮

出现这个结果是由各线程间的数据值没有可视性造成的，关键字 synchronized 具有增加可视性的作用，更改 Service.java 代码如下：

```java
package service;

public class Service {

    private boolean isContinueRun = true;

    public void runMethod() {
        String anyString = new String();
        while (isContinueRun == true) {
            synchronized (anyString) {
            }
        }
        System.out.println("停下来了！");
    }

    public void stopMethod() {
        isContinueRun = false;
```

            }
        }
再次运行程序后，线程正常退出，如图 2-78 所示。

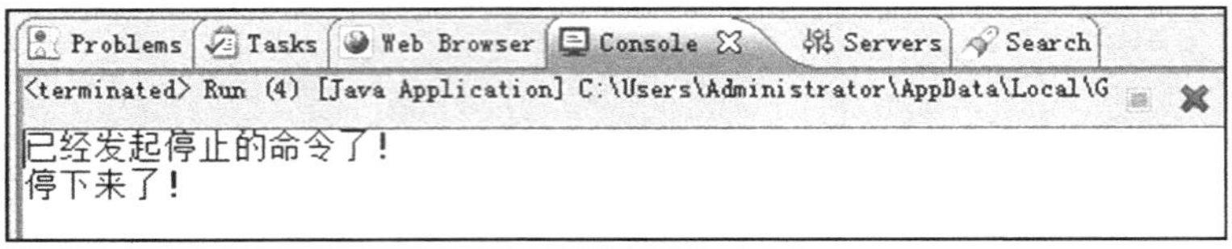

图 2-78　按钮变灰了

2.3.2　原子性的测试

在 32 位系统中，针对未使用 volatile 声明的 long 或 double 数据类型没有实现写原子性，如果想实现，则声明变量时添加 volatile。在 64 位系统中，原子性取决于具体的实现，在 X86 架构 64 位 JDK 版本中，写 double 或 long 是原子的。

另外，volatile 关键字最致命的缺点是不支持原子性，也就是多个线程对用 volatile 修饰的变量 i 执行 i-- 操作时，i-- 操作还会被分解成 3 步，造成非线程安全问题的出现。

1. 在 32 位系统中 long 或 double 数据类型写操作为非原子的

注意，本示例要在 32 位的 JDK 中进行测试。

创建测试用的项目 long_double_32_noATOMIC_test。

创建测试业务类代码如下：

```
package test;

public class MyService {
    public long i;
}
```

创建线程 A 代码如下：

```
package test;

public class MyThreadA extends Thread {
    private MyService service;

    public MyThreadA(MyService service) {
        super();
        this.service = service;
    }

    @Override
    public void run() {
        while (true) {
            service.i = 1;
```

            }
        }
    }

创建线程 B 代码如下:

```java
package test;

public class MyThreadB extends Thread {
    private MyService service;

    public MyThreadB(MyService service) {
        super();
        this.service = service;
    }

    @Override
    public void run() {
        while (true) {
            service.i = -1;
        }
    }
}
```

运行类代码如下:

```java
package test;

public class Test {
    public static void main(String[] args) throws InterruptedException {
        MyService service = new MyService();
        MyThreadA a = new MyThreadA(service);
        MyThreadB b = new MyThreadB(service);
        a.start();
        b.start();
        Thread.sleep(1000);
        System.out.println("long  1 二进制值是: " + Long.toBinaryString(1));
        System.out.println("long -1 二进制值是: " + Long.toBinaryString(-1));
        while (true) {
            long getValue = service.i;
            if (getValue != 1 && getValue != -1) {
                System.out.println(" i的值是: " + Long.toBinaryString(getValue) +
                        " 十进制是: " + getValue);
                System.exit(0);
            }
        }
    }
}
```

程序运行结果如下:

```
long  1 二进制值是: 1
```

```
long -1 二进制值是：11111111111111111111111111111111111111111111111111111111111111111111
i的值是：11111111111111111111111111111111 十进制是：4294967295
```

程序运行后变量 i 的值既不是 1 也不是 –1，是 4294967295，说明在 32 位系统中对 long 或 double 数据类型进行写操作是非原子的。

2. 使用 volatile 解决在 32 位系统中 long 或 double 数据类型写操作为非原子的

更改业务类代码如下：

```
package test;

public class MyService {
    volatile public long i;
}
```

在变量前加入 volatile 关键字，使在 32 位系统中对 long 或 double 数据类型进行写操作时为原子的，程序再次运行后控制台并未输出信息，说明 i 的值不是 1，就是 –1。

如下代码在 64 位系统中进行写操作是原子的，实验代码在项目 long_double_64_ATOMIC_test 中。

```
package test;

public class MyService {
    public long i;
}
```

3. 关键字 volatile int i++ 非原子的特性

使用多线程执行 volatile int i++ 赋值操作时为非原子的，下面用项目来进行测试。

创建项目 volatileTestThread，文件 MyThread.java 代码如下：

```
package extthread;

public class MyThread extends Thread {
    volatile public static int count;

    private static void addCount() {
        for (int i = 0; i < 100; i++) {
            count++;
        }
        System.out.println("count=" + count);
    }

    @Override
    public void run() {
        addCount();
    }
}
```

文件 Run.java 代码如下：

```java
package test.run;

import extthread.MyThread;

public class Run {
    public static void main(String[] args) {
        MyThread[] mythreadArray = new MyThread[100];
        for (int i = 0; i < 100; i++) {
            mythreadArray[i] = new MyThread();
        }

        for (int i = 0; i < 100; i++) {
            mythreadArray[i].start();
        }
    }
}
```

程序运行结果如图 2-79 所示。

运行结果不是 10000，说明在多线程环境下，volatile public static int count++ 操作是非原子的。

更改自定义线程类 MyThread.java 文件代码如下：

```java
package extthread;

public class MyThread extends Thread {
    volatile public static int count;
    //注意一定要添加static关键字
    //这样synchronized与static锁的内容就是MyThread.class类了
    //也就达到同步的效果了
    synchronized private static void addCount() {
        for (int i = 0; i < 100; i++) {
            count++;
        }
        System.out.println("count=" + count);
    }

    @Override
    public void run() {
        addCount();
    }
}
```

图 2-79　运行结果不是 10000

程序运行结果如图 2-80 所示。

在本示例中，如果在方法 private static void addCount() 前加入 synchronized 同步关键字，那么就没有必要再使用 volatile 关键字来声明 count 变量了。

关键字 volatile 使用的主要场合是在多个线程中可以感知实例变量被更改了，并且可以获得最新的值时，也就是可用于增加可见性/可视性。

关键字 volatile 提示线程每次从共享内存中去读取变量，而不是从私有内存中去读取，这样就保证了同步数据的可见性。但这里需要注意的是，如果修改实例变量中的数据，如 i++，即 i=i+1，则这样的操作其实并不是一个原子操作，也就是非线程安全的。表达式 i++ 的操作步骤分解如下：

1）从内存中取出 i 的值；
2）计算 i 的值；
3）将 i 的值写到内存中。

假如在第 2 步计算 i 的值时，另外一个线程也修改 i 的值，那么这个时候就会出现脏数据，解决的办法是使用 synchronized 关键字，这个知识点在前面已经介绍过了。所以，volatile 本身并不处理 int i++ 操作的原子性。

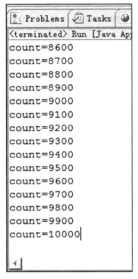

图 2-80　正确的程序运行结果

4. 使用 Atomic 原子类进行 i++ 操作实现原子性

除了在 i++ 操作时使用 synchronized 关键字实现同步外，还可以使用 AtomicInteger 原子类实现原子性。

原子操作是不能分割的整体，没有其他线程能够中断或检查处于原子操作中的变量。一个原子（atomic）类型就是一个原子操作可用的类型，它可以在没有锁（lock）的情况下做到线程安全（thread-safe）。

创建测试项目 AtomicIntegerTest，文件 AddCountThread.java 代码如下：

```java
package extthread;

import java.util.concurrent.atomic.AtomicInteger;

public class AddCountThread extends Thread {
    private AtomicInteger count = new AtomicInteger(0);

    @Override
    public void run() {
        for (int i = 0; i < 10000; i++) {
            System.out.println(count.incrementAndGet());
        }
    }
}
```

文件 Run.java 代码如下：

```java
package test;

import extthread.AddCountThread;
```

```
public class Run {

    public static void main(String[] args) {
        AddCountThread countService = new AddCountThread();

        Thread t1 = new Thread(countService);
        t1.start();

        Thread t2 = new Thread(countService);
        t2.start();

        Thread t3 = new Thread(countService);
        t3.start();

        Thread t4 = new Thread(countService);
        t4.start();

        Thread t5 = new Thread(countService);
        t5.start();

    }

}
```

程序运行结果如图 2-81 所示。

5. 出现逻辑混乱与解决

在有逻辑性的情况下，原子类的输出结果具有随机性。

创建测试用的项目 atomicIntergerNoSafe，类 MyService.java 代码如下：

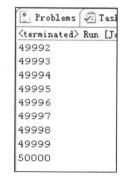

图 2-81　成功累加到 50000

```
package service;

import java.util.concurrent.atomic.AtomicLong;

public class MyService {

    public static AtomicLong aiRef = new AtomicLong();

    public void addNum() {
        System.out.println(Thread.currentThread().getName() + "加了100之后的值是:"
                + aiRef.addAndGet(100));
        aiRef.addAndGet(1);
    }

}
```

类 MyThread.java 代码如下：

```
package extthread;
```

```java
import service.MyService;

public class MyThread extends Thread {
    private MyService mySerivce;

    public MyThread(MyService mySerivce) {
        super();
        this.mySerivce = mySerivce;
    }

    @Override
    public void run() {
        mySerivce.addNum();
    }

}
```

类 Run.java 代码如下：

```java
package test.run;

import service.MyService;
import extthread.MyThread;

public class Run {

    public static void main(String[] args) {
        try {
            MyService service = new MyService();

            MyThread[] array = new MyThread[5];
            for (int i = 0; i < array.length; i++) {
                array[i] = new MyThread(service);
            }
            for (int i = 0; i < array.length; i++) {
                array[i].start();
            }
            Thread.sleep(1000);
            System.out.println(service.aiRef.get());
        } catch (InterruptedException e) {
            // TODO Auto-generated catch block
            e.printStackTrace();
        }
    }
}
```

取消选中 Limit console output 复选框，如图 2-82 所示。

程序运行结果如图 2-83 所示。

输出顺序出错了，应该每加 1 次 100 再加 1 次 1，出现这种情况是因为 addAndGet() 方法是原子的，但方法和方法之间的调用不是原子的，解决这种问题必须用同步了。

第 2 章 对象及变量的并发访问 ❖ 175

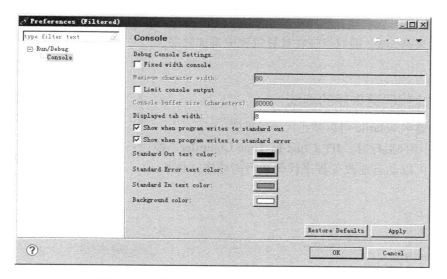

图 2-82 取消选中 Limit console output 复选框

图 2-83 结果值是正确的但顺序出错了

更改 MyService.java 文件代码如下：

```java
package service;

import java.util.concurrent.atomic.AtomicLong;

public class MyService {

    public static AtomicLong aiRef = new AtomicLong();

    synchronized public void addNum() {
        System.out.println(Thread.currentThread().getName() + "加了100之后的值是:"+
            aiRef.addAndGet(100));
        aiRef.addAndGet(1);
    }

}
```

程序运行结果如图 2-84 所示。

从运行结果可以看到，程序实现了依次加 100 再加 1，这就是我们想要得到的正确的计算过程，结果是 505，同时保证在计算过程中累加顺序也是正确的。

2.3.3 禁止代码重排序的测试

使用关键字 volatile 可以禁止代码重排序。

在 Java 程序运行时，JIT（Just-In-Time Compiler，即时编译器）可以动态地改变程序代码运行的顺序，例如，有如下代码：

图 2-84　结果正确

```
A代码-重耗时
B代码-轻耗时
C代码-重耗时
D代码-轻耗时
```

在多线程的环境中，JIT 有可能进行代码重排，重排序后的代码顺序有可能如下：

```
B代码-轻耗时
D代码-轻耗时
A代码-重耗时
C代码-重耗时
```

这样做的主要原因是 CPU 流水线是同时执行这 4 个指令的，那么轻耗时的代码在很大程度上先执行完，以让出 CPU 流水线资源给其他指令，所以代码重排序是为了追求更高的程序运行效率。

重排序发生在没有依赖关系时，例如，对于上面的 A、B、C、D 代码，B、C、D 代码不依赖 A 代码的结果，C、D 代码不依赖 A、B 代码的结果，D 代码不依赖 A、B、C 代码的结果，这种情况下就会发生重排序，如果代码之间有依赖关系，则代码不会重排序。

使用关键字 volatile 可以禁止代码重排序，例如，有如下代码：

```
A变量的操作
B变量的操作
volatile Z变量的操作
C变量的操作
D变量的操作
```

那么会有 4 种情况发生：

1）A、B 可以重排序。

2）C、D 可以重排序。

3）A、B 不可以重排到 Z 的后面。

4）C、D 不可以重排到 Z 的前面。

换言之，变量 Z 是一个"屏障"，Z 变量之前或之后的代码不可以跨越 Z 变量，这就是屏障的作用，关键字 synchronized 具有同样的特性。

1. 实现代码重排序的测试

虽然代码重排序后能提高程序的运行效率，但在有逻辑性的程序中就容易出现一些错误，下面通过示例验证一下。

创建测试用的项目 reorderTest。

创建类代码如下：

```
package test;

public class Test1 {
    private static long x = 0;
    private static long y = 0;
    private static long a = 0;
    private static long b = 0;
    private static long c = 0;
    private static long d = 0;
    private static long e = 0;
    private static long f = 0;

    private static long count = 0;

    public static void main(String[] args) throws InterruptedException {
        for (;;) {
            x = 0;
            y = 0;
            a = 0;
            b = 0;
            c = 0;
            d = 0;
            e = 0;
            f = 0;
            count++;
            Thread t1 = new Thread(new Runnable() {
                public void run() {
                    a = 1;
                    c = 101;
                    d = 102;
                    x = b;
                }
            });

            Thread t2 = new Thread(new Runnable() {
                public void run() {
                    b = 1;
                    e = 201;
                    f = 202;
                    y = a;
                }
            });
            t1.start();
```

```
            t2.start();
            t1.join();
            t2.join();
            String showString = "count=" + count + " " + x + "," + y + "";
            if (x == 0 && y == 0) {
                System.err.println(showString);
                break;
            } else {
                System.out.println(showString);
            }
        }
    }
}
```

程序运行后控制台输出最后的部分结果如下：

```
count=119630 0,1
count=119631 0,1
count=119632 0,1
count=119633 0,1
count=119634 0,1
count=119635 0,0
```

程序输出 x 和 y 都是 0，这时就出现了代码重排序，重排后的顺序如下：

```
x = b;
a = 1;
c = 101;
d = 102;
```

和

```
y = a;
b = 1;
e = 201;
f = 202;
```

结果是 x 和 y 均为 0。

2. 关键字 volatile 之前的代码可以重排

下面通过示例验证 volatile "之前" 的代码可以出现代码重排的效果，重排后的顺序如下：

```
x = b;
a = 1;
c = 101;
d = 102;
```

和

```
y = a;
b = 1;
```

```
e = 201;
f = 202;
```

结果是 x 和 y 均为 0。

程序代码如下：

```
package test;

public class Test2 {
    private static long x = 0;
    private static long y = 0;
    private static long a = 0;
    private static long b = 0;
    private static long c = 0;
    volatile private static long d = 0;
    private static long e = 0;
    volatile private static long f = 0;

    private static long count = 0;

    public static void main(String[] args) throws InterruptedException {
        for (;;) {
            x = 0;
            y = 0;
            a = 0;
            b = 0;
            c = 0;
            d = 0;
            e = 0;
            f = 0;
            count++;
            Thread t1 = new Thread(new Runnable() {
                public void run() {
                    a = 1;
                    c = 101;
                    x = b;
                    d = 102;
                }
            });

            Thread t2 = new Thread(new Runnable() {
                public void run() {
                    b = 1;
                    e = 201;
                    y = a;
                    f = 202;
                }
            });
            t1.start();
            t2.start();
            t1.join();
```

```
            t2.join();
            String showString = "count=" + count + " " + x + "," + y + "";
            if (x == 0 && y == 0) {
                System.err.println(showString);
                break;
            } else {
                System.out.println(showString);
            }
        }
    }
}
```

程序运行结果如下：

```
count=33853 0,1
count=33854 0,1
count=33855 0,1
count=33856 0,1
count=33857 0,1
count=33858 0,0
```

关键字 volatile 之前的代码发生了重排。

3. 关键字 volatile 之后的代码可以重排

下面通过示例验证 volatile 之后的代码可以出现代码重排的效果，重排后的顺序如下：

```
c = 101;
x = b;
a = 1;
d = 102;
```

和

```
e = 201;
y = a;
b = 1;
f = 202;
```

结果是 x 和 y 均为 0。

程序代码如下：

```
package test;

public class Test3 {
    private static long x = 0;
    private static long y = 0;
    private static long a = 0;
    private static long b = 0;
    volatile private static long c = 0;
    private static long d = 0;
```

```java
    volatile private static long e = 0;
    private static long f = 0;

    private static long count = 0;

    public static void main(String[] args) throws InterruptedException {
        for (;;) {
            x = 0;
            y = 0;
            a = 0;
            b = 0;
            c = 0;
            d = 0;
            e = 0;
            f = 0;
            count++;
            Thread t1 = new Thread(new Runnable() {
                public void run() {
                    c = 101;
                    a = 1;
                    d = 102;
                    x = b;
                }
            });

            Thread t2 = new Thread(new Runnable() {
                public void run() {
                    e = 201;
                    b = 1;
                    f = 202;
                    y = a;
                }
            });
            t1.start();
            t2.start();
            t1.join();
            t2.join();
            String showString = "count=" + count + " " + x + "," + y + "";
            if (x == 0 && y == 0) {
                System.err.println(showString);
                break;
            } else {
                System.out.println(showString);
            }
        }
    }
}
```

程序运行结果如下：

count=20898 0,1

```
count=20899 0,1
count=20900 0,1
count=20901 0,1
count=20902 0,0
```

关键字 volatile 之后的代码发生了重排。

4. 关键字 volatile 之前的代码不可以重排到 volatile 之后

下面通过示例验证 volatile 之前的代码不可以重排到 volatile 之后，所以 x 和 y 同时不为 0 的情况不会发生。

程序代码如下：

```java
package test;

public class Test4 {
    private static long x = 0;
    private static long y = 0;
    private static long a = 0;
    private static long b = 0;
    volatile private static long c = 0;
    volatile private static long d = 0;

    private static long count = 0;

    public static void main(String[] args) throws InterruptedException {
        for (;;) {
            x = 0;
            y = 0;
            a = 0;
            b = 0;
            c = 0;
            d = 0;
            count++;
            Thread t1 = new Thread(new Runnable() {
                public void run() {
                    x = b;
                    c = 101;
                    a = 1;
                }
            });

            Thread t2 = new Thread(new Runnable() {
                public void run() {
                    y = a;
                    d = 201;
                    b = 1;
                }
            });
            t1.start();
            t2.start();
```

```
            t1.join();
            t2.join();
            String showString = "count=" + count + " " + x + "," + y + "";
            if (x != 0 && y != 0) {
                System.err.println(showString);
                break;
            } else {
                System.out.println(showString);
            }
        }
    }
}
```

程序运行后，x 和 y 同时不为 0 的情况不会发生，所以关键字 volatile 之前的代码不可以重排到 volatile 之后。

5. 关键字 volatile 之后的代码不可以重排到 volatile 之前

下面通过示例验证 volatile 之后的代码不可以重排到 volatile 之前，所以 x 和 y 的值永远不可能同时为 0。

程序代码如下：

```
package test;

public class Test5 {
    private static long x = 0;
    private static long y = 0;
    private static long a = 0;
    private static long b = 0;
    volatile private static long c = 0;
    volatile private static long d = 0;

    private static long count = 0;

    public static void main(String[] args) throws InterruptedException {
        for (;;) {
            x = 0;
            y = 0;
            a = 0;
            b = 0;
            c = 0;
            d = 0;
            count++;
            Thread t1 = new Thread(new Runnable() {
                public void run() {
                    a = 1;
                    c = 101;
                    x = b;
                }
            });
```

```
            Thread t2 = new Thread(new Runnable() {
                public void run() {
                    b = 1;
                    d = 201;
                    y = a;
                }
            });
            t1.start();
            t2.start();
            t1.join();
            t2.join();
            String showString = "count=" + count + " " + x + "," + y + "";
            if (x == 0 && y == 0) {
                System.err.println(showString);
                break;
            } else {
                System.out.println(showString);
            }
        }
    }
}
```

程序运行后，x 和 y 的值永远不可能同时为 0，所以关键字 volatile 之后的代码不可以重排到 volatile 之前。

6. 关键字 synchronized 之前的代码不可以重排到 synchronized 之后

下面通过示例验证 synchronized 之前的代码不可以重排到 synchronized 之后，所以 x 和 y 同时不为 0 的情况不会发生。

程序代码如下：

```
package test;

public class Test6 {
    private static long x = 0;
    private static long y = 0;
    private static long a = 0;
    private static long b = 0;

    private static long count = 0;

    public static void main(String[] args) throws InterruptedException {
        for (;;) {
            x = 0;
            y = 0;
            a = 0;
            b = 0;
            count++;
            Thread t1 = new Thread(new Runnable() {
                public void run() {
```

```
                x = b;
                synchronized (this) {
                }
                a = 1;
            }
        });

        Thread t2 = new Thread(new Runnable() {
            public void run() {
                y = a;
                synchronized (this) {
                }
                b = 1;
            }
        });
        t1.start();
        t2.start();
        t1.join();
        t2.join();
        String showString = "count=" + count + " " + x + "," + y + "";
        if (x != 0 && y != 0) {
            System.err.println(showString);
            break;
        } else {
            System.out.println(showString);
        }
    }
  }
}
```

程序运行后，x 和 y 同时不为 0 的情况不会发生，所以关键字 synchronized 之前的代码不可以重排到 synchronized 之后。

7. 关键字 synchronized 之后的代码不可以重排到 synchronized 之前

下面通过示例验证 synchronized 之后的代码不可以重排到 synchronized 之前，所以 x 和 y 的值永远不可能同时为 0。

程序代码如下：

```
package test;

public class Test7 {
    private static long x = 0;
    private static long y = 0;
    private static long a = 0;
    private static long b = 0;

    private static long count = 0;

    public static void main(String[] args) throws InterruptedException {
```

```java
        for (;;) {
            x = 0;
            y = 0;
            a = 0;
            b = 0;
            count++;
            Thread t1 = new Thread(new Runnable() {
                public void run() {
                    a = 1;
                    synchronized (this) {
                    }
                    x = b;
                }
            });

            Thread t2 = new Thread(new Runnable() {
                public void run() {
                    b = 1;
                    synchronized (this) {
                    }
                    y = a;
                }
            });
            t1.start();
            t2.start();
            t1.join();
            t2.join();
            String showString = "count=" + count + " " + x + "," + y + "";
            if (x == 0 && y == 0) {
                System.err.println(showString);
                break;
            } else {
                System.out.println(showString);
            }
        }
    }
}
```

程序运行后，x 和 y 的值永远不可能同时为 0，所以关键字 synchronized 之后的代码不可以重排到 synchronized 之前。

8. 总结

关键字 synchronized 的主要作用是保证同一时刻，只有一个线程可以执行某一个方法，或是某一个代码块，synchronized 可以修饰方法及代码块。随着 JDK 的版本升级，synchronized 关键字在执行效率上得到很大提升。它包含三个特征。

1）可见性：synchronized 具有可见性。

2）原子性：使用 synchronized 实现了同步，同步实现了原子性，保证被同步的代码段在同一时间只有一个线程在执行。

3）禁止代码重排序：synchronized 禁止代码重排序。

关键字 volatile 的主要作用是让其他线程可以看到最新的值，volatile 只能修饰变量。它包含三个特征：

1）可见性：B 线程能马上看到 A 线程更改的数据。

2）原子性：在 32 位系统中，针对未使用 volatile 声明的 long 或 double 数据类型没有实现写原子性，如果想实现，则声明变量时添加 volatile，而在 64 位系统中，原子性取决于具体的实现，在 X86 架构 64 位 JDK 版本中，写 double 或 long 是原子的。另外，针对用 volatile 声明的 int i 变量进行 i++ 操作时是非原子的。

3）禁止代码重排序。

学习多线程与并发，要着重"外炼互斥，内修可见，内功有序"，这是掌握多线程、学习多线程和并发技术的重要知识点。

关键字 volatile 和 synchronized 的使用场景总结如下：

1）当想实现一个变量的值被更改时，让其他线程能取到最新的值时，就要对变量使用 volatile。

2）当多个线程对同一个对象中的同一个实例变量进行操作时，为了避免出现非线程安全问题，就要使用 synchronized。

2.4 本章小结

通过对本章的学习，我们对关键字 synchronized 的使用不再陌生，知道什么时候使用它，它解决哪些问题，这是开发的重点；学习完同步后就可以有效控制线程间处理数据的顺序性，以及对处理后的数据进行有效值的保证，更好地获知线程执行结果的预期性。

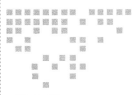

第 3 章
线程间通信

线程是操作系统中独立的个体，但这些个体如果不经过特殊的处理是不能成为一个整体的，线程间的通信是使这些独立的个体成为整体的必用方案之一，可以说，线程间进行通信后系统之间的交互性会更强大，在大大提高 CPU 利用率的同时也会使程序员对各线程任务的处理过程进行有效把控与监督。本章着重掌握的技术点如下：

- 如何使用 wait/notify 实现线程间的通信；
- 生产者 / 消费者模式的实现；
- join 方法的使用；
- ThreadLocal 类的使用。

3.1 wait/notify 机制

线程与线程之间不是独立的个体，它们彼此是可以互相通信和协作的。本节将介绍多个线程之间的通信。

3.1.1 不使用 wait/notify 机制实现线程间通信

创建新的项目，名称为 TwoThreadTransData，使用 sleep() 结合 while(true) 死循环法来实现多个线程间的通信。

类 MyList.java 代码如下：

```
package mylist;
import java.util.ArrayList;
import java.util.List;
```

```java
public class MyList {
    volatile private List list = new ArrayList();
    public void add() {
        list.add("高洪岩");
    }
    public int size() {
        return list.size();
    }
}
```

对变量：

```java
volatile private List list = new ArrayList();
```

添加 volatile 关键字，以实现 A 线程和 B 线程间的可视性。

线程类 ThreadA.java 代码如下：

```java
package extthread;
import mylist.MyList;
public class ThreadA extends Thread {
    private MyList list;
    public ThreadA(MyList list) {
        super();
        this.list = list;
    }
    @Override
    public void run() {
        try {
            for (int i = 0; i < 10; i++) {
                list.add();
                System.out.println("添加了" + (i + 1) + "个元素");
                Thread.sleep(1000);
            }
        } catch (InterruptedException e) {
            e.printStackTrace();
        }
    }
}
```

线程类 ThreadB.java 代码如下：

```java
package extthread;
import mylist.MyList;
public class ThreadB extends Thread {
    private MyList list;
    public ThreadB(MyList list) {
        super();
        this.list = list;
    }
    @Override
    public void run() {
        try {
```

```
            while (true) {
                if (list.size() == 5) {
                    System.out.println("==5了，线程b要退出了！");
                    throw new InterruptedException();
                }
            }
        } catch (InterruptedException e) {
            e.printStackTrace();
        }
    }
}
```

运行类 Test.java 代码如下：

```
package test;
import mylist.MyList;
import extthread.ThreadA;
import extthread.ThreadB;
public class Test {
    public static void main(String[] args) {
        MyList service = new MyList();
        ThreadA a = new ThreadA(service);
        a.setName("A");
        a.start();
        ThreadB b = new ThreadB(service);
        b.setName("B");
        b.start();
    }
}
```

程序运行结果如图 3-1 所示。

图 3-1　两个线程互相通信成功

虽然两个线程间实现了通信，但缺点是线程 ThreadB.java 不停地通过 while 语句轮询机制来检测某一个条件，这样会浪费 CPU 资源。

如果轮询的时间间隔很小，则更浪费 CPU 资源；如果轮询的时间间隔很大，则有可能取不到想要的数据。示例代码如下：

```
@Override
public void run() {
    try {
        while (true) {
            Thread.sleep(2000);
            if (list.size() == 5) {
                    System.out.println("eeeeeeeeeeeeeexit");
                    throw new InterruptedException();
            }
        }
    } catch (InterruptedException e) {
        e.printStackTrace();
    }
}
```

添加 Thread.sleep(2000) 代码后，B 线程并没有退出，所以需要引入一种机制——wait/notify（等待 / 通知）机制，以减少 CPU 资源浪费，还可以实现在多个线程间随时通信。

3.1.2　wait/notify 机制

wait/notify 机制在生活中比比皆是，例如在就餐时就会出现，如图 3-2 所示。

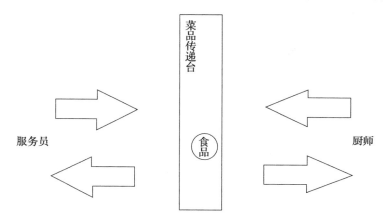

图 3-2　就餐的 wait/notify 机制

厨师和服务员的交互发生在"菜品传递台"上，在这期间需考虑以下几个问题。

1）厨师做完一个菜的时间未定，所以厨师将菜品放到"菜品传递台"上的时间也未定。

2）服务员取到菜的时间取决于厨师，所以服务员就有"等待"（wait）状态。

3）服务员如何取到菜呢，这取决于厨师，厨师将菜放在"菜品传递台"上，其实相当于一种通知（notify），这时服务员才可以拿到菜并交给就餐者。

在这个过程中出现了 wait/notify 机制。

需要说明的是，前面章节中多个线程之间也可以实现通信，原因是多个线程共同访问同一个变量，但那种通信机制不是 wait/notify 机制，两个线程完全是主动式地操作同一个共享变量，在花费读取时间的基础上，读到的值是不是想要的，并不能完全确定，例如，前面示例中添加了 "Thread.sleep(2000);" 代码，导致 B 线程不能退出，所以现在迫切需要引入 wait/notify 机制来满足上面的需求。

3.1.3 wait/notify 机制的原理

拥有相同锁的线程才可以实现 wait/notify 机制，所以后面的描述中都是假定操作同一个锁。

wait() 方法是 Object 类的方法，它的作用是使当前执行 wait() 方法的线程等待，在 wait() 所在的代码行处暂停执行，并释放锁，直到接到通知或被中断为止。在调用 wait() 之前，线程必须获得该对象的对象级别锁，即只能在同步方法或同步块中调用 wait() 方法。通过通知机制使某个线程继续执行 wait() 方法后面的代码时，对线程的选择是按照执行 wait() 方法的顺序确定的，并需要重新获得锁。如果调用 wait() 时没有持有适当的锁，则抛出 IllegalMonitorStateException，它是 RuntimeException 的一个子类，因此不需要 try-catch 语句捕捉异常。

notify() 方法要在同步方法或同步块中调用，即在调用前，线程必须获得锁，如果调用 notify() 时没有持有适当的锁，则会抛出 IllegalMonitorStateException。该方法用来通知那些可能等待该锁的其他线程，如果有多个线程等待，则按照执行 wait() 方法的顺序对处于 wait 状态的线程发出一次通知（notify），并使该线程重新获取锁。需要说明的是，执行 notify() 方法后，当前线程不会马上释放该锁，呈 wait 状态的线程也并不能马上获取该对象锁，要等到执行 notify() 方法的线程将程序执行完，也就是退出 synchronized 同步区域后，当前线程才会释放锁，而呈 wait 状态的线程才可以获取该对象锁。当第一个获得了该对象锁的 wait 线程运行完毕后，它会释放该对象锁，此时如果没有再次使用 notify 语句，那么其他呈 wait 状态的线程因为没有得到通知，会继续处于 wait 状态。

总结：wait() 方法使线程暂停运行，而 notify() 方法通知暂停的线程继续运行。

3.1.4 wait() 方法的基本使用

wait() 方法的作用是使当前线程暂停运行，并释放锁。

创建测试用的 Java 项目，名称为 test1，类 Test1.java 代码如下：

```
package test;

public class Test1 {
    public static void main(String[] args) {
        try {
            String newString = new String("");
```

```
            newString.wait();
        } catch (InterruptedException e) {
            e.printStackTrace();
        }
    }
}
```

程序运行结果如图 3-3 所示。

```
Exception in thread "main" java.lang.IllegalMonitorStateException
        at java.lang.Object.wait(Native Method)
        at java.lang.Object.wait(Object.java:485)
        at test.Test1.main(Test1.java:7)
```

图 3-3　出现异常

出现异常的原因是没有"对象监视器"，即没有锁。

继续创建 Test2.java 文件，代码如下：

```
package test;

public class Test2 {

    public static void main(String[] args) {
        try {
            String lock = new String();
            System.out.println("syn上面");
            synchronized (lock) {
                System.out.println("syn第一行");
                lock.wait();
                System.out.println("wait下的代码！");
            }
            System.out.println("syn下面的代码");
        } catch (InterruptedException e) {
            e.printStackTrace();
        }
    }

}
```

程序运行结果如图 3-4 所示。

线程不能永远等待下去，那样程序就停止不前，不能继续向下运行了，如何使呈 wait 状态的线程继续运行呢？答案就是使用 notify() 方法。

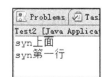

图 3-4　wait() 方法后面的代码不执行了

3.1.5　完整实现 wait/notify 机制

创建实验用的项目，名称为 test2，类 MyThread1.java 代码如下：

```java
package extthread;

public class MyThread1 extends Thread {
    private Object lock;

    public MyThread1(Object lock) {
        super();
        this.lock = lock;
    }

    @Override
    public void run() {
        try {
            synchronized (lock) {
                System.out.println("开始    wait time=" +
                        System.currentTimeMillis());
                lock.wait();
                System.out.println("结束    wait time=" +
                        System.currentTimeMillis());
            }
        } catch (InterruptedException e) {
            // TODO Auto-generated catch block
            e.printStackTrace();
        }
    }
}
```

类 MyThread2.java 代码如下：

```java
package extthread;

public class MyThread2 extends Thread {
    private Object lock;

    public MyThread2(Object lock) {
        super();
        this.lock = lock;
    }

    @Override
    public void run() {
        synchronized (lock) {
            System.out.println("开始notify time=" + System.currentTimeMillis());
            lock.notify();
            System.out.println("结束notify time=" + System.currentTimeMillis());
        }
    }
}
```

类 Test.java 代码如下：

```java
package test;

import extthread.MyThread1;
import extthread.MyThread2;

public class Test {
    public static void main(String[] args) {
        try {
            Object lock = new Object();

            MyThread1 t1 = new MyThread1(lock);
            t1.start();

            Thread.sleep(3000);

            MyThread2 t2 = new MyThread2(lock);
            t2.start();

        } catch (InterruptedException e) {
            e.printStackTrace();
        }
    }
}
```

程序运行结果如图 3-5 所示。

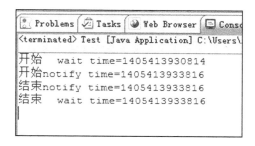

图 3-5　使用 wait / notify 方法的示例

从程序运行结果来看，3s 后线程被通知（notify）唤醒。

3.1.6　使用 wait/notify 机制实现 list.size() 等于 5 时的线程销毁

下面通过一个示例来演示如何使用 wait() 与 notify() 来实现前面 list.size() 值等于 5 时的线程销毁。创建新的项目 wait_notify_size5，类 MyList.java 代码如下：

```java
package extlist;

import java.util.ArrayList;
```

```java
import java.util.List;

public class MyList {

    private static List list = new ArrayList();

    public static void add() {
        list.add("anyString");
    }

    public static int size() {
        return list.size();
    }
}
```

类 ThreadA.java 代码如下：

```java
package extthread;

import extlist.MyList;

public class ThreadA extends Thread {

    private Object lock;

    public ThreadA(Object lock) {
        super();
        this.lock = lock;
    }

    @Override
    public void run() {
        try {
            synchronized (lock) {
                if (MyList.size() != 5) {
                    System.out.println("wait begin "
                            + System.currentTimeMillis());
                    lock.wait();
                    System.out.println("wait end   "
                            + System.currentTimeMillis());
                }
            }
        } catch (InterruptedException e) {
            e.printStackTrace();
        }
    }

}
```

类 ThreadB.java 代码如下：

```java
package extthread;

import extlist.MyList;

public class ThreadB extends Thread {
    private Object lock;

    public ThreadB(Object lock) {
        super();
        this.lock = lock;
    }

    @Override
    public void run() {
        try {
            synchronized (lock) {
                for (int i = 0; i < 10; i++) {
                    MyList.add();
                    if (MyList.size() == 5) {
                        lock.notify();
                        System.out.println("已发出通知！");
                    }
                    System.out.println("添加了" + (i + 1) + "个元素!");
                    Thread.sleep(1000);
                }
            }
        } catch (InterruptedException e) {
            e.printStackTrace();
        }
    }
}
```

类 Run.java 代码如下：

```java
package test;

import extthread.ThreadA;
import extthread.ThreadB;

public class Run {

    public static void main(String[] args) {

        try {
            Object lock = new Object();

            ThreadA a = new ThreadA(lock);
            a.start();

            Thread.sleep(50);
```

```
            ThreadB b = new ThreadB(lock);
            b.start();
        } catch (InterruptedException e) {
            e.printStackTrace();
        }

    }

}
```

程序运行结果如图 3-6 所示。

日志信息 wait end 在最后输出，这说明 notify() 方法执行后并不立即释放锁，这个知识点在后面章节会进行详细介绍。

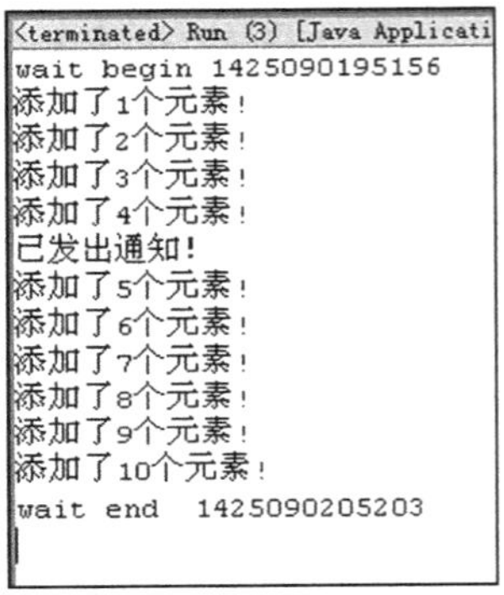

关键字 synchronized 可以将任何一个 Object 对象作为锁来看待，而 Java 为每个 Object 都实现了 wait() 和 notify() 方法，它们必须用在被 synchronized 同步的 Object 的临界区内。通过调用 wait() 方法可以使处于临界区内的线程进入等待状态，同时释放被同步对象的锁，而 notify 操作可以唤醒一个因调用了 wait 操作而处于 wait 状态中的线程，使其进入就绪状态，被重新唤醒的线程会试图重新获得临界区的控制权，也就是锁，并继续执行临界区内 wait 之后的代码。如果发出 notify 操作时没有处于 wait 状态中的线程，那么该命令会被忽略。

图 3-6 程序运行结果

wait() 方法可以使调用该方法的线程释放锁，然后从运行状态转换成 wait 状态，等待被唤醒。

notify() 方法按照执行 wait() 方法的顺序唤醒等待同一锁的"一个"线程，使其进入可运行状态，即 notify() 方法仅通知"一个"线程。

notifyAll() 方法执行后，会按照执行 wait() 方法相反的顺序依次唤醒全部的线程。

3.1.7　对业务代码进行封装

前面示例是在自定义的 MyThread 类中处理业务，而业务代码要尽量放在 Service 业务类中进行处理，这样代码更加标准。

创建测试用的项目 wait_notify_service。

创建类 MyList.java 代码如下：

```
package test;

import java.util.ArrayList;
import java.util.List;

public class MyList {
    volatile private List list = new ArrayList();
```

```java
    public void add() {
        list.add("anyString");
    }

    public int size() {
        return list.size();
    }

}
```

创建类 MyService.java 代码如下：

```java
package test;

public class MyService {

    private Object lock = new Object();
    private MyList list = new MyList();

    public void waitMethod() {
        try {
            synchronized (lock) {
                if (list.size() != 5) {
                    System.out.println(
                            "begin wait " + System.currentTimeMillis() + " "
                                    + Thread.currentThread().getName());
                    lock.wait();
                    System.out.println(
                            " end wait " + System.currentTimeMillis() + " "
                                    + Thread.currentThread().getName());
                }
            }
        } catch (InterruptedException e) {
            e.printStackTrace();
        }
    }

    public void notifyMethod() {
        try {
            synchronized (lock) {
                System.out.println("begin notify " + System.currentTimeMillis() + " "
                        + Thread.currentThread().getName());
                for (int i = 0; i < 10; i++) {
                    list.add();
                    if (list.size() == 5) {
                        lock.notify();
                        System.out.println("仅仅是发出通知而已，wait后面的代码并没有立即执行，因为锁没有释放");
                    }
                    System.out.println("add次数: " + (i + 1));
                    Thread.sleep(1000);
```

```
                }
                System.out.println(" end notify " + System.currentTimeMillis() + " "
                        + Thread.currentThread().getName());
            }
        } catch (InterruptedException e) {
            e.printStackTrace();
        }
    }

}
```

创建类 MyThreadA.java 和 MyThreadB.java 代码如下:

```
package test;

public class MyThreadA extends Thread {

    private MyService service;

    public MyThreadA(MyService service) {
        super();
        this.service = service;
    }

    @Override
    public void run() {
        service.waitMethod();
    }

}

package test;

public class MyThreadB extends Thread {
    private MyService service;

    public MyThreadB(MyService service) {
        super();
        this.service = service;
    }

    @Override
    public void run() {
        service.notifyMethod();
    }
}
```

创建类 Test.java 代码如下:

```
package test;

public class Test {
```

```
    public static void main(String[] args) throws InterruptedException {
        MyService service = new MyService();
        MyThreadA t1 = new MyThreadA(service);
        t1.start();
        Thread.sleep(5000);
        MyThreadB t2 = new MyThreadB(service);
        t2.start();
    }
}
```

控制台运行结果如下：

```
begin wait 1520220451697 Thread-0
begin notify 1520220456699 Thread-1
add次数：1
add次数：2
add次数：3
add次数：4
仅仅是发出通知而已，wait后面的代码并没有立即执行，因为锁没有释放
add次数：5
add次数：6
add次数：7
add次数：8
add次数：9
add次数：10
    end notify 1520220466700 Thread-1
    end wait 1520220466700 Thread-0
```

3.1.8 线程状态的切换

前面的章节已经介绍了与 Thread 有关的大部分 API，这些 API 可以改变线程对象的状态，如图 3-7 所示。

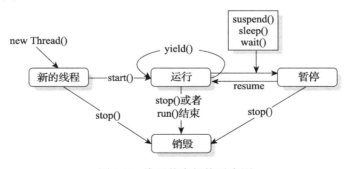

图 3-7　线程状态切换示意图

1）创建一个新的线程对象后，调用它的 start() 方法，系统会为此线程分配 CPU 资源，此时线程处于 runnable（可运行）状态，这是一个准备运行的阶段。如果线程抢占到 CPU 资源，则此线程就处于 running（运行）状态。

2）runnable 状态和 running 状态可相互切换，因为有可能线程运行一段时间后，其他高优先级的线程抢占了 CPU 资源，这时此线程就从 running 状态变成 runnable 状态。

线程进入 runnable 状态大体分为如下 4 种情况。

- 调用 sleep() 方法后经过的时间超过了指定的休眠时间；
- 线程成功获得了试图同步的监视器；
- 线程正在等待某个通知，其他线程发出了通知；
- 处于挂起状态的线程调用了 resume 恢复方法。

3）blocked 是阻塞的意思，例如，如果遇到了一个 I/O 操作，此时当前线程由 runnable 运行状态转成 blocked 阻塞状态，等待 I/O 操作的结果。这时操作系统会把宝贵的 CPU 时间片分配给其他线程，当 I/O 操作结束后，线程由 blocked 状态结束，进入 runnable 状态，线程会继续运行后面的任务。

出现阻塞的情况大体分为如下 5 种。

- 线程调用 sleep() 方法，主动放弃占用的处理器资源。
- 线程调用了阻塞式 I/O 方法，在该方法返回前，该线程被阻塞。
- 线程试图获得一个同步监视器，但该同步监视器正被其他线程所持有。
- 线程等待某个通知（notify）。
- 程序调用了 suspend() 方法将该线程挂起。此方法容易导致死锁，应尽量避免使用该方法。

4）run() 方法运行结束后进入销毁阶段，整个线程执行完毕。

3.1.9 wait() 方法：立即释放锁

执行 wait() 方法后，锁被立即释放。

创建实验用的项目，名称为 waitReleaseLock，类 Service.java 代码如下：

```java
package service;

public class Service {

    public void testMethod(Object lock) {
        try {
            synchronized (lock) {
                System.out.println("begin wait()");
                lock.wait();
                System.out.println("  end wait()");
            }
        } catch (InterruptedException e) {
            e.printStackTrace();
        }
    }
}
```

两个自定义线程类代码如图 3-8 所示。

```
package extthread;

import service.Service;

public class ThreadA extends Thread {
    private Object lock;

    public ThreadA(Object lock) {
        super();
        this.lock = lock;
    }

    @Override
    public void run() {
        Service service = new Service();
        service.testMethod(lock);
    }
}
```

```
package extthread;

import service.Service;

public class ThreadB extends Thread {
    private Object lock;

    public ThreadB(Object lock) {
        super();
        this.lock = lock;
    }

    @Override
    public void run() {
        Service service = new Service();
        service.testMethod(lock);
    }
}
```

图 3-8　两个自定义线程类代码

运行类 Test.java 代码如下：

```
package test;

import extthread.ThreadA;
import extthread.ThreadB;

public class Test {

    public static void main(String[] args) {

        Object lock = new Object();

        ThreadA a = new ThreadA(lock);
        a.start();

        ThreadB b = new ThreadB(lock);
        b.start();

    }
}
```

程序运行结果如图 3-9 所示。

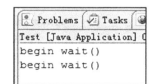

图 3-9　wait() 方法自动将锁释放

3.1.10　sleep() 方法：不释放锁

如果将 wait() 方法改成 sleep() 方法，就获得了同步的效果，因为 sleep() 方法不释放锁，如图 3-10 所示。

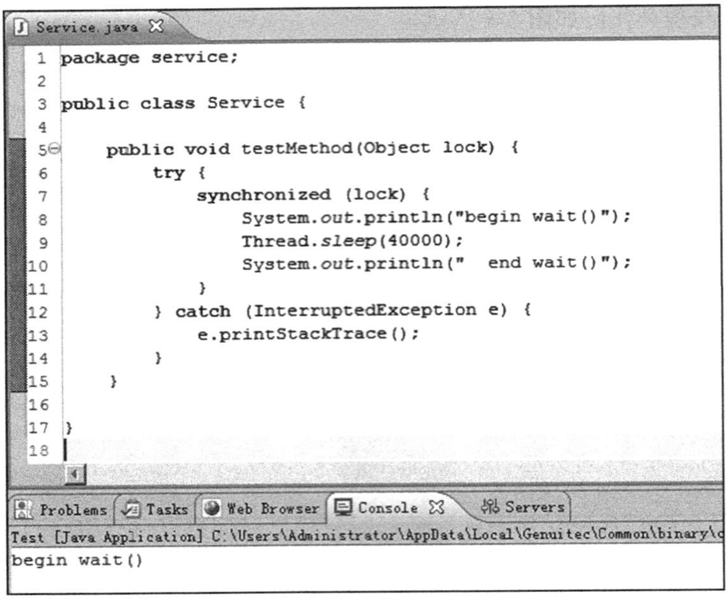

```
package service;

public class Service {

    public void testMethod(Object lock) {
        try {
            synchronized (lock) {
                System.out.println("begin wait()");
                Thread.sleep(40000);
                System.out.println("  end wait()");
            }
        } catch (InterruptedException e) {
            e.printStackTrace();
        }
    }
}
```

```
begin wait()
```

图 3-10 sleep() 方法不释放锁

3.1.11 notify() 方法：不立即释放锁

执行 notify() 方法后，不立即释放锁，下面通过示例来验证。

创建新的项目 notifyHoldLock，类 MyService.java 代码如下：

```java
package service;

public class MyService {

    private Object lock = new Object();

    public void waitMethod() {
        try {
            synchronized (lock) {
                System.out.println("begin wait() ThreadName=" +
                        Thread.currentThread().getName() + " time="
                        + System.currentTimeMillis());
                lock.wait();
                System.out.println("  end wait() ThreadName=" +
                        Thread.currentThread().getName() + " time="
                        + System.currentTimeMillis());
            }
        } catch (InterruptedException e) {
            e.printStackTrace();
        }
    }
}
```

```java
    public void notifyMethod() {
        try {
            synchronized (lock) {
                System.out.println("begin notify() ThreadName=" +
                        Thread.currentThread().getName() + " time="
                        + System.currentTimeMillis());
                lock.notify();
                Thread.sleep(5000);
                System.out.println("  end notify() ThreadName=" +
                        Thread.currentThread().getName() + " time="
                        + System.currentTimeMillis());
            }
        } catch (InterruptedException e) {
            e.printStackTrace();
        }
    }
}
```

类 MyThreadA.java 和 MyThreadB.java 代码如下：

```java
package extthread;

import service.MyService;

public class MyThreadA extends Thread {
    private MyService myService;

    public MyThreadA(MyService myService) {
        super();
        this.myService = myService;
    }

    @Override
    public void run() {
        myService.waitMethod();
    }
}
```

```java
package extthread;

import service.MyService;

public class MyThreadB extends Thread {
    private MyService myService;

    public MyThreadB(MyService myService) {
        super();
        this.myService = myService;
    }
```

```java
    @Override
    public void run() {
        myService.nctifyMethod();
    }

}
```

类 Test.java 代码如下：

```java
package test;

import extthread.MyThreadA;
import extthread.MyThreadB;
import service.MyService;

public class Test {

    public static void main(String[] args) throws InterruptedException {

        MyService myService = new MyService();

        MyThreadA a = new MyThreadA(myService);
        a.start();

        Thread.sleep(50);

        MyThreadB b = new MyThreadB(myService);
        b.start();

    }

}
```

程序运行结果如图 3-11 所示。

```
begin wait() ThreadName=Thread-0 time=1520221695524
begin notify() ThreadName=Thread-1 time=1520221695576
  end notify() ThreadName=Thread-1 time=1520221700576
  end wait() ThreadName=Thread-0 time=1520221700576
```

图 3-11　notify() 方法执行后锁不释放

通过对控制台输出的时间的分析，可以总结出：必须执行完 notify() 方法所在的同步 synchronized 代码块后才释放锁。

3.1.12　interrupt() 方法遇到 wait() 方法

当线程调用 wait() 方法后，再对该线程对象执行 interrupt() 方法会出现 Interrupted-Exception 异常。

创建测试用的项目 waitInterruptException，类 Service.java 代码如下：

```java
package service;

public class Service {

    public void testMethod(Object lock) {
        try {
            synchronized (lock) {
                System.out.println("begin wait()");
                lock.wait();
                System.out.println("  end wait()");
            }
        } catch (InterruptedException e) {
            e.printStackTrace();
            System.out.println("出现异常了，因为呈wait状态的线程被interrupt了！");
        }
    }

}
```

类 ThreadA.java 代码如下：

```java
package extthread;

import service.Service;

public class ThreadA extends Thread {

    private Object lock;

    public ThreadA(Object lock) {
        super();
        this.lock = lock;
    }

    @Override
    public void run() {
        Service service = new Service();
        service.testMethod(lock);
    }

}
```

运行类 Test.java 代码如下：

```java
package test;

import extthread.ThreadA;

public class Test {

    public static void main(String[] args) {
```

```java
        try {
            Object lock = new Object();

            ThreadA a = new ThreadA(lock);
            a.start();

            Thread.sleep(5000);

            a.interrupt();
        } catch (InterruptedException e) {
            e.printStackTrace();
        }

    }

}
```

程序运行结果如图 3-12 所示。

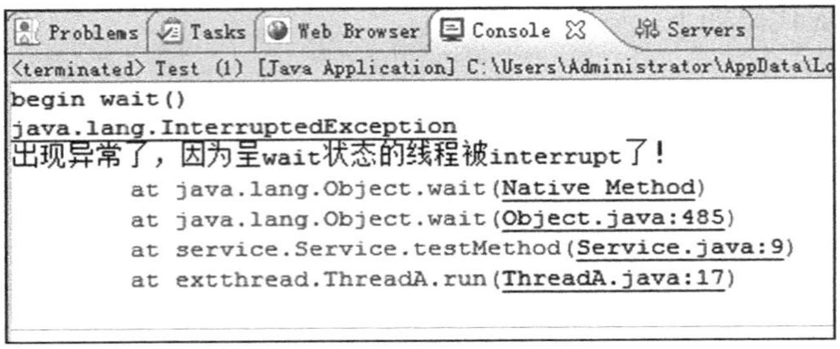

图 3-12 停止 wait 状态下的线程出现异常

通过上面的几个示例可以总结出如下 3 点。

1）执行完 notify() 方法后，按照执行 wait() 方法的顺序唤醒其他线程。notify() 所在的同步代码块执行完才会释放对象的锁，其他线程继续执行 wait() 之后的代码。

2）在执行同步代码块的过程中，遇到异常而导致线程终止，锁也会被释放。

3）在执行同步代码块的过程中，执行了锁所属对象的 wait() 方法，这个线程会释放对象锁，等待被唤醒。

3.1.13 notify() 方法：只通知一个线程

每调用一次 notify() 方法，只通知一个线程进行唤醒，唤醒的顺序与执行 wait() 方法的顺序一致。

创建测试用的项目 notifyOne，类 MyService.java 代码如下：

```java
package test;

public class MyService {
```

```java
    private Object lock = new Object();

    public void waitMethod() {
        try {
            synchronized (lock) {
                System.out.println("begin wait " + System.currentTimeMillis() + " "
                        + Thread.currentThread().getName());
                lock.wait();
                System.out.println("  end wait " + System.currentTimeMillis() + " "
                        + Thread.currentThread().getName());
            }
        } catch (InterruptedException e) {
            e.printStackTrace();
        }
    }

    public void notifyMethod() {
        synchronized (lock) {
            System.out.println("begin notify " + System.currentTimeMillis() + " "
                    + Thread.currentThread().getName());
            lock.notify();
            System.out.println("  end notify " + System.currentTimeMillis() + " "
                    + Thread.currentThread().getName());
        }
    }
}
```

执行 wait() 方法的线程 MyThreadA.java 源代码如下：

```java
package test;

public class MyThreadA extends Thread {

    private MyService service;

    public MyThreadA(MyService service) {
        super();
        this.service = service;
    }

    @Override
    public void run() {
        service.waitMethod();
    }
}
```

创建唤醒线程 MyThreadB.java，代码如下：

```java
package test;
```

```java
public class MyThreadB extends Thread {

    private MyService service;

    public MyThreadB(MyService service) {
        super();
        this.service = service;
    }

    @Override
    public void run() {
        service.notifyMethod();
    }

}
```

运行类 Test.java 代码如下：

```java
package test;

public class Test {
    public static void main(String[] args) throws InterruptedException {
        MyService service = new MyService();

        for (int i = 0; i < 10; i++) {
            MyThreadA t1 = new MyThreadA(service);
            t1.start();
        }

        Thread.sleep(1000);

        MyThreadB t1 = new MyThreadB(service);
        t1.start();
        Thread.sleep(500);
        MyThreadB t2 = new MyThreadB(service);
        t2.start();
        Thread.sleep(500);
        MyThreadB t3 = new MyThreadB(service);
        t3.start();
        Thread.sleep(500);
        MyThreadB t4 = new MyThreadB(service);
        t4.start();
        Thread.sleep(500);
        MyThreadB t5 = new MyThreadB(service);
        t5.start();
        // 一共唤醒5个线程
    }
}
```

程序运行结果如图 3-13 所示。

```
begin wait 1520234674318 Thread-0
begin wait 1520234674319 Thread-5
begin wait 1520234674319 Thread-4
begin wait 1520234674319 Thread-3
begin wait 1520234674319 Thread-2
begin wait 1520234674319 Thread-1
begin wait 1520234674320 Thread-6
begin wait 1520234674320 Thread-8
begin wait 1520234674320 Thread-7
begin wait 1520234674320 Thread-9
begin notify 1520234675376 Thread-10
  end notify 1520234675376 Thread-10
  end wait 1520234675376 Thread-0
begin notify 1520234675876 Thread-11
  end notify 1520234675876 Thread-11
  end wait 1520234675877 Thread-5
begin notify 1520234676376 Thread-12
  end notify 1520234676376 Thread-12
  end wait 1520234676377 Thread-4
begin notify 1520234676876 Thread-13
  end notify 1520234676876 Thread-13
  end wait 1520234676877 Thread-3
begin notify 1520234677376 Thread-14
  end notify 1520234677376 Thread-14
  end wait 1520234677377 Thread-2
```

图 3-13 仅有一个线程被唤醒

notify() 方法仅按照执行 wait() 方法的顺序依次唤醒一个线程，分别是 Thread-0、Thread-5、Thread-4、Thread-3、Thread-2。

3.1.14 notifyAll() 方法：通知所有线程

前面示例通过多次调用 notify() 方法实现 5 个线程被唤醒，但并不能保证系统中仅有 5 个线程，也就是 notify() 方法的调用次数小于线程对象的数量，那么会出现部分线程对象没有被唤醒的情况。为了唤醒全部线程可以使用 notifyAll() 方法。

 注意 notifyAll() 方法会按照执行 wait() 方法的倒序依次对其他线程进行唤醒。

创建测试用的项目 notifyAll，将 notifyOne 项目中的所有文件复制到 notifyAll 项目中，将 MyThreadB.java 类使用的方法更改为 notifyAll()，还要更改 Test.java 代码，如下所示：

```java
package test;

public class Test {
    public static void main(String[] args) throws InterruptedException {
        MyService service = new MyService();

        for (int i = 0; i < 10; i++) {
            MyThreadA t1 = new MyThreadA(service);
            t1.start();
```

```
            }
            Thread.sleep(1000);

            MyThreadB t1 = new MyThreadB(service);
            t1.start();
    }
}
```

程序运行结果如图 3-14 所示。

```
begin wait 1520237239994 Thread-0
begin wait 1520237239994 Thread-3
begin wait 1520237239994 Thread-2
begin wait 1520237239994 Thread-1
begin wait 1520237239995 Thread-8
begin wait 1520237239995 Thread-6
begin wait 1520237239995 Thread-5
begin wait 1520237239995 Thread-4
begin wait 1520237239995 Thread-9
begin wait 1520237239995 Thread-7
begin notify 1520237240994 Thread-10
  end notify 1520237240994 Thread-10
  end wait 1520237240994 Thread-7
  end wait 1520237240994 Thread-9
  end wait 1520237240994 Thread-4
  end wait 1520237240995 Thread-5
  end wait 1520237240995 Thread-6
  end wait 1520237240995 Thread-8
  end wait 1520237240995 Thread-1
  end wait 1520237240995 Thread-2
  end wait 1520237240995 Thread-3
  end wait 1520237240995 Thread-0
```

图 3-14 线程全部被唤醒并呈倒序

唤醒的顺序是正序、倒序、随机，取决于具体的 JVM 实现，不是所有的 JVM 在执行 notify() 时都是按调用 wait() 方法的正序进行唤醒的，也不是所有的 JVM 在执行 notifyAll() 时都是按调用 wait() 方法的倒序进行唤醒的，具体的唤醒顺序依赖于 JVM 的具体实现。

3.1.15 wait(long) 方法的基本使用

带一个参数的 wait(long) 方法的功能是等待某一时间内是否有线程对锁进行 notify() 通知唤醒，如果超过这个时间则线程自动唤醒，能继续向下运行的前提是再次持有锁。

创建测试用的项目 waitHasParamMethod，创建 MyRunnable.java 类代码如下：

```java
package myrunnable;

public class MyRunnable {
    static private Object lock = new Object();
    static private Runnable runnable1 = new Runnable() {
        @Override
        public void run() {
```

```
            try {
                synchronized (lock) {
                    System.out.println("wait begin timer="
                            + System.currentTimeMillis());
                    lock.wait(5000);
                    System.out.println("wait   end timer="
                            + System.currentTimeMillis());
                }
            } catch (InterruptedException e) {
                e.printStackTrace();
            }
        }
    };

    public static void main(String[] args) {
        Thread t = new Thread(runnable1);
        t.start();
    }

}
```

程序运行结果如图 3-15 所示。

图 3-15　5s 后自动被唤醒

也可以在 5s 内由其他线程进行唤醒。

代码更改如下：

```
package myrunnable;

public class MyRunnable {
    static private Object lock = new Object();
    static private Runnable runnable1 = new Runnable() {
        @Override
        public void run() {
            try {
                synchronized (lock) {
                    System.out.println("wait begin timer="
                            + System.currentTimeMillis());
                    lock.wait(5000);
                    System.out.println("wait   end timer="
                            + System.currentTimeMillis());
                }
            } catch (InterruptedException e) {
                e.printStackTrace();
```

```
                }
            }
        };

        static private Runnable runnable2 = new Runnable() {
            @Override
            public void run() {
                synchronized (lock) {
                    System.out.println("notify begin timer="
                            + System.currentTimeMillis());
                    lock.notify();
                    System.out.println("notify   end timer="
                            + System.currentTimeMillis());
                }
            }
        };

        public static void main(String[] args) throws InterruptedException {
            Thread t1 = new Thread(runnable1);
            t1.start();
            Thread.sleep(3000);
            Thread t2 = new Thread(runnable2);
            t2.start();
        }

}
```

程序运行结果如图 3-16 所示。

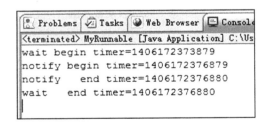

图 3-16 3s 后由其他线程唤醒

在输出日志中，wait begin 的时间尾数为 3879，在 3000ms 后，notify begin 6879 被执行，即在此时间点准备对呈 waiting 状态的线程进行唤醒。

3.1.16 wait(long) 方法自动向下运行需要重新持有锁

wait(long) 方法想要自动向下运行也要持有锁，如果没有锁，则一直在等待，直到持有锁为止。

创建测试用的项目 wait_time_backLock。创建 MyService.java 类代码如下：

```
package test;
```

```java
public class MyService {

    public void testMethod() {
        try {
            synchronized (this) {
                System.out.println("wait begin " +
                        Thread.currentThread().getName() + " "
                        + System.currentTimeMillis());
                wait(5000);
                System.out.println("wait   end " +
                        Thread.currentThread().getName() + " "
                        + System.currentTimeMillis());
            }
        } catch (InterruptedException e) {
            e.printStackTrace();
        }
    }

    synchronized public void longTimeSyn() {
        try {
            Thread.sleep(8000);
        } catch (InterruptedException e) {
            e.printStackTrace();
        }
    }

}
```

创建 MyThreadA.java 类代码如下：

```java
package test;

public class MyThreadA extends Thread {
    private MyService service;

    public MyThreadA(MyService service) {
        super();
        this.service = service;
    }

    @Override
    public void run() {
        service.testMethod();
    }
}
```

创建 MyThreadB.java 类代码如下：

```java
package test;

public class MyThreadB extends Thread {
```

```java
    private MyService service;

    public MyThreadB(MyService service) {
        super();
        this.service = service;
    }

    @Override
    public void run() {
        service.longTimeSyn();
    }
}
```

创建 Test.java 类代码如下：

```java
package test;

import java.io.IOException;

public class Test {
    public static void main(String[] args) throws IOException, InterruptedException {
        MyService service = new MyService();

        MyThreadA[] array = new MyThreadA[10];

        for (int i = 0; i < 10; i++) {
            array[i] = new MyThreadA(service);
        }

        for (int i = 0; i < 10; i++) {
            array[i].start();
        }

        MyThreadB b = new MyThreadB(service);
        b.start();
    }
}
```

程序运行结果如下：

```
wait begin Thread-0 1520403291508
wait begin Thread-9 1520403291508
wait begin Thread-8 1520403291508
wait begin Thread-7 1520403291508
wait begin Thread-6 1520403291508
wait begin Thread-5 1520403291508
wait begin Thread-4 1520403291508
wait begin Thread-3 1520403291508
wait begin Thread-2 1520403291508
wait begin Thread-1 1520403291508
wait   end Thread-7 1520403299509
wait   end Thread-8 1520403299510
```

```
wait    end Thread-1 1520403299510
wait    end Thread-2 1520403299511
wait    end Thread-9 1520403299512
wait    end Thread-5 1520403299512
wait    end Thread-3 1520403299512
wait    end Thread-6 1520403299512
wait    end Thread-4 1520403299513
wait    end Thread-0 1520403299513
```

由程序运行结果可以看出，end 输出的时间是在输出 begin 之后的 8s，并不是 5s。

3.1.17　通知过早问题与解决方法

如果通知过早，则会打乱程序正常的运行逻辑。

创建测试用的项目 firstNotify，类 MyRun.java 代码如下：

```java
package test;

public class MyRun {

    private String lock = new String("");

    private Runnable runnableA = new Runnable() {
        @Override
        public void run() {
            try {
                synchronized (lock) {
                    System.out.println("begin wait");
                    lock.wait();
                    System.out.println("end wait");

                }
            } catch (InterruptedException e) {
                e.printStackTrace();
            }
        }
    };

    private Runnable runnableB = new Runnable() {
        @Override
        public void run() {
            synchronized (lock) {
                System.out.println("begin notify");
                lock.notify();
                System.out.println("end notify");

            }
        }
    };
```

```java
public static void main(String[] args) {

    MyRun run = new MyRun();

    Thread a = new Thread(run.runnableA);
    a.start();

    Thread b = new Thread(run.runnableB);
    b.start();

}

}
```

程序运行结果如图 3-17 所示。

将 main() 方法中的代码更改如下：

```java
public static void main(String[] args) throws InterruptedException {

    MyRun run = new MyRun();

    Thread b = new Thread(run.runnableB);
    b.start();

    Thread.sleep(100);

    Thread a = new Thread(run.runnableA);
    a.start();

}
```

程序运行结果如图 3-18 所示。

图 3-17　程序正常运行结果

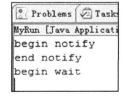

图 3-18　wait() 方法永远不会被通知

如果先通知 wait() 方法了，则 wait() 方法也就没有必要执行了，更改 MyRun.java 代码如下：

```java
package test;

public class MyRun {

    private String lock = new String("");
```

```java
    private boolean isFirstRunB = false;

    private Runnable runnableA = new Runnable() {
        @Override
        public void run() {
            try {
                synchronized (lock) {
                    while (isFirstRunB == false) {
                        System.out.println("begin wait");
                        lock.wait();
                        System.out.println("end wait");
                    }
                }
            } catch (InterruptedException e) {
                e.printStackTrace();
            }
        }
    };

    private Runnable runnableB = new Runnable() {
        @Override
        public void run() {
            synchronized (lock) {
                System.out.println("begin notify");
                lock.notify();
                System.out.println("end notify");
                isFirstRunB = true;
            }
        }
    };

    public static void main(String[] args) throws InterruptedException {

        MyRun run = new MyRun();

        Thread b = new Thread(run.runnableB);
        b.start();

        Thread.sleep(100);

        Thread a = new Thread(run.runnableA);
        a.start();

    }
}
```

程序运行结果如图 3-19 所示。

继续将上面程序中的 main() 方法代码更改如下：

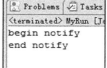

图 3-19　仅仅执行了 notify() 方法

```java
public static void main(String[] args) throws InterruptedException {

    MyRun run = new MyRun();

    Thread a = new Thread(run.runnableA);
    a.start();

    Thread.sleep(100);

    Thread b = new Thread(run.runnableB);
    b.start();

}
```

程序运行结果如图 3-20 所示。

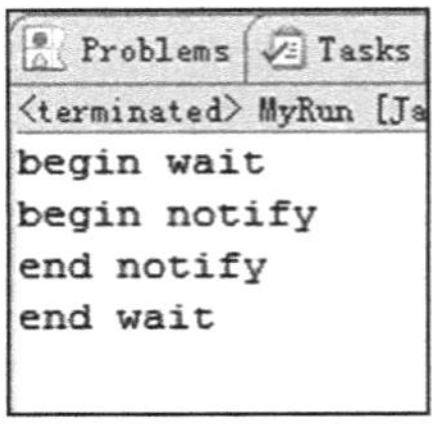

图 3-20 正确的结果

3.1.18 wait 条件发生变化与使用 while 的必要性

在使用 wait/notify 模式时，还需要注意一种情况，即 wait 条件发生变化，容易造成逻辑的混乱。

创建测试用的项目，名称为 waitOld，创建类 Add.java，代码如下：

```java
package entity;

//加法
public class Add {

    private String lock;

    public Add(String lock) {
        super();
        this.lock = lock;
    }

    public void add() {
        synchronized (lock) {
            ValueObject.list.add("anyString");
            lock.notifyAll();
        }
    }

}
```

创建类 Subtract.java，代码如下：

```java
package entity;
```

```java
//减法
public class Subtract {

    private String lock;

    public Subtract(String lock) {
        super();
        this.lock = lock;
    }

    public void subtract() {
        try {
            synchronized (lock) {
                if (ValueObject.list.size() == 0) {
                    System.out.println("wait begin ThreadName="
                            + Thread.currentThread().getName());
                    lock.wait();
                    System.out.println("wait   end ThreadName="
                            + Thread.currentThread().getName());
                }
                ValueObject.list.remove(0);
                System.out.println("list size=" +
                        ValueObject.list.size());
            }
        } catch (InterruptedException e) {
            e.printStackTrace();
        }
    }

}
```

类 ValueObject.java 代码如下：

```java
package entity;

import java.util.ArrayList;
import java.util.List;

public class ValueObject {

    public static List list = new ArrayList();

}
```

两个线程类代码如图 3-21 所示。

```
┌─ ThreadAdd.java ─────────────────┐  ┌─ ThreadSubtract.java ──────────────┐
│ package extthread;               │  │ package extthread;                 │
│                                  │  │                                    │
│ import entity.Add;               │  │ import entity.Subtract;            │
│                                  │  │                                    │
│ public class ThreadAdd extends Thread {│ public class ThreadSubtract extends Thread {│
│     private Add p;               │  │     private Subtract r;            │
│                                  │  │                                    │
│     public ThreadAdd(Add p) {    │  │     public ThreadSubtract(Subtract r) {│
│         super();                 │  │         super();                   │
│         this.p = p;              │  │         this.r = r;                │
│     }                            │  │     }                              │
│                                  │  │                                    │
│     @Override                    │  │     @Override                      │
│     public void run() {          │  │     public void run() {            │
│         p.add();                 │  │         r.subtract();              │
│     }                            │  │     }                              │
│ }                                │  │ }                                  │
└──────────────────────────────────┘  └────────────────────────────────────┘
```

图 3-21　两个线程类代码

类 Run.java 代码如下：

```java
package test;

import entity.Add;
import entity.Subtract;
import extthread.ThreadAdd;
import extthread.ThreadSubtract;

public class Run {

    public static void main(String[] args) throws InterruptedException {

        String lock = new String("");

        Add add = new Add(lock);
        Subtract subtract = new Subtract(lock);

        ThreadSubtract subtract1Thread = new ThreadSubtract(subtract);
        subtract1Thread.setName("subtract1Thread");
        subtract1Thread.start();

        ThreadSubtract subtract2Thread = new ThreadSubtract(subtract);
        subtract2Thread.setName("subtract2Thread");
        subtract2Thread.start();

        Thread.sleep(1000);

        ThreadAdd addThread = new ThreadAdd(add);
        addThread.setName("addThread");
        addThread.start();
```

}
}

程序运行结果如图 3-22 所示。

```
wait begin ThreadName=subtract1Thread
wait begin ThreadName=subtract2Thread
wait   end ThreadName=subtract2Thread
list size=0
wait   end ThreadName=subtract1Thread
Exception in thread "subtract1Thread" java.lang.IndexOutOfBoundsException: Index: 0, Size: 0
    at java.util.ArrayList.RangeCheck(ArrayList.java:547)
    at java.util.ArrayList.remove(ArrayList.java:387)
    at entity.Subtract.subtract(Subtract.java:23)
    at extthread.ThreadSubtract.run(ThreadSubtract.java:16)
```

图 3-22　程序运行结果报异常

出现异常的原因是有两个实现删除 remove() 操作的线程，它们在 "Thread.sleep(1000);" 之前都执行了 wait() 方法，呈等待状态，当加操作的线程在 1s 之后被运行时，通知了所有呈等待状态的减操作线程，那么第一个实现减操作的线程能正确地删除 list 中索引为 0 的数据，但第二个实现减操作的线程则出现索引溢出的异常，因为 list 中仅仅添加了一个数据，也只能删除一个数据，没有第二个数据可供删除，所以出现了 java.lang.IndexOutOfBoundsException 异常。

如何解决这种情况呢？更改 Subtract.java 中的 subtract 方法代码如下：

```java
public void subtract() {
    try {
        synchronized (lock) {
            while (ValueObject.list.size() == 0) {
                System.out.println("wait begin ThreadName="
                        + Thread.currentThread().getName());
                lock.wait();
                System.out.println("wait   end ThreadName="
                        + Thread.currentThread().getName());
            }
            ValueObject.list.remove(0);
            System.out.println("list size=" + ValueObject.list.size());
        }
    } catch (InterruptedException e) {
        e.printStackTrace();
    }
}
```

程序运行结果如图 3-23 所示。

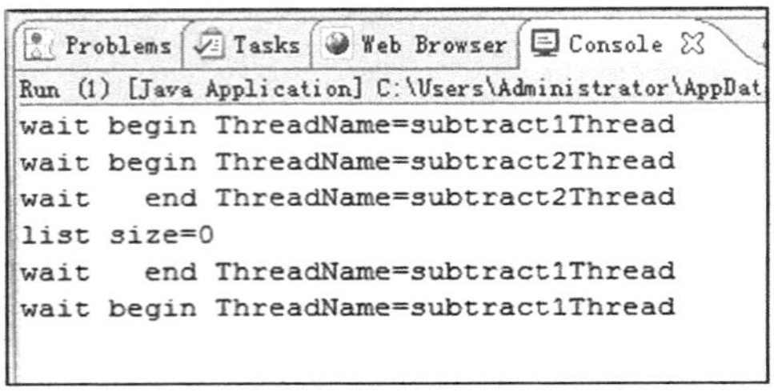

```
Problems  Tasks  Web Browser  Console
Run (1) [Java Application] C:\Users\Administrator\AppDat
wait begin ThreadName=subtract1Thread
wait begin ThreadName=subtract2Thread
wait   end ThreadName=subtract2Thread
list size=0
wait   end ThreadName=subtract1Thread
wait begin ThreadName=subtract1Thread
```

图 3-23 不再出现异常

3.1.19 生产者/消费者模式的实现

wait / notify 模式最经典的案例就是生产者 / 消费者模式，但此模式在使用上有几种"变形"，还有一些注意事项，但原理都是基于 wait/notify 的。

1. 一生产与一消费：操作值

创建名称为 p_r_test 的 Java 项目，创建生产者 P.java 类，代码如下：

```java
package entity;

//生产者
public class P {

    private String lock;

    public P(String lock) {
        super();
        this.lock = lock;
    }

    public void setValue() {
        try {
            synchronized (lock) {
                if (!ValueObject.value.equals("")) {
                    lock.wait();
                }
                String value = System.currentTimeMillis() + "_"
                        + System.nanoTime();
                System.out.println("set的值是" + value);
                ValueObject.value = value;
                lock.notify();
            }
        } catch (InterruptedException e) {
            e.printStackTrace();
```

            }
        }
    }

创建消费者类 C.java,代码如下:

```java
package entity;

//消费者
public class C {

    private String lock;

    public C(String lock) {
        super();
        this.lock = lock;
    }

    public void getValue() {
        try {
            synchronized (lock) {
                if (ValueObject.value.equals("")) {
                    lock.wait();
                }
                System.out.println("get的值是" + ValueObject.value);
                ValueObject.value = "";
                lock.notify();
            }
        } catch (InterruptedException e) {
            e.printStackTrace();
        }
    }

}
```

创建存储值的对象 ValueObject.java,代码如下:

```java
package entity;

public class ValueObject {

    public static String value = "";

}
```

创建两个线程对象,一个是生产者线程,另一个是消费者线程,代码如图 3-24 所示。

```
ThreadP.java
package extthread;

import entity.P;

public class ThreadP extends Thread {

    private P p;

    public ThreadP(P p) {
        super();
        this.p = p;
    }

    @Override
    public void run() {
        while (true) {
            p.setValue();
        }
    }
}
```

```
ThreadC.java
package extthread;

import entity.C;

public class ThreadC extends Thread {

    private C r;

    public ThreadC(C r) {
        super();
        this.r = r;
    }

    @Override
    public void run() {
        while (true) {
            r.getValue();
        }
    }
}
```

图 3-24　线程代码

运行类 Run.java 代码如下：

```
package test;

import entity.P;
import entity.C;
import extthread.ThreadP;
import extthread.ThreadC;

public class Run {

    public static void main(String[] args) {

        String lock = new String("");
        P p = new P(lock);
        C r = new C(lock);

        ThreadP pThread = new ThreadP(p);
        ThreadC rThread = new ThreadC(r);

        pThread.start();
        rThread.start();
    }

}
```

程序运行结果如图 3-25 所示。

本示例是一个生产者和一个消费者进行数据的交互，在控制台中输出的日志 get 和 set 是交替运行的。

图 3-25　程序运行结果（一部分）

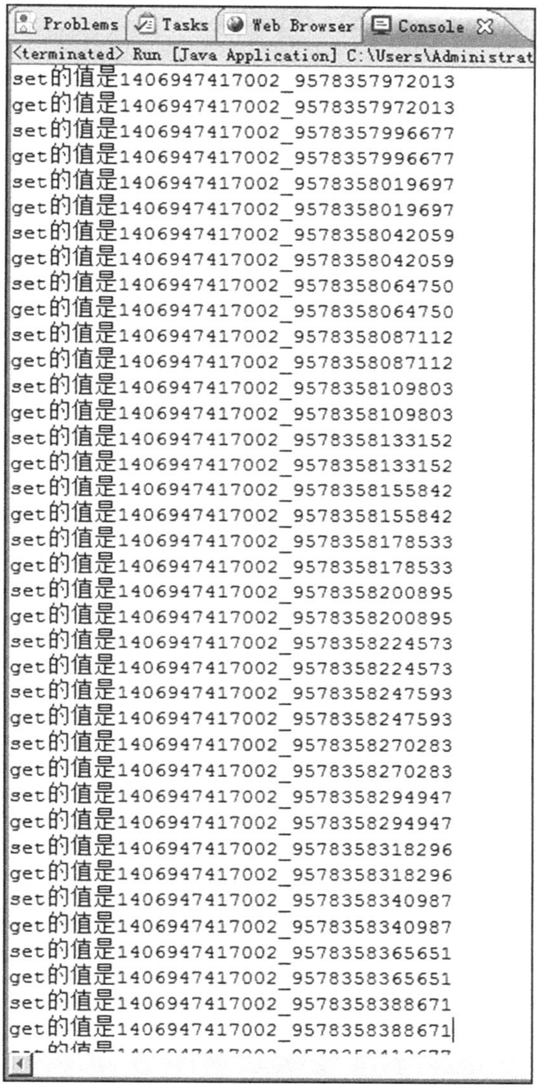

但如果在此示例的基础上设计出多个生产者和多个消费者，那么在运行的过程中极有可能出现"假死"的情况，也就是所有的线程都呈 waiting 状态，关于这个示例请看下面的内容。

2. 多生产与多消费：操作值（假死）

"假死"的现象其实就是线程进入 waiting 状态，如果全部线程都进入 waiting 状态，则程序就不再执行任何业务功能了，整个项目呈停止状态，这在使用生产者/消费者模式时经常遇到。

创建 Java 项目 p_c_allWait，类 P.java 代码如下：

```java
package entity;

//生产者
public class P {

    private String lock;

    public P(String lock) {
        super();
        this.lock = lock;
    }

    public void setValue() {
        try {
            synchronized (lock) {
                while (!ValueObject.value.equals("")) {
                    System.out.println("生产者 "
                            + Thread.currentThread().getName() + " WAITING了★");
                    lock.wait();
                }
                System.out.println("生产者 " + Thread.currentThread().getName()
                        + " RUNNABLE了");
                String value = System.currentTimeMillis() + "_"
                        + System.nanoTime();
                ValueObject.value = value;
                lock.notify();
            }
        } catch (InterruptedException e) {
            e.printStackTrace();
        }
    }

}
```

类 C.java 代码如下：

```java
package entity;

//消费者
public class C {

    private String lock;

    public C(String lock) {
        super();
        this.lock = lock;
    }

    public void getValue() {
```

```
        try {
            synchronized (lock) {
                while (ValueObject.value.equals("")) {
                    System.out.println("消费者 "
                            + Thread.currentThread().getName() + " WAITING了☆");
                    lock.wait();
                }
                System.out.println("消费者 " + Thread.currentThread().getName()
                        + " RUNNABLE了");
                ValueObject.value = "";
                lock.notify();
            }
        } catch (InterruptedException e) {
            e.printStackTrace();
        }
    }
}
```

由于此示例涉及多生产与多消费，所以不再使用 if 语句，而使用 while 语句作为 wait 条件的判断。

线程类和工具类代码如图 3-26 所示。

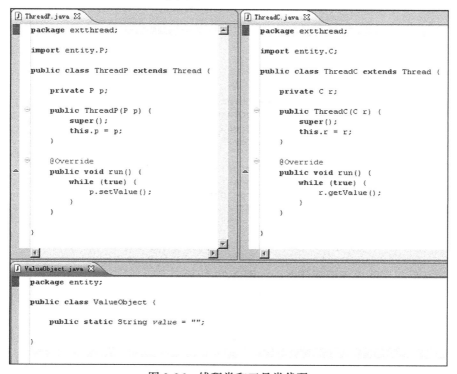

图 3-26　线程类和工具类代码

类 Run.java 代码如下：

```java
package test;

import entity.P;
import entity.C;
import extthread.ThreadP;
import extthread.ThreadC;

public class Run {

    public static void main(String[] args) throws InterruptedException {
        String lock = new String("");
        P p = new P(lock);
        C r = new C(lock);

        ThreadP[] pThread = new ThreadP[2];
        ThreadC[] rThread = new ThreadC[2];

        for (int i = 0; i < 2; i++) {
            pThread[i] = new ThreadP(p);
            pThread[i].setName("生产者" + (i + 1));

            rThread[i] = new ThreadC(r);
            rThread[i].setName("消费者" + (i + 1));

            pThread[i].start();
            rThread[i].start();
        }

        Thread.sleep(5000);
        Thread[] threadArray = new Thread[Thread.currentThread()
                .getThreadGroup().activeCount()];
        Thread.currentThread().getThreadGroup()
                .enumerate(threadArray);

        for (int i = 0; i < threadArray.length; i++) {
            System.out.println(threadArray[i].getName() + " "
                    + threadArray[i].getState());
        }
    }
}
```

程序运行后很可能出现假死状态，如图 3-27 所示。

从运行结果来看，呈假死状态就是所有的线程都呈 waiting 状态，不执行任何任务了，在代码中已经使用了 wait/notify 机制，为什么还会出现这种情况呢？

在代码中确实已经通过 wait/notify 进行通信了，但不保证 notify 唤醒的是异类，也许是同类，如"生产者"唤醒"生产者"，或"消费者"唤醒"消费者"这样的情况，如果这样的情况积少成多，就会导致所有的线程都不能继续运行下去，大家都在等待，都呈

waiting 状态，程序最后就呈"假死"状态，不能继续运行下去了。

那么出现假死的具体过程是怎样的呢？将控制台中的日志复制到 EditPlus 中以显示行号，如图 3-28 所示。

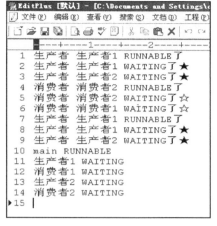

图 3-27　右上角按钮呈红色表示正在运行状态　　　图 3-28　显示行号的 EditPlus 工具

分析 EditPlus 工具中的日志执行过程，其实完全可以得出结论，分析的步骤与行号对应：

1）生产者 1 进行生产，生产完毕后发出通知（但此通知属于"通知过早"），并释放锁，准备进入下一次的 while 循环。

2）生产者 1 进入下一次 while 循环，迅速再次持有锁，发现产品并没有被消费，所以生产者 1 呈等待状态★。

3）生产者 2 被 start() 启动，生产者 2 发现产品还没有被消费，所以生产者 2 也呈等待状态★。

4）消费者 2 被 start() 启动，消费者 2 持有锁，将产品消费并发出通知（发出的通知唤醒了第 7 行生产者 1），运行结束后释放锁，等待消费者 2 进入下次循环。

5）消费者 2 进入下一次 while 循环，并快速持有锁，发现产品并未生产，所以释放锁并呈等待状态★。

6）消费者 1 被 start() 启动，快速持有锁，发现产品并未生产，所以释放锁并呈等待状态★。

7）由于消费者 2 在第 4 行已经将产品进行消费，唤醒了第 7 行的生产者 1 顺利生产后释放锁并发出通知（此通知唤醒了第 9 行的生产者 2），生产者 1 准备进入下一次 while 循环。

8）这时生产者 1 进入下一次 while 循环，再次持有锁，发现产品还并未消费，所以生产者 1 也呈等待状态了★。

9）由于第 7 行的生产者 1 唤醒了生产者 2，生产者 2 发现产品还并未被消费，所以生产者 2 呈等待状态★。

出现★符号就代表本线程进入等待状态，需要额外注意这样的执行结果。

出现假死的极大的原因是有可能连续唤醒同类，怎么解决这样的问题？不光唤醒同类，异类也一同唤醒就解决了，解决的办法请看后面的内容。

3. 多生产与多消费：操作值（解决假死）

创建 p_c_allWait_fix 项目，将 p_c_allWait 项目中的所有源代码复制到 p_c_allWait_fix 项目中。解决"假死"问题的方法很简单，将 P.java 和 C.java 文件中的 notify() 改成 notifyAll() 方法即可，它的原理就是通知的线程不光包括同类，也包括异类，这样就不至于出现假死的状态了，程序会一直运行下去。

4. 一生产与一消费：操作栈

本示例是使用生产者向堆栈 List 对象中放入数据，使用消费者从 List 堆栈中取出数据，List 最大容量是 1，实验环境只有一个生产者与一个消费者。

创建项目 stack_1，类 MyStack.java 代码如下：

```java
package entity;

import java.util.ArrayList;
import java.util.List;

public class MyStack {
    private List list = new ArrayList();

    synchronized public void push() {
        try {
            if (list.size() == 1) {
                this.wait();
            }
            list.add("anyString=" + Math.random());
            this.notify();
            System.out.println("push=" + list.size());
        } catch (InterruptedException e) {
            e.printStackTrace();
        }
    }

    synchronized public String pop() {
        String returnValue = "";
        try {
            if (list.size() == 0) {
                System.out.println("pop操作中的: "
                        + Thread.currentThread().getName() + " 线程呈wait状态");
                this.wait();
            }
            returnValue = "" + list.get(0);
            list.remove(0);
            this.notify();
            System.out.println("pop=" + list.size());
        } catch (InterruptedException e) {
            e.printStackTrace();
        }
        return returnValue;
    }
}
```

生产者和消费者线程代码如图 3-29 所示。

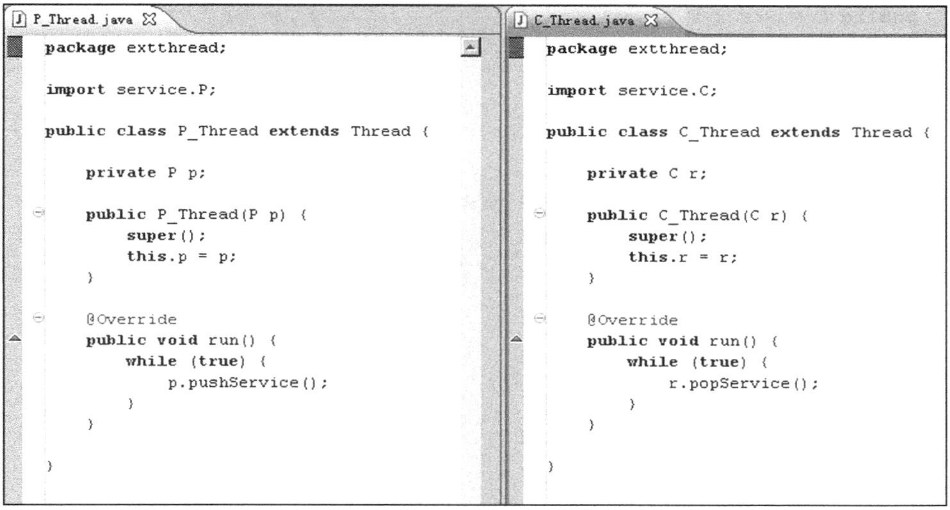

```
P_Thread.java
package extthread;

import service.P;

public class P_Thread extends Thread {

    private P p;

    public P_Thread(P p) {
        super();
        this.p = p;
    }

    @Override
    public void run() {
        while (true) {
            p.pushService();
        }
    }
}
```

```
C_Thread.java
package extthread;

import service.C;

public class C_Thread extends Thread {

    private C r;

    public C_Thread(C r) {
        super();
        this.r = r;
    }

    @Override
    public void run() {
        while (true) {
            r.popService();
        }
    }
}
```

图 3-29　生产者和消费者线程代码

生产者 P.java 服务代码如下：

```java
package service;

import entity.MyStack;

public class P {

    private MyStack myStack;

    public P(MyStack myStack) {
        super();
        this.myStack = myStack;
    }

    public void pushService() {
        myStack.push();
    }
}
```

消费者 C.java 服务代码如下：

```java
package service;

import entity.MyStack;

public class C {

    private MyStack myStack;
```

```java
    public C(MyStack myStack) {
        super();
        this.myStack = myStack;
    }

    public void popService() {
        System.out.println("pop=" + myStack.pop());
    }
}
```

运行类 Run.java 代码如下：

```java
package test.run;

import service.P;
import service.C;
import entity.MyStack;
import extthread.P_Thread;
import extthread.C_Thread;

public class Run {
    public static void main(String[] args) {
        MyStack myStack = new MyStack();

        P p = new P(myStack);
        C r = new C(myStack);

        P_Thread pThread = new P_Thread(p);
        C_Thread rThread = new C_Thread(r);
        pThread.start();
        rThread.start();
    }
}
```

程序运行结果如图 3-30 所示。

通过使用一生产者一消费者模式，容器 size() 的值不会大于 1，这也是本示例想要实现的效果，值在 0 和 1 之间进行交替，即生产和消费这两个过程在交替执行。

5. 一生产与多消费：操作栈（解决 wait 条件改变与假死）

本示例是使用一个生产者向堆栈 List 对象中放入数据，使用多个消费者从 List 堆栈中取出数据，List 最大容量还是 1。

创建新的项目 stack_2_old，将项目 stack_1 中的所有代码内容复制到 stack_2_old 项目中，更改 Run.java 代码如下：

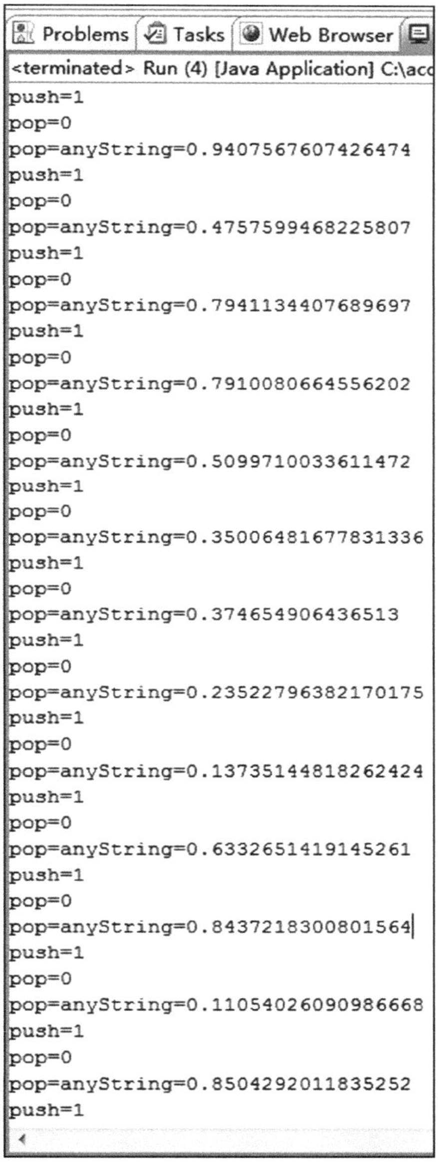

```
Problems | Tasks | Web Browser
<terminated> Run (4) [Java Application] C:\acc
push=1
pop=0
pop=anyString=0.9407567607426474
push=1
pop=0
pop=anyString=0.4757599468225807
push=1
pop=0
pop=anyString=0.7941134407689697
push=1
pop=0
pop=anyString=0.7910080664556202
push=1
pop=0
pop=anyString=0.5099710033611472
push=1
pop=0
pop=anyString=0.35006481677831336
push=1
pop=0
pop=anyString=0.374654906436513
push=1
pop=0
pop=anyString=0.23522796382170175
push=1
pop=0
pop=anyString=0.13735144818262424
push=1
pop=0
pop=anyString=0.6332651419145261
push=1
pop=0
pop=anyString=0.8437218300801564
push=1
pop=0
pop=anyString=0.11054026090986668
push=1
pop=0
pop=anyString=0.8504292011835252
push=1
```

图 3-30　程序运行结果（size() 不会大于 1 正常输出）

```java
package test.run;

import service.C;
import service.P;
import entity.MyStack;
import extthread.C_Thread;
import extthread.P_Thread;
```

```java
public class Run {
    public static void main(String[] args) throws InterruptedException {
        MyStack myStack = new MyStack();

        P p = new P(myStack);

        C r1 = new C(myStack);
        C r2 = new C(myStack);
        C r3 = new C(myStack);
        C r4 = new C(myStack);
        C r5 = new C(myStack);

        P_Thread pThread = new P_Thread(p);
        pThread.start();

        C_Thread cThread1 = new C_Thread(r1);
        C_Thread cThread2 = new C_Thread(r2);
        C_Thread cThread3 = new C_Thread(r3);
        C_Thread cThread4 = new C_Thread(r4);
        C_Thread cThread5 = new C_Thread(r5);
        cThread1.start();
        cThread2.start();
        cThread3.start();
        cThread4.start();
        cThread5.start();
    }

}
```

程序运行后在某些情况下出现异常，如图 3-31 所示。

```
push=1
pop=0
pop=anyString=0.5881162916547992
pop操作中的: Thread-3 线程呈wait状态
pop操作中的: Thread-4 线程呈wait状态
pop操作中的: Thread-1 线程呈wait状态
pop操作中的: Thread-2 线程呈wait状态
pop操作中的: Thread-5 线程呈wait状态
push=1
pop=0
pop=anyString=0.5787081529092245
pop操作中的: Thread-3 线程呈wait状态
Exception in thread "Thread-4" java.lang.IndexOutOfBoundsException: Index: 0, Size: 0
        at java.util.ArrayList.RangeCheck(ArrayList.java:547)
        at java.util.ArrayList.get(ArrayList.java:322)
        at entity.MyStack.pop(MyStack.java:30)
        at service.C.popService(C.java:15)
        at extthread.C_Thread.run(C_Thread.java:17)
```

图 3-31　出现异常索引溢出

出现此问题是因为在 MyStack.java 类中使用了 if 语句来做条件判断,代码如下:

```java
synchronized public String pop() {
    String returnValue = "";
    try {
        if (list.size() == 0) {
            System.out.println("pop操作中的: "
                    + Thread.currentThread().getName() + " 线程呈wait状态");
            this.wait();
        }
        returnValue = "" + list.get(0);
        list.remove(0);
        this.notify();
        System.out.println("pop=" + list.size());
    } catch (InterruptedException e) {
        e.printStackTrace();
    }
    return returnValue;
}
```

条件发生改变时并没有得到及时响应,所以多个呈 wait 状态的线程被唤醒,继而执行 list.remove(0) 代码而出现异常,解决这个问题的办法是将 if 语句改成 while 语句。

新建项目 stack_2_new,将 stack_2_old 中的全部代码复制到 stack_2_new 项目中,并且更改 MyStack.java 类代码如下:

```java
package entity;

import java.util.ArrayList;
import java.util.List;

public class MyStack {
    private List list = new ArrayList();

    synchronized public void push() {
        try {
            while (list.size() == 1) {
                this.wait();
            }
            list.add("anyString=" + Math.random());
            this.notify();
            System.out.println("push=" + list.size());
        } catch (InterruptedException e) {
            e.printStackTrace();
        }
    }

    synchronized public String pop() {
        String returnValue = "";
        try {
            while (list.size() == 0) {
```

```
                System.out.println("pop操作中的: "
                        + Thread.currentThread().getName() + " 线程呈wait状态");
                this.wait();
            }
            returnValue = "" + list.get(0);
            list.remove(0);
            this.notify();
            System.out.println("pop=" + list.size());
        } catch (InterruptedException e) {
            e.printStackTrace();
        }
        return returnValue;
    }
}
```

运行项目没有出现执行异常，但出现了"假死"情况，如图3-32所示。

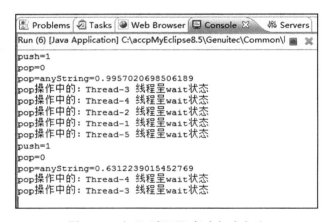

图 3-32　出现"假死"但有解决办法

解决的办法是使用notifyAll()方法。

创建全新的项目 stack_2_new_final，将 stack_2_new 项目中的所有源代码复制到 stack_2_new_final 项目中，将 MyStack.java 类中的两处调用 notify() 方法改成调用 notifyAll() 方法，程序运行后不再出现"假死"，永远并且正常地运行下去。

6. 多生产与一消费：操作栈

本示例是使用生产者向堆栈 List 对象中放入数据，使用消费者从 List 堆栈中取出数据，List 最大容量还是1，实验环境是多个生产者与一个消费者。

创建项目 stack_3，将 stack_2_new_final 项目中 src 下的所有包及类复制到 stack_3 项目中。

只需要将 Run.java 代码进行如下更改：

```java
package test.run;

import service.C;
import service.P;
import entity.MyStack;
import extthread.C_Thread;
import extthread.P_Thread;

public class Run {
    public static void main(String[] args) throws InterruptedException {
        MyStack myStack = new MyStack();

        P p1 = new P(myStack);
        P p2 = new P(myStack);
        P p3 = new P(myStack);
        P p4 = new P(myStack);
        P p5 = new P(myStack);
        P p6 = new P(myStack);

        P_Thread pThread1 = new P_Thread(p1);
        P_Thread pThread2 = new P_Thread(p2);
        P_Thread pThread3 = new P_Thread(p3);
        P_Thread pThread4 = new P_Thread(p4);
        P_Thread pThread5 = new P_Thread(p5);
        P_Thread pThread6 = new P_Thread(p6);
        pThread1.start();
        pThread2.start();
        pThread3.start();
        pThread4.start();
        pThread5.start();
        pThread6.start();

        C c1 = new C(myStack);
        C_Thread cThread = new C_Thread(c1);
        cThread.start();

    }

}
```

程序运行结果如图 3-33 所示。

7. 多生产与多消费：操作栈

本示例是使用生产者向堆栈 List 对象中放入数据，使用消费者从 List 堆栈中取出数据，List 最大容量是 1，实验环境是多个生产者与多个消费者。

创建项目 stack_4，将 stack_3 项目中 src 下的所有包及类复制到 stack_4 项目中。

只需要将 Run.java 代码更改如下：

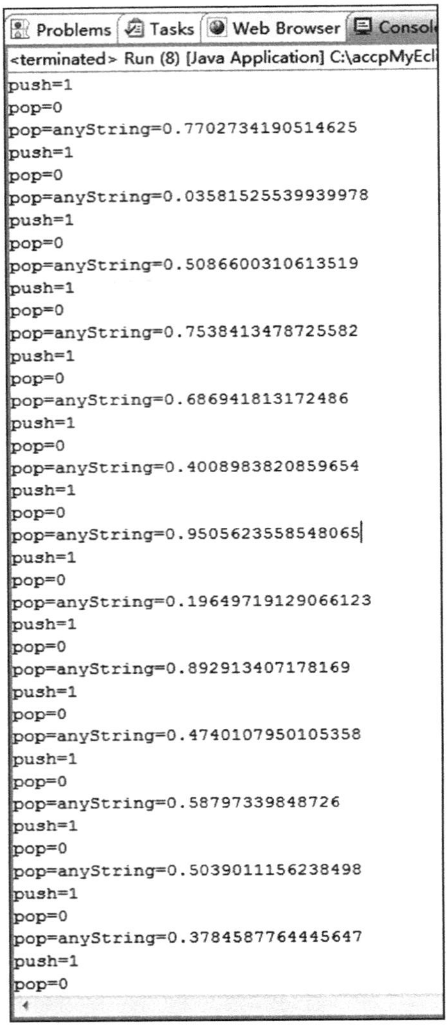

```
Problems  Tasks  Web Browser  Console
<terminated> Run (8) [Java Application] C:\accpMyEcli
push=1
pop=0
pop=anyString=0.7702734190514625
push=1
pop=0
pop=anyString=0.03581525539939978
push=1
pop=0
pop=anyString=0.5086600310613519
push=1
pop=0
pop=anyString=0.7538413478725582
push=1
pop=0
pop=anyString=0.686941813172486
push=1
pop=0
pop=anyString=0.4008983820859654
push=1
pop=0
pop=anyString=0.9505623558548065
push=1
pop=0
pop=anyString=0.19649719129066123
push=1
pop=0
pop=anyString=0.892913407178169
push=1
pop=0
pop=anyString=0.4740107950105358
push=1
pop=0
pop=anyString=0.58797339848726
push=1
pop=0
pop=anyString=0.5039011156238498
push=1
pop=0
pop=anyString=0.3784587764445647
push=1
pop=0
```

图 3-33　程序运行结果（多生产与一消费）

```java
package test.run;

import service.C;
import service.P;
import entity.MyStack;
import extthread.C_Thread;
import extthread.P_Thread;

public class Run {
    public static void main(String[] args) throws InterruptedException {
        MyStack myStack = new MyStack();
```

```java
        P p1 = new P(myStack);
        P p2 = new P(myStack);
        P p3 = new P(myStack);
        P p4 = new P(myStack);
        P p5 = new P(myStack);
        P p6 = new P(myStack);

        P_Thread pThread1 = new P_Thread(p1);
        P_Thread pThread2 = new P_Thread(p2);
        P_Thread pThread3 = new P_Thread(p3);
        P_Thread pThread4 = new P_Thread(p4);
        P_Thread pThread5 = new P_Thread(p5);
        P_Thread pThread6 = new P_Thread(p6);
        pThread1.start();
        pThread2.start();
        pThread3.start();
        pThread4.start();
        pThread5.start();
        pThread6.start();

        C r1 = new C(myStack);
        C r2 = new C(myStack);
        C r3 = new C(myStack);
        C r4 = new C(myStack);
        C r5 = new C(myStack);
        C r6 = new C(myStack);
        C r7 = new C(myStack);
        C r8 = new C(myStack);

        C_Thread cThread1 = new C_Thread(r1);
        C_Thread cThread2 = new C_Thread(r2);
        C_Thread cThread3 = new C_Thread(r3);
        C_Thread cThread4 = new C_Thread(r4);
        C_Thread cThread5 = new C_Thread(r5);
        C_Thread cThread6 = new C_Thread(r6);
        C_Thread cThread7 = new C_Thread(r7);
        C_Thread cThread8 = new C_Thread(r8);

        cThread1.start();
        cThread2.start();
        cThread3.start();
        cThread4.start();
        cThread5.start();
        cThread6.start();
        cThread7.start();
        cThread8.start();

    }

}
```

程序运行结果如图 3-34 所示。

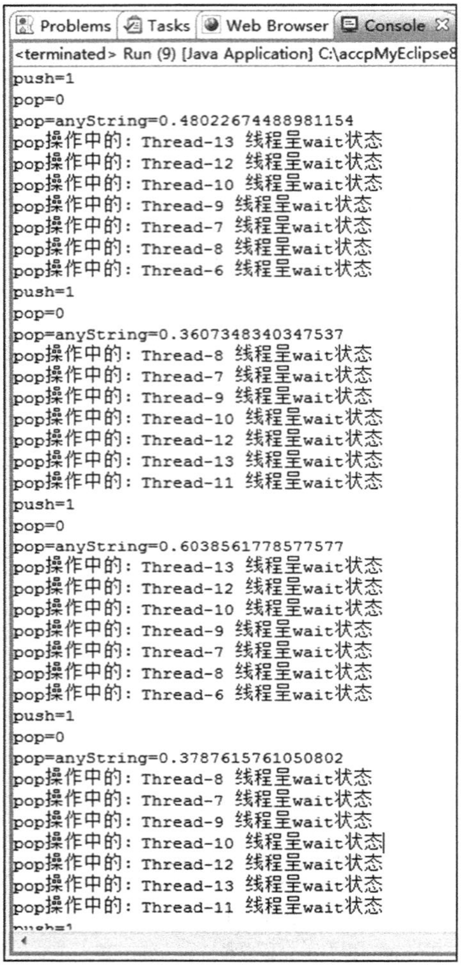

图 3-34　程序运行结果（多生产与多消费）

从程序运行结果来看，list 对象的 size() 并没有超过 1。

想要实现任意数量的几对几生产与消费的示例，可使用 while 结合 notifyAll() 的方法，这种组合具有通用性。

8. 允许连续生产多个：允许连续消费多个

前面章节代码执行的过程都是生产－消费的模式，本案例要实现连续的生产或连续的消费，运行过程可以是生产－生产－生产－消费－消费－生产这样的过程。

下面要实现的是多个生产与多个消费向一个 Box 容器中连续地放入或取出，容器的最大容量不能超过 50。

创建测试用的项目 Stack_50。创建容器 Box 类代码如下:

```java
package tools;

import java.util.ArrayList;
import java.util.List;

public class Box {

    private static List list = new ArrayList();

    synchronized public void add() {
        if (size() < 50) {
            list.add("anyString");
            System.out.println("线程: " + Thread.currentThread().getName() + "执行"
                    + "add()方法, size大小为: " + size());
        }
    }

    synchronized public int size() {
        return list.size();
    }

    synchronized public Object popFirst() {
        Object value = list.remove(0);
        System.out.println("线程: " + Thread.currentThread().getName() + "执行"
                + "popFirst()方法, size大小为: " + size());
        return value;
    }
}
```

创建生产者业务类代码如下:

```java
package set;

import tools.Box;

public class SetService {
    private Box box;

    public SetService(Box box) {
        super();
        this.box = box;
    }

    public void setMethod() {
        try {
            synchronized (this) {
                while (box.size() == 50) {
                    System.out.println("●●●●●");
```

```java
                    this.wait();
                }
            }
            Thread.sleep(300);
            box.add();
        } catch (InterruptedException e) {
            e.printStackTrace();
        }
    }

    public void checkBoxStatus() {
        try {
            while (true) {
                synchronized (this) {
                    if (box.size() < 50) {
                        this.notifyAll();
                    }
                }
                System.out.println("set checkboxBox = " + box.size());
                Thread.sleep(1000);
            }
        } catch (InterruptedException e) {
            e.printStackTrace();
        }
    }
}
```

创建生产者线程类代码如下：

```java
package set;

public class SetValueThread extends Thread {
    private SetService service;

    public SetValueThread(SetService service) {
        super();
        this.service = service;
    }

    @Override
    public void run() {
        while (true) {
            service.setMethod();
        }
    }
}
```

创建生产者容器大小测试类代码如下：

```java
package set;
```

```
public class SetCheckThread extends Thread {
    private SetService service;

    public SetCheckThread(SetService service) {
        super();
        this.service = service;
    }

    @Override
    public void run() {
        service.checkBoxStatus();
    }
}
```

创建消费者业务类代码如下:

```
package get;

import tools.Box;

public class GetService {
    private Box box;

    public GetService(Box box) {
        super();
        this.box = box;
    }

    public void getMethod() {
        try {
            synchronized (this) {
                while (box.size() == 0) {
                    System.out.println("○○○○○");
                    this.wait();
                }
                box.popFirst();
            }
            Thread.sleep(300);
        } catch (InterruptedException e) {
            e.printStackTrace();
        }
    }

    public void checkBoxStatus() {
        try {
            while (true) {
                synchronized (this) {
                    if (box.size() > 0) {
                        this.notifyAll();
                    }
                }
```

```
                    System.out.println("get checkboxBox = " + box.size());
                    Thread.sleep(1000);
                }
            } catch (InterruptedException e) {
                e.printStackTrace();
            }
        }
    }
}
```

创建消费者线程类代码如下：

```
package get;

public class GetValueThread extends Thread {
    private GetService service;

    public GetValueThread(GetService service) {
        super();
        this.service = service;
    }

    @Override
    public void run() {
        while (true) {
            service.getMethod();
        }
    }
}
```

创建消费者容器大小测试类代码如下：

```
package get;

public class GetCheckThread extends Thread {
    private GetService service;

    public GetCheckThread(GetService service) {
        super();
        this.service = service;
    }

    @Override
    public void run() {
        service.checkBoxStatus();
    }
}
```

创建运行类代码如下：

```
package test;
```

```
import get.GetCheckThread;
import get.GetService;
import get.GetValueThread;
import set.SetCheckThread;
import set.SetService;
import set.SetValueThread;
import tools.Box;

public class Test1 {

    public static void main(String[] args) throws InterruptedException {
        Box box = new Box();
        SetService setService = new SetService(box);
        for (int i = 0; i < 2; i++) {
            SetValueThread setValueThread1 =
                new SetValueThread(setService);
                    setValueThread1.start();
        }

        Thread.sleep(50);
        SetCheckThread setCheckThread = new SetCheckThread(setService);
        setCheckThread.start();

        /////

        Thread.sleep(10000);

        GetService getService = new GetService(box);
        for (int i = 0; i < 10; i++) {
            GetValueThread getValueThread1 = new GetValueThread(getService);
            getValueThread1.start();
        }
        Thread.sleep(50);
        GetCheckThread getCheckThread = new GetCheckThread(getService);
        getCheckThread.start();
    }

}
```

程序运行后控制台输出的部分结果如下：

```
set checkboxBox = 0
线程：Thread-0执行add()方法，size大小为：1
线程：Thread-1执行add()方法，size大小为：2
线程：Thread-1执行add()方法，size大小为：3
线程：Thread-0执行add()方法，size大小为：4
线程：Thread-1执行add()方法，size大小为：5
线程：Thread-0执行add()方法，size大小为：6
set checkboxBox = 6
线程：Thread-1执行add()方法，size大小为：7
线程：Thread-0执行add()方法，size大小为：8
线程：Thread-1执行add()方法，size大小为：9
```

```
线程：Thread-0执行add()方法，size大小为: 10
线程：Thread-1执行add()方法，size大小为: 11
线程：Thread-0执行add()方法，size大小为: 12
set checkboxBox = 12
线程：Thread-1执行add()方法，size大小为: 13
线程：Thread-0执行add()方法，size大小为: 14
线程：Thread-1执行add()方法，size大小为: 15
线程：Thread-0执行add()方法，size大小为: 16
线程：Thread-0执行add()方法，size大小为: 17
线程：Thread-1执行add()方法，size大小为: 18
线程：Thread-1执行add()方法，size大小为: 19
线程：Thread-0执行add()方法，size大小为: 20
set checkboxBox = 20
线程：Thread-1执行add()方法，size大小为: 21
线程：Thread-0执行add()方法，size大小为: 22
线程：Thread-1执行add()方法，size大小为: 23
线程：Thread-0执行add()方法，size大小为: 24
线程：Thread-1执行add()方法，size大小为: 25
线程：Thread-0执行add()方法，size大小为: 26
set checkboxBox = 26
线程：Thread-0执行add()方法，size大小为: 27
线程：Thread-1执行add()方法，size大小为: 28
线程：Thread-1执行add()方法，size大小为: 29
线程：Thread-0执行add()方法，size大小为: 30
线程：Thread-0执行add()方法，size大小为: 31
线程：Thread-1执行add()方法，size大小为: 32
set checkboxBox = 32
线程：Thread-1执行add()方法，size大小为: 33
线程：Thread-0执行add()方法，size大小为: 34
线程：Thread-0执行add()方法，size大小为: 35
线程：Thread-1执行add()方法，size大小为: 36
线程：Thread-0执行add()方法，size大小为: 37
线程：Thread-1执行add()方法，size大小为: 38
线程：Thread-0执行add()方法，size大小为: 39
线程：Thread-1执行add()方法，size大小为: 40
set checkboxBox = 40
线程：Thread-0执行add()方法，size大小为: 41
线程：Thread-1执行add()方法，size大小为: 42
线程：Thread-1执行add()方法，size大小为: 43
线程：Thread-0执行add()方法，size大小为: 44
线程：Thread-0执行add()方法，size大小为: 45
线程：Thread-1执行add()方法，size大小为: 46
set checkboxBox = 46
线程：Thread-1执行add()方法，size大小为: 47
线程：Thread-0执行add()方法，size大小为: 48
线程：Thread-0执行add()方法，size大小为: 49
线程：Thread-1执行add()方法，size大小为: 50
●●●●●
●●●●●
set checkboxBox = 50
set checkboxBox = 50
```

```
set checkboxBox = 50
线程：Thread-3执行popFirst()方法，size大小为：49
线程：Thread-5执行popFirst()方法，size大小为：48
线程：Thread-4执行popFirst()方法，size大小为：47
线程：Thread-6执行popFirst()方法，size大小为：46
线程：Thread-7执行popFirst()方法，size大小为：45
线程：Thread-8执行popFirst()方法，size大小为：44
线程：Thread-9执行popFirst()方法，size大小为：43
线程：Thread-10执行popFirst()方法，size大小为：42
线程：Thread-11执行popFirst()方法，size大小为：41
线程：Thread-12执行popFirst()方法，size大小为：40
get checkboxBox = 40
线程：Thread-4执行popFirst()方法，size大小为：39
线程：Thread-5执行popFirst()方法，size大小为：38
线程：Thread-8执行popFirst()方法，size大小为：37
线程：Thread-7执行popFirst()方法，size大小为：36
线程：Thread-6执行popFirst()方法，size大小为：35
线程：Thread-3执行popFirst()方法，size大小为：34
线程：Thread-12执行popFirst()方法，size大小为：33
线程：Thread-10执行popFirst()方法，size大小为：32
线程：Thread-11执行popFirst()方法，size大小为：31
线程：Thread-9执行popFirst()方法，size大小为：30
线程：Thread-8执行popFirst()方法，size大小为：29
线程：Thread-7执行popFirst()方法，size大小为：28
线程：Thread-3执行popFirst()方法，size大小为：27
线程：Thread-6执行popFirst()方法，size大小为：26
线程：Thread-4执行popFirst()方法，size大小为：25
线程：Thread-5执行popFirst()方法，size大小为：24
线程：Thread-12执行popFirst()方法，size大小为：23
线程：Thread-10执行popFirst()方法，size大小为：22
线程：Thread-9执行popFirst()方法，size大小为：21
线程：Thread-11执行popFirst()方法，size大小为：20
线程：Thread-4执行popFirst()方法，size大小为：19
线程：Thread-6执行popFirst()方法，size大小为：18
线程：Thread-7执行popFirst()方法，size大小为：17
线程：Thread-3执行popFirst()方法，size大小为：16
线程：Thread-8执行popFirst()方法，size大小为：15
线程：Thread-5执行popFirst()方法，size大小为：14
线程：Thread-11执行popFirst()方法，size大小为：13
线程：Thread-9执行popFirst()方法，size大小为：12
线程：Thread-12执行popFirst()方法，size大小为：11
线程：Thread-10执行popFirst()方法，size大小为：10
set checkboxBox = 10
get checkboxBox = 10
线程：Thread-3执行popFirst()方法，size大小为：9
线程：Thread-5执行popFirst()方法，size大小为：8
线程：Thread-8执行popFirst()方法，size大小为：7
线程：Thread-6执行popFirst()方法，size大小为：6
线程：Thread-7执行popFirst()方法，size大小为：5
线程：Thread-4执行popFirst()方法，size大小为：4
线程：Thread-12执行popFirst()方法，size大小为：3
```

线程：Thread-9执行popFirst()方法，size大小为：2
线程：Thread-10执行popFirst()方法，size大小为：1
线程：Thread-11执行popFirst()方法，size大小为：0
线程：Thread-1执行add()方法，size大小为：1
线程：Thread-0执行add()方法，size大小为：2
线程：Thread-6执行popFirst()方法，size大小为：1
线程：Thread-5执行popFirst()方法，size大小为：0
○○○○○
○○○○○
○○○○○
○○○○○
○○○○○
○○○○○
○○○○○
线程：Thread-0执行add()方法，size大小为：1
线程：Thread-1执行add()方法，size大小为：2
线程：Thread-6执行popFirst()方法，size大小为：1
线程：Thread-5执行popFirst()方法，size大小为：0
线程：Thread-0执行add()方法，size大小为：1
线程：Thread-1执行add()方法，size大小为：2
set checkboxBox = 2
get checkboxBox = 2
线程：Thread-9执行popFirst()方法，size大小为：1
线程：Thread-10执行popFirst()方法，size大小为：0
○○○○○
○○○○○
○○○○○
○○○○○
○○○○○
○○○○○
○○○○○
线程：Thread-1执行add()方法，size大小为：1
线程：Thread-0执行add()方法，size大小为：2

运行结果比较多，但从控制台输出的结果来看，由于消费者数量比生产者多，所以多次输出○○○○○，说明消费者执行了wait()等待。

3.1.20 通过管道进行线程间通信——字节流

Java 语言提供了各种各样的输入/输出流，使我们能够很方便地对数据进行操作，其中管道流（pipe stream）是一种特殊的流，用于在不同线程间直接传送数据。一个线程发送数据到输出管道，另一个线程从输入管道中读数据。通过使用管道，实现不同线程间的通信，而无须借助于类似临时文件之类的东西。

Java JDK 提供了 4 个类来使线程间可以进行通信，即 PipedInputStream 和 PipedOutputStream、PipedReader 和 PipedWriter。

创建测试用的项目 pipeInputOutput。

类 WriteData.java 代码如下：

```java
package service;

import java.io.IOException;
import java.io.PipedOutputStream;

public class WriteData {

    public void writeMethod(PipedOutputStream out) {
        try {
            System.out.println("write :");
            for (int i = 0; i < 300; i++) {
                String outData = "" + (i + 1);
                out.write(outData.getBytes());
                System.out.print(outData);
            }
            System.out.println();
            out.close();
        } catch (IOException e) {
            e.printStackTrace();
        }
    }
}
```

类 ReadData.java 代码如下：

```java
package service;

import java.io.IOException;
import java.io.PipedInputStream;

public class ReadData {

    public void readMethod(PipedInputStream input) {
        try {
            System.out.println("read  :");
            byte[] byteArray = new byte[20];
            int readLength = input.read(byteArray);
            while (readLength != -1) {
                String newData = new String(byteArray, 0, readLength);
                System.out.print(newData);
                readLength = input.read(byteArray);
            }
            System.out.println();
            input.close();
        } catch (IOException e) {
            e.printStackTrace();
        }
    }
}
```

两个自定义线程类代码如图 3-35 所示。

```
package extthread;

import java.io.PipedOutputStream;

import service.WriteData;

public class ThreadWrite extends Thread {

    private WriteData write;
    private PipedOutputStream out;

    public ThreadWrite(WriteData write, PipedOutputStream out) {
        super();
        this.write = write;
        this.out = out;
    }

    @Override
    public void run() {
        write.writeMethod(out);
    }
}
```

```
package extthread;

import java.io.PipedInputStream;

import service.ReadData;

public class ThreadRead extends Thread {

    private ReadData read;
    private PipedInputStream input;

    public ThreadRead(ReadData read, PipedInputStream input) {
        super();
        this.read = read;
        this.input = input;
    }

    @Override
    public void run() {
        read.readMethod(input);
    }
}
```

图 3-35　两个自定义线程类代码

类 Run.java 代码如下：

```
package test;

import java.io.IOException;
import java.io.PipedInputStream;
import java.io.PipedOutputStream;

import service.ReadData;
import service.WriteData;
import extthread.ThreadRead;
import extthread.ThreadWrite;

public class Run {

    public static void main(String[] args) {

        try {
            WriteData writeData = new WriteData();
            ReadData readData = new ReadData();

            PipedInputStream inputStream = new PipedInputStream();
            PipedOutputStream outputStream = new PipedOutputStream();

            // inputStream.connect(outputStream);
            outputStream.connect(inputStream);

            ThreadRead threadRead = new ThreadRead(readData, inputStream);
            threadRead.start();

            Thread.sleep(2000);
```

```
                ThreadWrite threadWrite = new ThreadWrite(writeData, outputStream);
                threadWrite.start();
            } catch (IOException e) {
                e.printStackTrace();
            } catch (InterruptedException e) {
                e.printStackTrace();
            }
        }
    }
```

代码 inputStream.connect(outputStream) 或 outputStream.connect(inputStream) 的作用是使两个管道之间建立通信连接，这样才可以对数据进行输出与输入。

程序运行结果如图 3-36 所示。

```
<terminated> Run (2) [Java Application] C:\Program Files\Genuitec\Common\binary\com.sun.java.
read   :
write :
1234567891011121314151617181920212223242526272829303132333435363738394
1234567891011121314151617181920212223242526272829303132333435363738394
```

图 3-36 程序运行结果（从 1 开始）

从程序运行结果来看，两个线程通过管道流成功进行数据的传输。

但在此示例中，首先读取线程 new ThreadRead(inputStream) 启动，由于当时没有数据被写入，所以线程阻塞在 " int readLength = in.read(byteArray);" 代码中，直到有数据被写入，才继续向下运行。

3.1.21 通过管道进行线程间通信——字符流

在管道中还可以传递字符流。创建测试用的项目 pipeReaderWriter。

类 WriteData.java 代码如下：

```java
package service;

import java.io.IOException;
import java.io.PipedWriter;

public class WriteData {

    public void writeMethod(PipedWriter out) {
        try {
            System.out.println("write :");
            for (int i = 0; i < 300; i++) {
                String outData = "" + (i + 1);
                out.write(outData);
                System.out.print(outData);
```

```
            }
            System.out.println();
            out.close();
        } catch (IOException e) {
            e.printStackTrace();
        }
    }
}
```

类 ReadData.java 代码如下:

```
package service;

import java.io.IOException;
import java.io.PipedReader;

public class ReadData {

    public void readMethod(PipedReader input) {
        try {
            System.out.println("read  :");
            char[] byteArray = new char[20];
            int readLength = input.read(byteArray);
            while (readLength != -1) {
                String newData = new String(byteArray, 0, readLength);
                System.out.print(newData);
                readLength = input.read(byteArray);
            }
            System.out.println();
            input.close();
        } catch (IOException e) {
            e.printStackTrace();
        }
    }
}
```

两个自定义线程类代码如图 3-37 所示。

```
1  package extthread;
2
3  import java.io.PipedWriter;
4
5  import service.WriteData;
6
7  public class ThreadWrite extends Thread {
8
9      private WriteData write;
10     private PipedWriter out;
11
12     public ThreadWrite(WriteData write, PipedWriter out) {
13         super();
14         this.write = write;
15         this.out = out;
16     }
17
18     @Override
19     public void run() {
20         write.writeMethod(out);
21     }
22
23 }
24
```

```
1  package extthread;
2
3  import java.io.PipedReader;
4
5  import service.ReadData;
6
7  public class ThreadRead extends Thread {
8
9      private ReadData read;
10     private PipedReader input;
11
12     public ThreadRead(ReadData read, PipedReader input) {
13         super();
14         this.read = read;
15         this.input = input;
16     }
17
18     @Override
19     public void run() {
20         read.readMethod(input);
21     }
22
23 }
```

图 3-37 两个自定义线程类代码

类 Run.java 代码如下：

```java
package test;

import java.io.IOException;
import java.io.PipedReader;
import java.io.PipedWriter;

import service.ReadData;
import service.WriteData;
import extthread.ThreadRead;
import extthread.ThreadWrite;

public class Run {

    public static void main(String[] args) {

        try {
            WriteData writeData = new WriteData();
            ReadData readData = new ReadData();

            PipedReader inputStream = new PipedReader();
            PipedWriter outputStream = new PipedWriter();

            // inputStream.connect(outputStream);
            outputStream.connect(inputStream);

            ThreadRead threadRead = new ThreadRead(readData, inputStream);
            threadRead.start();

            Thread.sleep(2000);

            ThreadWrite threadWrite = new ThreadWrite(writeData, outputStream);
            threadWrite.start();
        } catch (IOException e) {
            e.printStackTrace();
        } catch (InterruptedException e) {
            e.printStackTrace();
        }

    }

}
```

程序运行结果如图 3-38 所示。

```
read :
write :
12345678910111213141516171819202122232425262728293031323334353637383940414243444546474849505152535455565758596061626364656667…
12345678910111213141516171819202122232425262728293031323334353637383940414243444546474849505152535455565758596061626364656667…
```

图 3-38　程序运行结果（从 1 开始）

运行结果和前一个示例基本一样,此示例演示了在两个线程中通过管道进行字符数据的传输。

3.1.22 实现 wait/notify 的交叉备份

本节将继续学习 wait/notify 相关的知识点,实现创建 20 个线程,其中 10 个线程用于将数据备份到 A 数据库中,另外 10 个线程用于将数据备份到 B 数据库中,并且备份 A 数据库和 B 数据库是交叉的效果。

先创建出 20 个线程,效果如图 3-39 所示。

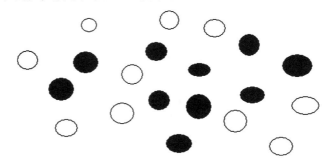

图 3-39 创建 20 个线程

通过一些方法将这 20 个线程的运行效果变成有序的,如图 3-40 所示。

图 3-40 具有交叉间隔行的效果

使用的技术还是 wait/notify,使用组合 while 循环 +notifyAll() 方法。

创建测试用的项目,名称为 wait_notify_insert_test。

创建 DBTools.java 类代码如下:

```java
package service;

public class DBTools {

    volatile private boolean prevIsA = false;

    synchronized public void backupA() {
        try {
            while (prevIsA == true) {
                wait();
            }
            for (int i = 0; i < 5; i++) {
                System.out.println("★★★★★");
            }
            prevIsA = true;
```

```java
                notifyAll();
            } catch (InterruptedException e) {
                e.printStackTrace();
            }
        }

        synchronized public void backupB() {
            try {
                while (prevIsA == false) {
                    wait();
                }
                for (int i = 0; i < 5; i++) {
                    System.out.println("☆☆☆☆☆");
                }
                prevIsA = false;
                notifyAll();
            } catch (InterruptedException e) {
                e.printStackTrace();
            }
        }
    }
```

变量 prevIsA 的主要作用是确保备份"★★★★★"数据库 A 首先执行，然后与"☆☆☆☆☆"数据库 B 交替进行备份。

创建两个线程工具类，如图 3-41 所示。

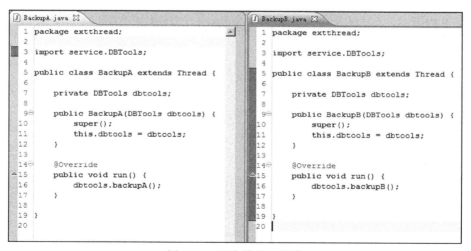

图 3-41　两个线程工具类

运行类 Run.java 代码如下：

```java
package test.run;

import service.DBTools;
import extthread.BackupA;
```

```
import extthread.BackupB;

public class Run {

    public static void main(String[] args) {
        DBTools dbtools = new DBTools();
        for (int i = 0; i < 20; i++) {
            BackupB output = new BackupB(dbtools);
            output.start();
            BackupA input = new BackupA(dbtools);
            input.start();
        }
    }

}
```

程序运行后部分输出效果如下:

```
★★★★★
★★★★★
★★★★★
★★★★★
★★★★★
☆☆☆☆☆
☆☆☆☆☆
☆☆☆☆☆
☆☆☆☆☆
☆☆☆☆☆
★★★★★
★★★★★
★★★★★
★★★★★
★★★★★
☆☆☆☆☆
☆☆☆☆☆
☆☆☆☆☆
☆☆☆☆☆
☆☆☆☆☆
★★★★★
★★★★★
★★★★★
★★★★★
★★★★★
☆☆☆☆☆
☆☆☆☆☆
☆☆☆☆☆
☆☆☆☆☆
☆☆☆☆☆
```

可见运行结果是交替输出。

交替输出的原理是使用如下代码作为标记,以实现 A 线程和 B 线程交替备份的效果。

```
volatile private boolean prevIsA = false;
```

3.2 join() 方法的使用

在很多情况下，主线程创建并启动子线程，如果子线程要进行大量的耗时运算，主线程往往将早于子线程结束之前结束，这时如果主线程想等待子线程执行完成之后再结束，例如，当子线程处理一个数据，主线程要取得这个数据中的值时，就要用到 join() 方法了。方法 join() 的作用是等待线程对象销毁。

3.2.1 学习 join() 方法前的铺垫

在介绍 join() 方法之前，先来看一个示例。

创建测试用的 java 项目，名称为 joinTest1，类 MyThread.java 代码如下：

```java
package extthread;

public class MyThread extends Thread {

    @Override
    public void run() {
        try {
            int secondValue = (int) (Math.random() * 10000);
            System.out.println(secondValue);
            Thread.sleep(secondValue);
        } catch (InterruptedException e) {
            // TODO Auto-generated catch block
            e.printStackTrace();
        }
    }

}
```

类 Test.java 代码如下：

```java
package test;

import extthread.MyThread;

public class Test {

    public static void main(String[] args) {

        MyThread threadTest = new MyThread();
        threadTest.start();

        // Thread.sleep(?)
```

```
            System.out.println("main线程想实现当threadTest对象执行完毕后我再继续向下执行,");
            System.out.println("但上面代码中的sleep()中的值应该写多少呢？");
            System.out.println("答案是：不能确定");
        }

}
```

程序运行结果如图 3-42 所示。

> main线程想实现当threadTest对象执行完毕后我再继续向下执行,
> 但上面代码中的sleep()中的值应该写多少呢？
> 答案是：不能确定
> 7528

图 3-42   sleep() 方法中的值不能确定

join() 方法可以解决这个问题，新建 java 项目 joinTest2，类 MyThread.java 代码如下：

```java
package extthread;

public class MyThread extends Thread {

    @Override
    public void run() {
        try {
            int secondValue = (int) (Math.random() * 10000);
            System.out.println(secondValue);
            Thread.sleep(secondValue);
        } catch (InterruptedException e) {
            // TODO Auto-generated catch block
            e.printStackTrace();
        }
    }

}
```

类 Test.java 代码如下：

```java
package test;

import extthread.MyThread;

public class Test {

    public static void main(String[] args) {
        try {
            MyThread threadTest = new MyThread();
            threadTest.start();
            threadTest.join();

            System.out.println("我想当threadTest对象执行完毕后我再执行，我做到了");
        } catch (InterruptedException e) {
```

```
            e.printStackTrace();
        }
    }
}
```

程序运行结果如图 3-43 所示。

图 3-43　程序运行结果

join() 方法的作用是使所属的线程对象 x 正常执行 run() 方法中的任务，而使当前线程 z 进行无限期的阻塞，等待线程 x 销毁后再继续执行线程 z 后面的代码，具有串联执行的效果。

join() 方法具有使线程排队运行的效果，有些类似同步的运行效果，但是 join() 方法与 synchronized 的区别是 join() 方法在内部使用 wait() 方法进行等待，而 synchronized 关键字使用锁作为同步。

3.2.2　join() 方法和 interrupt() 方法出现异常

在使用 join() 方法的过程中，如果当前线程对象被中断，则当前线程出现异常。

创建测试用的项目 joinException，类 ThreadA.java 代码如下：

```java
package extthread;

public class ThreadA extends Thread {
    @Override
    public void run() {
        for (int i = 0; i < Integer.MAX_VALUE; i++) {
            String newString = new String();
            Math.random();
        }
    }
}
```

类 ThreadB.java 代码如下：

```java
package extthread;

public class ThreadB extends Thread {

    @Override
    public void run() {
        try {
            ThreadA a = new ThreadA();
```

```
            a.start();
            a.join();

            System.out.println("线程B在run end处打印了");
        } catch (InterruptedException e) {
            System.out.println("线程B在catch处打印了");
            e.printStackTrace();
        }
    }

}
```

类 ThreadC.java 代码如下：

```
package extthread;

public class ThreadC extends Thread {

    private ThreadB threadB;

    public ThreadC(ThreadB threadB) {
        super();
        this.threadB = threadB;
    }

    @Override
    public void run() {
        threadB.interrupt();
    }

}
```

类 Run.java 代码如下：

```
package test.run;

import extthread.ThreadB;
import extthread.ThreadC;

public class Run {

    public static void main(String[] args) {
        try {
            ThreadB b = new ThreadB();
            b.start();

            Thread.sleep(500);

            ThreadC c = new ThreadC(b);
            c.start();
        } catch (InterruptedException e) {
```

```
            e.printStackTrace();
        }
    }
}
```

程序运行结果如图 3-44 所示。

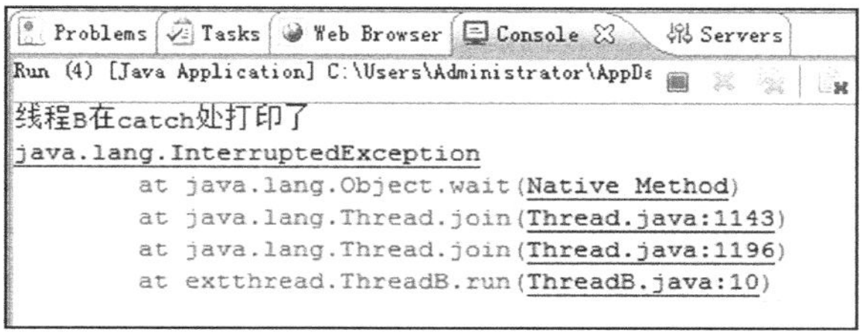

图 3-44 出现异常

join() 方法与 interrupt() 方法如果彼此遇到，则出现异常，不管先后顺序。进程按钮还呈红色，原因是线程 ThreadA 还在继续运行，并未出现异常，是正常执行的状态。

3.2.3 join(long) 方法的使用

x.join(long) 方法中的参数用于设定等待的时间，不管 x 线程是否执行完毕，时间到了并且重新获得了锁，则当前线程会继续向后运行。如果没有重新获得锁，则一直在尝试，直到获得锁为止。

创建测试用的项目 joinLong，类 MyThread.java 代码如下：

```java
package extthread;

public class MyThread extends Thread {
    @Override
    public void run() {
        try {
            System.out.println("run begin Timer=" + System.currentTimeMillis());
            Thread.sleep(5000);
            System.out.println("run   end Timer=" + System.currentTimeMillis());
        } catch (InterruptedException e) {
            e.printStackTrace();
        }
    }
}
```

类 Test.java 代码如下：

```java
package test;

import extthread.MyThread;
```

```
public class Test {
    public static void main(String[] args) {
        try {
            MyThread threadTest = new MyThread();
            threadTest.start();

            System.out.println("   main begin time=" + System.
                currentTimeMillis());
            threadTest.join(2000);// 只等2s
            // Thread.sleep(2000);
            System.out.println("   main   end time=" + System.
                currentTimeMillis());
        } catch (InterruptedException e) {
            e.printStackTrace();
        }
    }
}
```

程序运行结果如图 3-45 所示。

从运行结果可以发现，run() 方法执行了 5s，main() 方法暂停了 2s。

但将 main() 方法中的代码，改成使用 sleep(2000) 方法时，运行的效果还是 main 线程等待了 2s，如图 3-46 所示。

```
    main begin time=1520388684343
run begin Timer=1520388684343
    main   end time=1520388686343
run   end Timer=1520388689343
```

图 3-45　运行结果是等待了 2s

```
    main begin time=1520388775072
run begin Timer=1520388775072
    main   end time=1520388777072
run   end Timer=1520388780072
```

图 3-46　使用方法 sleep(2000) 也等待了 2s

那使用 join(2000) 和使用 sleep(2000) 有什么区别呢？上面的示例中在运行效果上并没有区别，其实区别主要来自于这两个方法在同步的处理上。

3.2.4　join(long) 方法与 sleep(long) 方法的区别

join(long) 方法的功能在内部是使用 wait(long) 方法来进行实现的，所以 join(long) 方法具有释放锁的特点。

join(long) 方法源代码如下：

```
public final synchronized void join(long millis)
        throws InterruptedException {
    long base = System.currentTimeMillis();
    long now = 0;

    if (millis < 0) {
        throw new IllegalArgumentException("timeout value is negative");
    }
```

```
        if (millis == 0) {
            while (isAlive()) {
                wait(0);
            }
        } else {
            while (isAlive()) {
                long delay = millis - now;
                if (delay <= 0) {
                    break;
                }
                wait(delay);
                now = System.currentTimeMillis() - base;
            }
        }
    }
```

从源代码中可以了解到，当执行 wait(long) 方法后当前线程的锁被释放，那么其他线程就可以调用此线程中的同步方法了。

而 Thread.sleep(long) 方法却不释放锁，下面通过示例来演示 Thread.sleep(long) 方法具有不释放锁的特点。

创建测试用的项目 join_sleep_1，类 ThreadA.java 代码如下：

```
package extthread;

public class ThreadA extends Thread {

    private ThreadB b;

    public ThreadA(ThreadB b) {
        super();
        this.b = b;
    }

    @Override
    public void run() {
        try {
            synchronized (b) {
                b.start();
                Thread.sleep(6000);
                // Thread.sleep()不释放锁！
            }
        } catch (InterruptedException e) {
            e.printStackTrace();
        }
    }
}
```

类 ThreadB.java 代码如下：

```
package extthread;
```

```java
public class ThreadB extends Thread {

    @Override
    public void run() {
        try {
            System.out.println("  b run  begin timer="
                    + System.currentTimeMillis());
            Thread.sleep(5000);
            System.out.println("  b run    end timer="
                    + System.currentTimeMillis());
        } catch (InterruptedException e) {
            e.printStackTrace();
        }
    }

    synchronized public void bService() {
        System.out.println("打印了bService timer=" + System.currentTimeMillis());
    }

}
```

类 ThreadC.java 代码如下:

```java
package extthread;

public class ThreadC extends Thread {

    private ThreadB threadB;

    public ThreadC(ThreadB threadB) {
        super();
        this.threadB = threadB;
    }

    @Override
    public void run() {
        threadB.bService();
    }
}
```

类 Run.java 代码如下:

```java
package test.run;

import extthread.ThreadA;
import extthread.ThreadB;
import extthread.ThreadC;

public class Run {

    public static void main(String[] args) {
```

```
    try {
        ThreadB b = new ThreadB();

        ThreadA a = new ThreadA(b);
        a.start();

        Thread.sleep(1000);

        ThreadC c = new ThreadC(b);
        c.start();
    } catch (InterruptedException e) {
        e.printStackTrace();
    }
}
```

}

程序运行结果如图 3-47 所示。

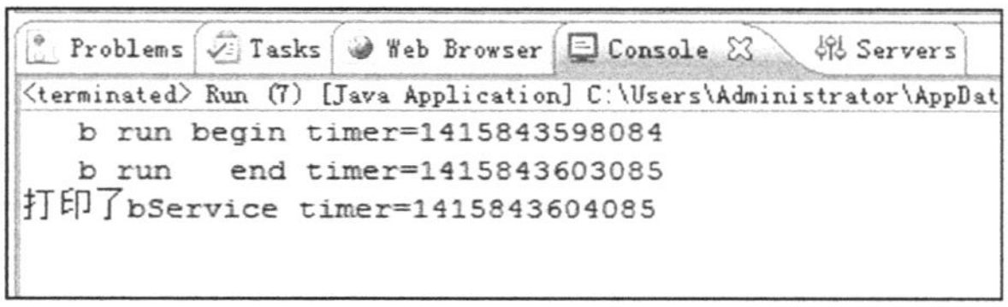

图 3-47　线程 ThreadA 不释放 ThreadB 的锁

由于线程 ThreadA 使用 Thread.sleep(6000) 方法一直持有 ThreadB 对象的锁，时间达到 6s，所以线程 ThreadC 只有在 ThreadA 时间到达 6s 后释放 ThreadB 的锁时，才可以调用 ThreadB 中的同步方法 synchronized public void bService()。

上面示例证明 Thread.sleep(long) 方法不释放锁。

下面通过示例，验证 join() 方法释放锁的特点。

创建实验用的项目 join_sleep_2，将 join_sleep_1 中的所有代码复制到 join_sleep_2 项目中，更改 ThreadA.java 类代码如下：

```
package extthread;

public class ThreadA extends Thread {

    private ThreadB b;

    public ThreadA(ThreadB b) {
        super();
        this.b = b;
    }

    @Override
    public void run() {
```

```
    try {
        synchronized (b) {
            b.start();
            b.join();// 执行join()方法的一瞬间，b锁立即释放
            for (int i = 0; i < Integer.MAX_VALUE; i++) {
                String newString = new String();
                Math.random();
            }
        }
    } catch (InterruptedException e) {
        e.printStackTrace();
    }
}
```

程序运行结果如图 3-48 所示。

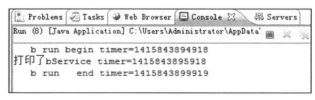

图 3-48　join() 方法释放锁

由于线程 ThreadA 释放了 ThreadB 的锁，所以线程 ThreadC 可以调用 ThreadB 中的同步方法 synchronized public void bService()。

此示例再次说明 join(long) 方法具有释放锁的特点。

另外需要注意一点，join() 无参方法或 join(time) 有参方法一旦执行，说明源代码中的 wait(time) 已经被执行，也就证明锁被立即释放，仅仅在指定的 join(time) 时间后当前线程会继续向下运行。

3.2.5　join() 方法后面的代码提前运行——出现意外

针对前面章节中的代码进行测试的过程中，还可以延伸出 "陷阱式" 结果，如果稍加不注意，就会掉进 "坑" 里。

创建测试用的项目 joinMoreTest，类 ThreadA.java 代码如下：

```
package extthread;

public class ThreadA extends Thread {
    private ThreadB b;

    public ThreadA(ThreadB b) {
        super();
        this.b = b;
    }
```

```
    @Override
    public void run() {
        try {
            synchronized (b) {
                System.out.println("begin A ThreadName="
                        + Thread.currentThread().getName() + "  "
                        + System.currentTimeMillis());
                Thread.sleep(500);
                System.out.println("  end A ThreadName="
                        + Thread.currentThread().getName() + "  "
                        + System.currentTimeMillis());
            }
        } catch (InterruptedException e) {
            e.printStackTrace();
        }
    }
}
```

类 ThreadB.java 代码如下：

```
package extthread;

public class ThreadB extends Thread {
    @Override
    synchronized public void run() {
        try {
            System.out.println("begin B ThreadName="
                    + Thread.currentThread().getName() + "  "
                    + System.currentTimeMillis());
            Thread.sleep(500);
            System.out.println("  end B ThreadName="
                    + Thread.currentThread().getName() + "  "
                    + System.currentTimeMillis());
        } catch (InterruptedException e) {
            e.printStackTrace();
        }
    }
}
```

一定要注意，ThreadB.java 中的 run() 方法是 synchronized 同步的。

创建 Run1.java 代码如下：

```
package test.run;

import extthread.ThreadA;
import extthread.ThreadB;

public class Run1 {

    public static void main(String[] args) {
        try {
```

```java
            ThreadB b = new ThreadB();
            ThreadA a = new ThreadA(b);
            a.start();
            b.start();
            b.join(200);
            System.out.println("              main end "
                    + System.currentTimeMillis());
        } catch (InterruptedException e) {
            e.printStackTrace();
        }
    }
}
```

程序运行后，在控制台有时会出现截然不同的输出，如图 3-49～图 3-51 所示。

```
begin A ThreadName=Thread-1  1520394585265
  end A ThreadName=Thread-1  1520394585765
                   main end 1520394585765
begin B ThreadName=Thread-0  1520394585765
  end B ThreadName=Thread-0  1520394586265
```

图 3-49　运行结果 1

```
begin B ThreadName=Thread-0  1520394725161
  end B ThreadName=Thread-0  1520394725661
begin A ThreadName=Thread-1  1520394725661
  end A ThreadName=Thread-1  1520394726161
                   main end 1520394726161
```

图 3-50　运行结果 2

```
begin A ThreadName=Thread-1  1520394761485
  end A ThreadName=Thread-1  1520394761985
begin B ThreadName=Thread-0  1520394761985
                   main end 1520394761985
  end B ThreadName=Thread-0  1520394762485
```

图 3-51　运行结果 3

为什么会出现截然不同的运行结果呢？3.2.6 节将给出答案。

3.2.6　join() 方法后面的代码提前运行——解释意外

为了查看 join() 方法在 Run1.java 类中执行的时机，创建 RunFirst.java 类文件，代码如下：

```java
package test.run;

import extthread.ThreadA;
import extthread.ThreadB;

public class RunFirst {

    public static void main(String[] args) {
        ThreadB b = new ThreadB();
        ThreadA a = new ThreadA(b);
        a.start();
        b.start();
        System.out.println("    main end=" + System.currentTimeMillis());
    }

}
```

程序第 1 次运行结果如图 3-52 所示。程序第 2 次运行结果如图 3-53 所示。

图 3-52　程序第 1 次运行结果

图 3-53　程序第 2 次运行结果

多次运行 RunFirst.java 文件后，可以发现一个规律：main end 大部分是第一个输出的，但这不代表其每次都是第一个输出的。可以得出一个结论：当将 System.out.println() 方法当作 join(2000) 时，join(2000) 大部分是先运行的，也就是先抢到 ThreadB 的锁，然后快速进行释放。但由于线程执行 run() 方法的随机性，所以 ThreadA、ThreadB 及 main 线程都有可能先获得锁，不过为了在解释上的方便性，都假定 join() 方法先获得锁。

执行 Run1.java 文件后会出现一些不同的运行结果，在此示例中出现了 3 个线程、2 个锁，3 个线程分别是 ThreadA 线程、ThreadB 线程和 main 线程，2 个锁分别是 ThreadB 对象锁和 PrintStream 打印锁。

先来看看有可能出现的运行结果 A，如图 3-54 所示。

图 3-54　运行结果 A

说明：

1）b.join(200) 方法先抢到 B 锁，执行 JDK 源代码内部的 wait(200) 方法后立即将 B 锁进行释放。

2）ThreadA 和 ThreadB 开始抢 ThreadB 对象锁，而 ThreadA 抢到了锁，输出 ThreadA begin 并且运行 sleep(500)。

3）在运行 sleep(500) 之后 ThreadA 输出 ThreadA end 并释放锁。

4）当在 ThreadA 中的 sleep(500) 执行第 200ms 时，join(200) 的源代码 wait(200) 时间已到，想要继续执行 JDK 中的源代码时必须有 ThreadB 对象锁，但 ThreadB 对象锁此时被 ThreadA 持有，ThreadA 还要持有 300ms，在这 300ms 的过程中，main 线程和 ThreadB 一起争抢 ThreadB 对象锁。300ms 过后 ThreadA 释放锁，而 main 线程和 ThreadB 还会一起争抢 ThreadB 对象锁，main 线程抢到了锁，导致 join(200) 方法执行完毕继续运行，然后 main 线程和 ThreadB 都要执行 print() 方法，main 线程和 ThreadB 线程由原来抢 ThreadB 对象锁转而变为抢 PrintStream 锁，这时 main 线程抢到了 PrintStream 锁，优先输出 main end。

5）ThreadB 得到锁 PrintStream 输出 ThreadB begin。

6）500ms 之后再输出 ThreadB end。

再来看看有可能出现的运行结果 B，如图 3-55 所示。

```
begin B ThreadName=Thread-0   1520394725161
  end B ThreadName=Thread-0   1520394725661
begin A ThreadName=Thread-1   1520394725661
  end A ThreadName=Thread-1   1520394726161
                     main end 1520394726161
```

图 3-55　运行结果 B

说明：

1）b.join(200) 方法先抢到 B 锁，执行 JDK 源代码内部的 wait(200) 方法后立即将 B 锁进行释放。

2）ThreadA 和 ThreadB 开始抢 ThreadB 对象锁，而 ThreadA 抢到了锁，输出 ThreadA begin 并且运行 sleep(500)。

3）在运行 sleep(500) 之后 ThreadA 输出 ThreadA end 并释放锁。

4）当在 ThreadA 中的 sleep(500) 执行第 200ms 时，join(200) 的源代码 wait(200) 时间已到，想要继续执行 JDK 中的源代码时必须有 ThreadB 对象锁，但 ThreadB 对象锁此时被 ThreadA 持有，ThreadA 还要持有 300ms，在这 300ms 的过程中，main 线程和 ThreadB 一起争抢 ThreadB 对象锁。300ms 过后 ThreadA 释放锁，而 main 线程和 ThreadB 还会一起争抢 ThreadB 对象锁，可惜这次 main 线程并没有抢到 ThreadB 对象锁，而 ThreadB 抢到了锁。

5）ThreadB 抢到锁输出 ThreadB begin。

6）500ms 之后再输出 ThreadB end，并释放锁。

7）ThreadB 释放锁后，wait(200) 重新获得锁，发现时间已过，结束 join(200) 的运行，继续运行，在最后输出 main end。

再来看看有可能出现的运行结果 C，如图 3-56 所示。

```
begin A ThreadName=Thread-1   1520394761485
  end A ThreadName=Thread-1   1520394761985
begin B ThreadName=Thread-0   1520394761985
                     main end 1520394761985
  end B ThreadName=Thread-0   1520394762485
```

图 3-56　运行结果 C

说明：

1）b.join(200) 方法先抢到 B 锁，执行 JDK 源代码内部的 wait(200) 方法后立即将 B 锁进行释放。

2）ThreadA 和 ThreadB 开始抢 ThreadB 对象锁，而 ThreadA 抢到了锁，输出 ThreadA begin 并且运行 sleep(500)。

3）在运行 sleep(500) 之后，ThreadA 输出 ThreadA end 并释放锁。

4）当在 ThreadA 中的 sleep(500) 执行第 200ms 时，join(200) 的源代码 wait(200) 时间已到，想要继续执行 JDK 中的源代码时必须有 ThreadB 对象锁，但 ThreadB 对象锁此时被 ThreadA 持有，ThreadA 还要持有 300ms，在这 300ms 的过程中，main 线程和 ThreadB 一起争抢 ThreadB 对象锁。300ms 过后 ThreadA 释放锁，而 main 线程和 ThreadB 还会一起争抢 ThreadB 对象锁，main 线程抢到了锁导致 join(200) 方法执行完毕继续运行并释放 ThreadB 锁，然后 main 线程和 ThreadB 都要执行 print() 方法，main 线程和 ThreadB 线程由原来的抢 ThreadB 对象锁转而变为抢 PrintStream 锁，这时 ThreadB 抢到了 PrintStream 对象锁并输出 ThreadB begin 后释放锁，在执行 ThreadB 线程中的 sleep(500) 的同时 main 线程接着抢到了 PrintStream 锁，继续输出 main end。

5）500ms 过后 ThreadB 继续输出 ThreadB end。

3.2.7　join(long millis, int nanos) 方法的使用

public final synchronized void join(long millis, int nanos) 方法的作用是等待该线程终止的时间最长为 millis 毫秒 + nanos 纳秒。如果参数 nanos < 0 或者 nanos > 999999，则出现异常"nanosecond timeout value out of range"。

创建测试用的代码如下：

```java
public class Test1 {
    public static void main(String[] args) throws InterruptedException {
        // 秒
        // 毫秒
        // 微秒
        // 纳秒
        long beginTime = System.currentTimeMillis();
        Thread.currentThread().join(2000, 999999);
        long endTime = System.currentTimeMillis();
        System.out.println(endTime - beginTime);
    }
}
```

程序运行结果如下：

2001

join() 方法耗时 2001ms。

3.3　类 ThreadLocal 的使用

变量值的共享可以使用 public static 变量的形式实现，所有的线程都使用同一个 public static 变量，那如何实现每一个线程都有自己的变量呢？JDK 提供的 ThreadLocal 可用于解决这样的问题。

类 ThreadLocal 的主要作用是将数据放入当前线程对象中的 Map 中，这个 Map 是 Thread 类的实例变量。类 ThreadLocal 自己不管理、不存储任何数据，它只是数据和 Map 之间的桥梁，用于将数据放入 Map 中，执行流程如下：数据→ ThreadLocal → currentThread() → Map。

执行后每个线程中的 Map 存有自己的数据，Map 中的 key 存储的是 ThreadLocal 对象，value 就是存储的值。每个 Thread 中的 Map 值只对当前线程可见，其他线程不可以访问当前线程对象中 Map 的值。当前线程销毁，Map 随之销毁，Map 中的数据如果没有被引用、没有被使用，则随时 GC 收回。

线程、Map、数据之间的关系可以做以下类比：

人（Thread）随身带有兜子（Map），兜子（Map）里面有东西（value），这样，Thread 随身也有自己的数据了，随时可以访问自己的数据了。

由于 Map 中的 key 不可以重复，所以一个 ThreadLocal 对象对应一个 value，内存存储结构如图 3-57 所示。

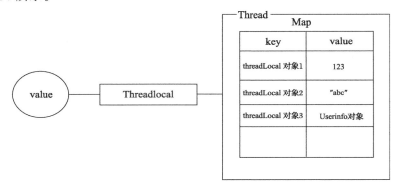

图 3-57　内存存储结构

3.3.1　get() 方法与 null

如果从未在 Thread 中的 Map 存储 ThreadLocal 对象对应的 value 值，则 get() 方法返回 null。
创建名称为 ThreadLocal11 的项目，类 Run.java 代码如下：

```
package test;

public class Run {
    public static ThreadLocal tl = new ThreadLocal();

    public static void main(String[] args) {
        if (tl.get() == null) {
            System.out.println("从未放过值");
            tl.set("我的值");
        }
        System.out.println(tl.get());
```

```
            System.out.println(tl.get());
    }
}
```

程序运行结果如图 3-58 所示。

从程序运行结果来看，第一次调用 tl 对象的 get() 方法时返回的值是 null，通过调用 set() 方法赋值后顺利取出值并输出到控制台上。类 ThreadLocal 解决的是变量在不同线程的隔离性，也就是不同线程拥有自己的值，不同线程的值是可以通过 ThreadLocal 类进行保存的。

图 3-58　程序运行结果

3.3.2　类 ThreadLocal 存取数据流程分析

下面通过以下测试程序来从 JDK 源代码的角度分析一下 ThreadLocal 类执行存取操作的流程：

```
public class Test {
    public static void main(String[] args) throws IOException, InterruptedException {
        ThreadLocal local = new ThreadLocal();
        local.set("我是任意的值");
        System.out.println(local.get());
    }
}
```

1）执行 ThreadLocal.set(" 我是任意的值 ") 代码时，ThreadLocal 代码如下：

```
public void set(T value) {
    Thread t = Thread.currentThread();//对象t就是main线程
    ThreadLocalMap map = getMap(t);//从main线程中获得ThreadLocalMap
    if (map != null)//不是第一次调用set方法时，map值不是null
        map.set(this, value);
    else
        createMap(t, value);//是第一次调用set方法时，执行createMap()方法
}//此源代码在ThreadLocal.java类中
```

2）代码 ThreadLocalMap map = getMap(t) 中的 getMap(t) 的源代码如下：

```
ThreadLocalMap getMap(Thread t) {//参数t就是前面传入的main线程
    return t.threadLocals;
    //返回main线程中threadLocals变量对应的ThreadLocalMap对象
}//此源代码在ThreadLocal.java类中
```

3）声明变量 t.threadLocals 的源代码如下：

```
public class Thread implements Runnable {
    ThreadLocal.ThreadLocalMap threadLocals = null;//默认值为null
    ……   //此源代码在Thread.java类中
```

对象 threadLocals 数据类型就是 ThreadLocal.ThreadLocalMap，变量 threadLocals 是

Thread 类中的实例变量。

4）取得 Thread 中的 ThreadLocal.ThreadLocalMap 后，第一次向其存放数据时会调用 createMap() 方法来创建 ThreadLocal.ThreadLocalMap 对象，因为 ThreadLocal.ThreadLocalMap threadLocals 对象默认值是 null，执行 createMap() 方法的源代码如图 3-59 所示。此源代码在 ThreadLocal.java 类中。

```
199   public void set(T value) {
200       Thread t = Thread.currentThread();
201       ThreadLocalMap map = getMap(t);
202       if (map != null)
203           map.set(this, value);
204       else
205           createMap(t, value);
206   }
```

图 3-59 执行 createMap() 方法的源代码

5）createMap() 方法的功能是创建一个新的 ThreadLocalMap，并向这个新的 ThreadLocalMap 存储数据，ThreadLocalMap 中的 key 就是当前的 ThreadLocal 对象，值就是传入的 value。createMap() 方法源代码如下：

```
void createMap(Thread t, T firstValue) {
    t.threadLocals = new ThreadLocalMap(this, firstValue);
}//此源代码在ThreadLocal.java类中
```

在实例化 ThreadLocalMap 的时候，向构造方法传入 this 和 firstValue，参数 this 就是当前 ThreadLocal 对象，firstValue 就是调用 ThreadLocal 对象 set() 方法传入的参数值。

6）new ThreadLocalMap(this, firstValue) 构造方法的源代码如图 3-60 所示。此源代码在 ThreadLocal.java 类中。

```
365   ThreadLocalMap(ThreadLocal<?> firstKey, Object firstValue) {
366       table = new Entry[INITIAL_CAPACITY];
367       int i = firstKey.threadLocalHashCode & (INITIAL_CAPACITY
368       table[i] = new Entry(firstKey, firstValue);
369       size = 1;
370       setThreshold(INITIAL_CAPACITY);
371   }
```
核心代码

图 3-60 类 ThreadLocalMap 有参构造方法的源代码

在源代码中可以发现，ThreadLocal 对象与 firstValue 被封装进 Entry 对象中，并放入 table[] 数组中。

7）table[] 数组的源代码如下：

```
static class ThreadLocalMap {
……
private Entry[] table;
……//此源代码在ThreadLocal.java类中
```

变量 table 就是 Entry[] 数组类型。

经过上面 7 个步骤，成功将 value 通过 ThreadLocal 放入当前线程 currentThread() 中的 ThreadLocalMap 对象中。

下面看看 get() 的执行流程。当执行"System.out.println(local.get());"代码时，ThreadLocal.get() 源代码如下：

```
public T get() {
    Thread t = Thread.currentThread();//t就是main线程
    ThreadLocalMap map = getMap(t);//从main线程中获得Map
    if (map != null) {//进入此分支，因为map不是null
        //执行getEntry()以this作为key，获得对应的Entry对象
        ThreadLocalMap.Entry e = map.getEntry(this);
        if (e != null) {//进入此分支，因为Entry对象不为null
            @SuppressWarnings("unchecked")
            T result = (T)e.value;//从Entry对象中取得value并返回
            return result;
        }
    }
    return setInitialValue();
}//此源代码在ThreadLocal.java类中
```

上面的流程就是 get() 方法的执行过程。

上面就是 set() 和 get() 的执行流程，流程比较麻烦，为什么不能直接向 Thread 类中的 ThreadLocalMap 对象存取数据呢？这是不能实现的，原因请看如图 3-61 所示的源代码。

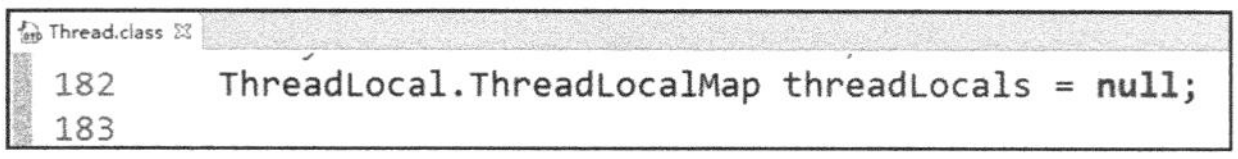

```
182        ThreadLocal.ThreadLocalMap threadLocals = null;
183
```

图 3-61　变量 threadLocals 默认是包级访问

变量 threadLocals 默认是包级访问，所以不能直接从外部访问该变量，只有同包中的类可以访问 threadLocals 变量，而 ThreadLocal 和 Thread 恰好在同一个包中，源代码如图 3-62 所示。

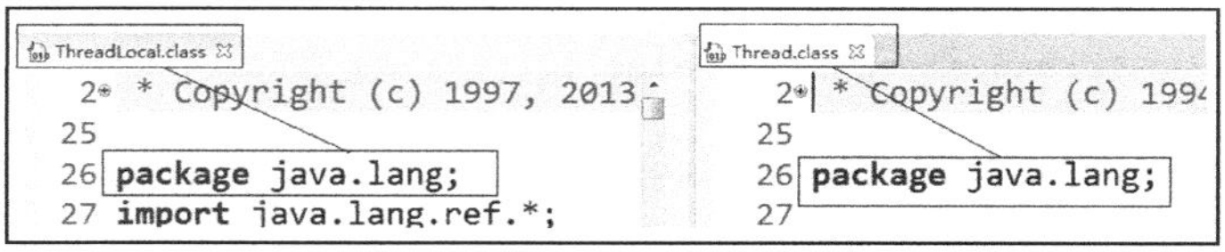

```
 2* * Copyright (c) 1997, 2013         2* * Copyright (c) 1994
25                                      25
26  package java.lang;                   26  package java.lang;
27  import java.lang.ref.*;              27
```

图 3-62　同在 lang 包下 ThreadLocal 可以访问 Thread 中的 ThreadLocalMap

由于在同一个 lang 包下，所以外部代码通过 ThreadLocal 可以访问 Thread 类中的"秘密对象"——ThreadLocalMap。

3.3.3　验证线程变量的隔离性

下面通过示例来验证通过 ThreadLocal 可向每个线程中存储自己的私有数据。

创建测试用的项目 ThreadLocalTest，类 Tools.java 代码如下：

```java
package test;

public class Tools {
    public static ThreadLocal tl = new ThreadLocal();
}
```

两个自定义线程类代码如下：

```java
package test;

public class MyThreadA extends Thread {
    @Override
    public void run() {
        try {
            for (int i = 0; i < 10; i++) {
                Tools.tl.set("A " + (i + 1));
                System.out.println("A get " + Tools.tl.get());
                int sleepValue = (int) (Math.random() * 1000);
                Thread.sleep(sleepValue);
            }
        } catch (InterruptedException e) {
            e.printStackTrace();
        }
    }
}
```

```java
package test;

public class MyThreadB extends Thread {
    @Override
    public void run() {
        try {
            for (int i = 0; i < 10; i++) {
                Tools.tl.set("B " + (i + 1));
                System.out.println("    B get " + Tools.tl.get());
                int sleepValue = (int) (Math.random() * 1000);
                Thread.sleep(sleepValue);
            }
        } catch (InterruptedException e) {
            e.printStackTrace();
        }
    }
}
```

类 Test.java 代码如下：

```java
package test;

import java.io.IOException;
```

```java
public class Test {
    public static void main(String[] args) throws IOException, InterruptedException {
        MyThreadA a = new MyThreadA();
        MyThreadB b = new MyThreadB();
        a.start();
        b.start();
        for (int i = 0; i < 10; i++) {
            Tools.tl.set("main " + (i + 1));
            System.out.println("          main get " + Tools.tl.get());
            int sleepValue = (int) (Math.random() * 1000);
            Thread.sleep(sleepValue);
        }
    }
}
```

程序运行结果如下：

```
              main get main 1
A get A 1
    B get B 1
              main get main 2
    B get B 2
A get A 2
              main get main 3
A get A 3
    B get B 3
A get A 4
              main get main 4
A get A 5
    B get B 4
              main get main 5
    B get B 5
              main get main 6
A get A 6
    B get B 6
              main get main 7
A get A 7
A get A 8
              main get main 8
    B get B 7
A get A 9
    B get B 8
    B get B 9
              main get main 9
A get A 10
              main get main 10
    B get B 10
```

程序运行结果表明通过 ThreadLocal 可向每一个线程存储自己的私有数据，虽然 3 个线程都向 tl 对象中 set() 数据值，但每个线程还是能取出自己的数据，不能取出别的线程的。

创建新的项目 s5 来再次验证数据的隔离性。类 Tools.java 代码如下：

```java
package test;

public class Tools {
    public static ThreadLocal tl = new ThreadLocal();
}
```

创建两个线程类代码如下：

```java
package test;

public class MyThreadA extends Thread {
    @Override
    public void run() {
        try {
            for (int i = 0; i < 10; i++) {
                if (Tools.tl.get() == null) {
                    Tools.tl.set("A " + (i + 1));
                }
                System.out.println("A get " + Tools.tl.get());
                int sleepValue = (int) (Math.random() * 1000);
                Thread.sleep(sleepValue);
            }
        } catch (InterruptedException e) {
            e.printStackTrace();
        }
    }
}
```

```java
package test;

public class MyThreadB extends Thread {
    @Override
    public void run() {
        try {
            for (int i = 1; i < 10; i++) {
                if (Tools.tl.get() == null) {
                    Tools.tl.set("B " + (i + 1));
                }
                System.out.println("    B get " + Tools.tl.get());
                int sleepValue = (int) (Math.random() * 1000);
                Thread.sleep(sleepValue);
            }
        } catch (InterruptedException e) {
            e.printStackTrace();
        }
    }
}
```

类 Test.java 代码如下：

```java
package test;
```

```
import java.io.IOException;

public class Test {
    public static void main(String[] args) throws IOException, InterruptedException {
        MyThreadA a = new MyThreadA();
        MyThreadB b = new MyThreadB();
        a.start();
        b.start();
        for (int i = 2; i < 10; i++) {
            if (Tools.tl.get() == null) {
                Tools.tl.set("main " + (i + 1));
            }
            System.out.println("              main get " + Tools.tl.get());
            int sleepValue = (int) (Math.random() * 1000);
            Thread.sleep(sleepValue);
        }
    }
}
```

程序运行结果如下：

```
              main get main 3
    B get B 2
A get A 1
              main get main 3
A get A 1
    B get B 2
A get A 1
    B get B 2
A get A 1
A get A 1
    B get B 2
              main get main 3
A get A 1
A get A 1
    B get B 2
A get A 1
              main get main 3
A get A 1
              main get main 3
    B get B 2
    B get B 2
A get A 1
    B get B 2
              main get main 3
    B get B 2
              main get main 3
              main get main 3
```

在 for 循环中使用 if 语句来判断当前线程的 ThreadLocalMap 中是否有数据，如果有则不再重复 set，所以线程 a 存取值 1，线程 b 存取值 2，线程 main 存取值 3。

3.3.4 解决 get() 方法返回 null 的问题

在第一次调用 Threadlocal 类的 get() 方法时，返回值是 null，怎样实现第一次调用 get() 不返回 null 呢？也就是说，怎样使其具有默认值的效果呢？

创建名称为 ThreadLocal22 的项目，继承 ThreadLocal 类产生 ThreadLocalExt.java 类，代码如下：

```java
package ext;

public class ThreadLocalExt extends ThreadLocal {
    @Override
    protected Object initialValue() {
        return "我是默认值 第一次get不再为null";
    }
}
```

覆盖 initialValue() 方法具有初始值，因为 ThreadLocal.java 中的 initialValue() 方法默认返回值是 null，所以要在子类中进行重写，源代码如下：

```java
protected T initialValue() {
    return null;
}
```

运行类 Run.java 代码如下：

```java
package test;

import ext.ThreadLocalExt;

public class Run {
    public static ThreadLocalExt tl = new ThreadLocalExt();

    public static void main(String[] args) {
        if (tl.get() == null) {
            System.out.println("从未放过值");
            tl.set("我的值");
        }
        System.out.println(tl.get());
        System.out.println(tl.get());
    }
}
```

程序运行结果如图 3-63 所示。

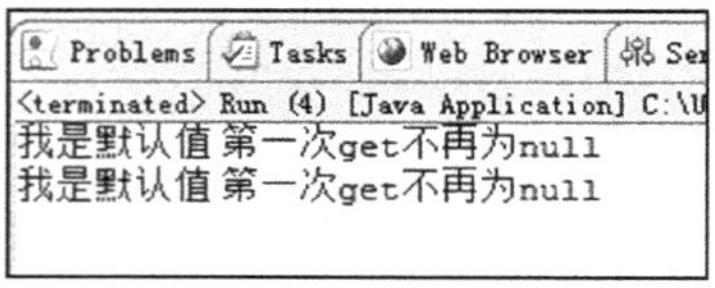

图 3-63 get() 方法返回 null 问题得以解决

此示例仅仅证明 main 线程有自己的值，那其他线程是否会有自己的初始值呢？

3.3.5 验证重写 initialValue() 方法的隔离性

创建名称为 ThreadLocal33 的项目，类 Tools.java 代码如下：

```java
package tools;

import ext.ThreadLocalExt;

public class Tools {
    public static ThreadLocalExt tl = new ThreadLocalExt();
}
```

类 ThreadLocalExt.java 代码如下：

```java
package ext;

import java.util.Date;

public class ThreadLocalExt extends ThreadLocal {
    @Override
    protected Object initialValue() {
        return new Date().getTime();
    }
}
```

类 ThreadA.java 代码如下：

```java
package extthread;

import tools.Tools;

public class ThreadA extends Thread {

    @Override
    public void run() {
        try {
            for (int i = 0; i < 10; i++) {
                System.out.println("在ThreadA线程中取值=" + Tools.tl.get());
                Thread.sleep(100);
            }
        } catch (InterruptedException e) {
            // TODO Auto-generated catch block
            e.printStackTrace();
        }
    }

}
```

类 Run.java 代码如下：

```
package test;

import tools.Tools;
import extthread.ThreadA;

public class Run {
    public static void main(String[] args) {
        try {
            for (int i = 0; i < 10; i++) {
                System.out.println("         在Main线程中取值=" + Tools.tl.get());
                Thread.sleep(100);
            }
            Thread.sleep(5000);
            ThreadA a = new ThreadA();
            a.start();
        } catch (InterruptedException e) {
            e.printStackTrace();
        }
    }
}
```

程序运行结果如图 3-64 所示。

图 3-64　程序运行结果

程序运行结果表明子线程和父线程各有自己所拥有的值。

3.4　类 InheritableThreadLocal 的使用

使用类 InheritableThreadLocal 可使子线程继承父线程的值。

3.4.1 类 ThreadLocal 不能实现值继承

创建测试用的项目 ThreadLocalNoExtends。

创建类 Tools.java 代码如下：

```java
package test;

public class Tools {
    public static ThreadLocal tl = new ThreadLocal();
}
```

创建类 ThreadA.java 代码如下：

```java
package test;

public class ThreadA extends Thread {
    @Override
    public void run() {
        try {
            for (int i = 0; i < 10; i++) {
                System.out.println("在ThreadA线程中取值=" + Tools.tl.get());
                Thread.sleep(100);
            }
        } catch (InterruptedException e) {
            e.printStackTrace();
        }
    }
}
```

创建类 Test.java 代码如下：

```java
package test;

public class Test {

    public static void main(String[] args) {
        try {
            for (int i = 0; i < 10; i++) {
                if (Tools.tl.get() == null) {
                    Tools.tl.set("此值是main线程放入的！");
                }
                System.out.println("        在Main线程中取值=" + Tools.tl.get());
                Thread.sleep(100);
            }
            Thread.sleep(5000);
            ThreadA a = new ThreadA();
            a.start();
        } catch (InterruptedException e) {
            e.printStackTrace();
        }
```

        }
    }
程序运行结果如下：

```
在Main线程中取值=此值是main线程放入的！
在Main线程中取值=此值是main线程放入的！
在Main线程中取值=此值是main线程放入的！
在Main线程中取值=此值是main线程放入的！
在Main线程中取值=此值是main线程放入的！
在Main线程中取值=此值是main线程放入的！
在Main线程中取值=此值是main线程放入的！
在Main线程中取值=此值是main线程放入的！
在Main线程中取值=此值是main线程放入的！
在Main线程中取值=此值是main线程放入的！
在ThreadA线程中取值=null
在ThreadA线程中取值=null
在ThreadA线程中取值=null
在ThreadA线程中取值=null
在ThreadA线程中取值=null
在ThreadA线程中取值=null
在ThreadA线程中取值=null
在ThreadA线程中取值=null
在ThreadA线程中取值=null
在ThreadA线程中取值=null
```

main 线程创建了 ThreadA 线程，所以 main 线程是 ThreadA 线程的父线程，从运行结果可以发现，main 线程中的值并没有继承给 ThreadA，所以 ThreadLocal 并不具有值继承特性，这时就要使用 InheritableThreadLocal 类进行替换了。

3.4.2 使用 InheritableThreadLocal 体现值继承特性

使用 InheritableThreadLocal 类可以让子线程从父线程继承值。

创建测试用的项目 InheritableThreadLocal1。

类 Tools.java 代码如下：

```java
package tools;

public class Tools {
    public static InheritableThreadLocal tl = new
                    InheritableThreadLocal();
}
```

类 ThreadA.java 代码如下：

```java
package extthread;

import tools.Tools;
```

```java
public class ThreadA extends Thread {
    @Override
    public void run() {
        try {
            for (int i = 0; i < 10; i++) {
                System.out.println("在ThreadA线程中取值=" + Tools.tl.get());
                Thread.sleep(100);
            }
        } catch (InterruptedException e) {
            e.printStackTrace();
        }
    }
}
```

类 Test.java 代码如下：

```java
package test;

import extthread.ThreadA;
import tools.Tools;

public class Test {

    public static void main(String[] args) {
        try {
            for (int i = 0; i < 10; i++) {
                if (Tools.tl.get() == null) {
                    Tools.tl.set("此值是main线程放入的！");
                }
                System.out.println("         在Main线程中取值=" + Tools.tl.get());
                Thread.sleep(100);
            }
            Thread.sleep(5000);
            ThreadA a = new ThreadA();
            a.start();
        } catch (InterruptedException e) {
            e.printStackTrace();
        }
    }
}
```

程序运行结果如下：

```
在Main线程中取值=此值是main线程放入的！
在Main线程中取值=此值是main线程放入的！
在Main线程中取值=此值是main线程放入的！
在Main线程中取值=此值是main线程放入的！
在Main线程中取值=此值是main线程放入的！
在Main线程中取值=此值是main线程放入的！
在Main线程中取值=此值是main线程放入的！
```

```
            在Main线程中取值=此值是main线程放入的!
            在Main线程中取值=此值是main线程放入的!
            在Main线程中取值=此值是main线程放入的!
在ThreadA线程中取值=此值是main线程放入的!
在ThreadA线程中取值=此值是main线程放入的!
在ThreadA线程中取值=此值是main线程放入的!
在ThreadA线程中取值=此值是main线程放入的!
在ThreadA线程中取值=此值是main线程放入的!
在ThreadA线程中取值=此值是main线程放入的!
在ThreadA线程中取值=此值是main线程放入的!
在ThreadA线程中取值=此值是main线程放入的!
在ThreadA线程中取值=此值是main线程放入的!
在ThreadA线程中取值=此值是main线程放入的!
```

子 ThreadA 线程获取的值是从父线程 main 继承的。

3.4.3 值继承特性在源代码中的执行流程

使用 InheritableThreadLocal 的确可以实现值继承特性,那么 JDK 的源代码是如何实现这个特性的呢?下面按步骤来分析一下。

1)下面分析一下类 InheritableThreadLocal 的源代码,如下:

```java
public class InheritableThreadLocal<T> extends ThreadLocal<T> {
    protected T childValue(T parentValue) {
        return parentValue;
    }

    ThreadLocalMap getMap(Thread t) {
        return t.inheritableThreadLocals;
    }

    void createMap(Thread t, T firstValue) {
        t.inheritableThreadLocals = new ThreadLocalMap(this, firstValue);
    }
}
```

InheritableThreadLocal 类的源代码中存在 3 个方法,这 3 个方法都是对父类 ThreadLocal 中的同名方法进行重写,但在源代码中并没有使用 @Override 进行标识,所以在初期分析时,流程是比较复杂的。

2)在 main() 方法中使用 main 线程执行 InheritableThreadLocal.set() 方法,源代码如下:

```java
public static void main(String[] args) {
    try {
        for (int i = 0; i < 10; i++) {
            if (Tools.tl.get() == null) {
                Tools.tl.set("此值是main线程放入的!");//在此处执行
            }
```

调用 InheritableThreadLocal 对象中的 set() 方法其实就是调用 ThreadLocal.java 类中的 set() 方法，因为 InheritableThreadLocal 并没有重写 set() 方法。

3）下面分析一下 ThreadLocal.java 类中的 set() 方法，源代码如下：

```java
public void set(T value) {
    Thread t = Thread.currentThread();
    ThreadLocalMap map = getMap(t);
    if (map != null)
        map.set(this, value);
    else
        createMap(t, value);
}
```

执行 ThreadLocal.java 类中的 set() 方法时，有两个方法已经被 InheritableThreadLocal 类重写了，分别是 getMap(t) 和 createMap(t, value)，一定要留意，所以在执行这两个方法时，调用的是 InheritableThreadLocal 类中重写的 getMap(t) 和 createMap(t, value) 这两个方法。重写的这两个方法在 InheritableThreadLocal 类中的源代码如下：

```java
public class InheritableThreadLocal<T> extends ThreadLocal<T> {
    ThreadLocalMap getMap(Thread t) {
        return t.inheritableThreadLocals;
    }
    void createMap(Thread t, T firstValue) {
        t.inheritableThreadLocals = new ThreadLocalMap(this, firstValue);
    }
}
```

4）通过查看 InheritableThreadLocal 类中 getMap(Thread t) 和 createMap(Thread t, T firstValue) 方法的源代码可以明确一个重要的知识点，那就是不再向 Thread 类中的 ThreadLocal.ThreadLocalMap threadLocals 存入数据了，而是向 ThreadLocal.ThreadLocal-Map inheritableThreadLocals 存入数据，这两个对象在 Thread.java 类中的声明如下：

```java
public class Thread implements Runnable {
    ThreadLocal.ThreadLocalMap threadLocals = null;
    ThreadLocal.ThreadLocalMap inheritableThreadLocals = null;
```

上面的分析步骤已经明确一个知识点，就是线程 main 向 inheritableThreadLocals 对象存入数据，对象 inheritableThreadLocals 就是存储数据的容器，那么子线程如何实现从父线程中的 inheritableThreadLocals 对象继承值呢？

5）这个实现的思路就是在创建子线程 ThreadA 时，子线程主动引用父线程 main 中的 inheritableThreadLocals 对象值，源代码如下：

```java
public class Thread implements Runnable {
    private void init(ThreadGroup g, Runnable target, String name,
                      long stackSize, AccessControlContext acc,
                      boolean inheritThreadLocals) {
        ......
```

```
        if (inheritThreadLocals && parent.inheritableThreadLocals != null)
            this.inheritableThreadLocals = ThreadLocal.createInheritedMap(parent.
                inheritableThreadLocals);
        ......
    }
```

方法 init(ThreadGroup g, Runnable target, String name, long stackSize, AccessControl-Context acc, boolean inheritThreadLocals) 是被 Thread 的构造方法调用的，所以在 new ThreadA() 时，Thread.java 源代码内部会自动调用 init(ThreadGroup g, Runnable target, String name, long stackSize, AccessControlContext acc, boolean inheritThreadLocals) 方法。

在 init(ThreadGroup g, Runnable target, String name, long stackSize, AccessControlContext acc, boolean inheritThreadLocals) 方法中，最后一个参数 inheritThreadLocals 代表当前线程对象是否会从父线程继承值，每一次都会继承值，因为这个值被永远传入 true，传入 true 的源代码在 init(ThreadGroup g, Runnable target, String name, long stackSize) 方法中，源代码如下：

```
    private void init(ThreadGroup g, Runnable target, String name,
            long stackSize) {
        init(g, target, name, stackSize, null, true);
    }
```

也就是方法 init(ThreadGroup g, Runnable target, String name, long stackSize) 调用方法 init(ThreadGroup g, Runnable target, String name, long stackSize, AccessControlContext acc, boolean inheritThreadLocals)。

对最后一个参数永远传入 true，所以最后一个参数 inheritThreadLocals 永远为 true。

5）在执行 init(ThreadGroup g, Runnable target, String name, long stackSize, Access-ControlContext acc, boolean inheritThreadLocals) 方法中的 if 语句时：

```
        if (inheritThreadLocals && parent.inheritableThreadLocals != null)
```

如果运算符 && 左边的表达式 inheritThreadLocals 值为 true，那么开始计算 && 右边的表达式 parent.inheritableThreadLocals != null。

当向 main 线程中的 inheritableThreadLocals 存放数据时，对象 inheritableThreadLocals 的值并不是 null，所以 && 运算符两端的结果都为 true，那么程序继续运行，对当前线程的 inheritableThreadLocals 对象变量进行赋值，代码如下：

```
    this.inheritableThreadLocals = 
    ThreadLocal.createInheritedMap(parent.inheritableThreadLocals);
```

6）代码 this.inheritableThreadLocals 中的 this 就是当前 ThreadA.java 类的对象，执行 createInheritedMap() 的目的是创建一个新的 ThreadLocalMap 对象，然后将新的 ThreadLocalMap 对象赋值给 ThreadA 对象中的 inheritableThreadLocals 变量，方法 create-InheritedMap() 源代码如下：

```java
static ThreadLocalMap createInheritedMap(ThreadLocalMap parentMap) {
    return new ThreadLocalMap(parentMap);
}
```

7）下面继续分析 new ThreadLocalMap(parentMap) 构造方法中的核心源代码，如下：

```java
private ThreadLocalMap(ThreadLocalMap parentMap) {
    Entry[] parentTable = parentMap.table;
    int len = parentTable.length;
    setThreshold(len);
    table = new Entry[len];//新建Entry[]数组

    for (int j = 0; j < len; j++) {
        Entry e = parentTable[j];
        if (e != null) {
            @SuppressWarnings("unchecked")
            ThreadLocal<Object> key = (ThreadLocal<Object>) e.get();
            if (key != null) {
                Object value = key.childValue(e.value);
                Entry c = new Entry(key, value);
                //实例化新的Entry对象
                int h = key.threadLocalHashCode & (len - 1);
                while (table[h] != null)
                    h = nextIndex(h, len);
                table[h] = c;//将父线程中的数据复制到新数组中
                size++;
            }
        }
    }
}
```

在构造方法的完整源代码算法中可以发现，子线程将父线程中的 table 对象以复制的方式赋值给子线程的 table 数组，这个过程是在创建 Thread 类对象时发生的，也就说明当子线程对象创建完毕后，子线程中的数据就是主线程中旧的数据，主线程使用新的数据时，子线程还是使用旧的数据，因为主子线程使用两个 Entry[] 对象数组各自存储自己的值。此知识点在后面章节中会进行详细讲解。

3.4.4 父线程有最新的值，子线程仍是旧值

创建测试用的项目 InheritableThreadLocal101。

创建类 Tools.java 代码如下：

```java
package tools;

public class Tools {
    public static InheritableThreadLocal tl = new InheritableThreadLocal();
}
```

创建类 ThreadA.java 代码如下：

```java
package extthread;

import tools.Tools;

public class ThreadA extends Thread {
    @Override
    public void run() {
        try {
            for (int i = 0; i < 10; i++) {
                System.out.println("在ThreadA线程中取值=" + Tools.tl.get());
                Thread.sleep(1000);
            }
        } catch (InterruptedException e) {
            e.printStackTrace();
        }
    }
}
```

创建类 Test.java 代码如下:

```java
package test;

import extthread.ThreadA;
import tools.Tools;

public class Test {
    public static void main(String[] args) throws InterruptedException {
        if (Tools.tl.get() == null) {
            Tools.tl.set("此值是main线程放入的!");
        }
        System.out.println("        在Main线程中取值=" + Tools.tl.get());
        Thread.sleep(100);
        ThreadA a = new ThreadA();
        a.start();
        Thread.sleep(5000);
        Tools.tl.set("此值是main线程newnewnewnewnew放入的!");
    }
}
```

程序运行后, 子线程仍持有旧的数据, 输出结果如下:

```
        在Main线程中取值=此值是main线程放入的!
在ThreadA线程中取值=此值是main线程放入的!
在ThreadA线程中取值=此值是main线程放入的!
在ThreadA线程中取值=此值是main线程放入的!
在ThreadA线程中取值=此值是main线程放入的!
在ThreadA线程中取值=此值是main线程放入的!
在ThreadA线程中取值=此值是main线程放入的!
在ThreadA线程中取值=此值是main线程放入的!
在ThreadA线程中取值=此值是main线程放入的!
```

在ThreadA线程中取值=此值是main线程放入的！
在ThreadA线程中取值=此值是main线程放入的！

3.4.5 子线程有最新的值，父线程仍是旧值

创建测试用的项目 InheritableThreadLocal102。

创建类 Tools.java 代码如下：

```java
package tools;

public class Tools {
    public static InheritableThreadLocal tl = new InheritableThreadLocal();
}
```

创建类 ThreadA.java 代码如下：

```java
package extthread;

import tools.Tools;

public class ThreadA extends Thread {
    @Override
    public void run() {
        try {
            for (int i = 0; i < 10; i++) {
                System.out.println("在ThreadA线程中取值=" + Tools.tl.get());
                Thread.sleep(1000);
                if (i == 5) {
                    Tools.tl.set("我是ThreadA的newnewnewnew最新的值！");
                    System.out.println("ThreadA已经存在最的值----------------");
                }
            }
        } catch (InterruptedException e) {
            e.printStackTrace();
        }
    }
}
```

创建类 Test.java 代码如下：

```java
package test;

import extthread.ThreadA;
import tools.Tools;

public class Test {
    public static void main(String[] args) throws InterruptedException {
        if (Tools.tl.get() == null) {
            Tools.tl.set("此值是main线程放入的！");
        }
```

```java
        System.out.println("          在Main线程中取值=" + Tools.tl.get());
        Thread.sleep(100);
        ThreadA a = new ThreadA();
        a.start();
        Thread.sleep(3000);
        for (int i = 0; i < 10; i++) {
            System.out.println("main end get value=" + Tools.tl.get());
            Thread.sleep(1000);
        }
    }
}
```

程序运行后，子线程仍持有旧的数据，输出结果如下：

```
          在Main线程中取值=此值是main线程放入的!
在ThreadA线程中取值=此值是main线程放入的!
在ThreadA线程中取值=此值是main线程放入的!
在ThreadA线程中取值=此值是main线程放入的!
在ThreadA线程中取值=此值是main线程放入的!
main end get value=此值是main线程放入的!
在ThreadA线程中取值=此值是main线程放入的!
main end get value=此值是main线程放入的!
main end get value=此值是main线程放入的!
在ThreadA线程中取值=此值是main线程放入的!
ThreadA已经存在最的值----------------
main end get value=此值是main线程放入的!
在ThreadA线程中取值=我是ThreadA的newnewnewnew最新的值!
在ThreadA线程中取值=我是ThreadA的newnewnewnew最新的值!
main end get value=此值是main线程放入的!
在ThreadA线程中取值=我是ThreadA的newnewnewnew最新的值!
main end get value=此值是main线程放入的!
在ThreadA线程中取值=我是ThreadA的newnewnewnew最新的值!
main end get value=此值是main线程放入的!
main end get value=此值是main线程放入的!
main end get value=此值是main线程放入的!
main end get value=此值是main线程放入的!
```

main 线程存储的值永远是旧的数据。

3.4.6 子线程可以感应对象属性值的变化

前面示例都是在主、子线程中使用 String 数据类型做继承特性的演示，当子线程从父线程继承可变对象数据类型时，子线程可以取到最新对象中的属性值。

创建测试用的项目 InheritableThreadLocal103。

创建类 Userinfo.java 代码如下：

```java
package entity;

public class Userinfo {
```

```java
    private String username;

    public String getUsername() {
        return username;
    }

    public void setUsername(String username) {
        this.username = username;
    }
}
```

创建类 Tools.java 代码如下：

```java
package tools;

import entity.Userinfo;

public class Tools {
    public static InheritableThreadLocal<Userinfo> tl = new InheritableThreadLocal<>();
}
```

创建类 ThreadA.java 代码如下：

```java
package extthread;

import entity.Userinfo;
import tools.Tools;

public class ThreadA extends Thread {
    @Override
    public void run() {
        try {
            for (int i = 0; i < 10; i++) {
                Userinfo userinfo = Tools.tl.get();
                System.out.println("在ThreadA线程中取值=" + userinfo.getUsername()
                    + " " + userinfo.hashCode());
                Thread.sleep(1000);
            }
        } catch (InterruptedException e) {
            e.printStackTrace();
        }
    }
}
```

创建类 Test.java 代码如下：

```java
package test;

import entity.Userinfo;
```

```
import extthread.ThreadA;
import tools.Tools;

public class Test {
    public static void main(String[] args) throws InterruptedException {
        Userinfo userinfo = new Userinfo();
        System.out.println("A userinfo " + userinfo.hashCode());
        userinfo.setUsername("中国");
        if (Tools.tl.get() == null) {
            Tools.tl.set(userinfo);
        }
        System.out.println("            在Main线程中取值=" + Tools.tl.get().
            getUsername() + " " + Tools.tl.get().hashCode());
        Thread.sleep(100);
        ThreadA a = new ThreadA();
        a.start();
        Thread.sleep(5000);
        Tools.tl.get().setUsername("美国");
    }
}
```

程序运行结果就是 ThreadA 取到 userinfo 对象的最新属性值 "美国"，运行结果如下：

```
A userinfo 366712642
        在Main线程中取值=中国 366712642
在ThreadA线程中取值=中国 366712642
在ThreadA线程中取值=中国 366712642
在ThreadA线程中取值=中国 366712642
在ThreadA线程中取值=中国 366712642
在ThreadA线程中取值=中国 366712642
在ThreadA线程中取值=美国 366712642
在ThreadA线程中取值=美国 366712642
在ThreadA线程中取值=美国 366712642
在ThreadA线程中取值=美国 366712642
在ThreadA线程中取值=美国 366712642
```

如果在 main() 方法的最后重新放入一个新的 Userinfo 对象，则 ThreadA 线程输出的结果永远是"中国"，因为 ThreadA 永远引用的是"中国"对应的 Userinfo 对象，并不是"美国"对应的 Userinfo 对象，仍符合"父线程有最新的值，子线程仍是旧值"，代码如下：

```
package test;

import entity.Userinfo;
import extthread.ThreadA;
import tools.Tools;

public class Test2 {
    public static void main(String[] args) throws InterruptedException {
        Userinfo userinfo = new Userinfo();
        System.out.println("A userinfo " + userinfo.hashCode());
```

```
            userinfo.setUsername("中国");
            if (Tools.tl.get() == null) {
                Tools.tl.set(userinfo);
            }
            System.out.println("             在Main线程中取值=" + Tools.tl.get().
                    getUsername() + " " + Tools.tl.get().hashCode());
            Thread.sleep(100);
            ThreadA a = new ThreadA();
            a.start();
            Thread.sleep(5000);
            Userinfo userinfo2 = new Userinfo();
            userinfo2.setUsername("美国");
            System.out.println("B userinfo " + userinfo2.hashCode());
            Tools.tl.set(userinfo2);
        }

}
```

程序运行结果如下：

```
A userinfo 366712642
           在Main线程中取值=中国 366712642
在ThreadA线程中取值=中国 366712642
在ThreadA线程中取值=中国 366712642
在ThreadA线程中取值=中国 366712642
在ThreadA线程中取值=中国 366712642
在ThreadA线程中取值=中国 366712642
在ThreadA线程中取值=中国 366712642
B userinfo 1829164700
在ThreadA线程中取值=中国 366712642
在ThreadA线程中取值=中国 366712642
在ThreadA线程中取值=中国 366712642
在ThreadA线程中取值=中国 366712642
```

3.4.7　重写 childValue() 方法实现对继承的值进行加工

在继承的同时还可以对值进行进一步的加工。创建测试用的项目 InheritableThread-Local2，将 InheritableThreadLocal1 项目中的所有类复制到 InheritableThreadLocal2 项目中。

更改类 InheritableThreadLocalExt.java 代码如下：

```
package ext;

import java.util.Date;

public class InheritableThreadLocalExt extends InheritableThreadLocal {
    @Override
    protected Object initialValue() {
        return new Date().getTime();
    }

    @Override
```

```java
    protected Object childValue(Object parentValue) {
        return parentValue + " 我在子线程加的~!";
    }
}
```

程序运行结果如图 3-65 所示。

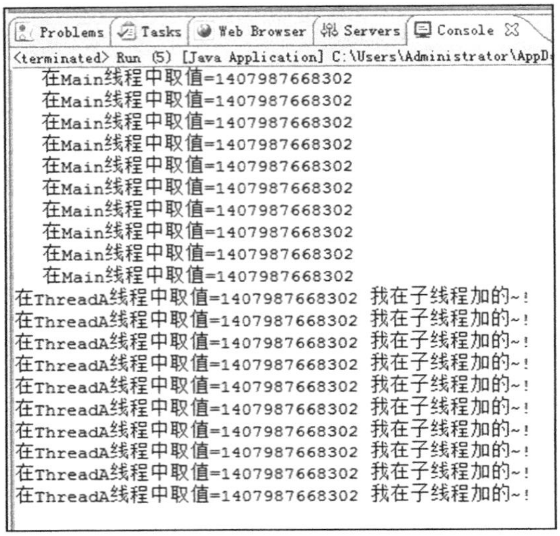

图 3-65　成功继承并修改

通过重写 childValue() 方法，子线程可以对父线程继承的值进行加工修改。

在子线程的任意时刻执行 InheritableThreadLocalExt.set() 方法可使子线程具有最新的值。另外，通过重写 childValue() 方法也会使子线程具有最新的值，那么这两点有什么区别？区别就是子线程可以在任意的时间执行 InheritableThreadLocalExt.set() 方法任意次，使自身具有最新的值，而重写 childValue() 方法实现子线程具有最新的值是只有在创建子线程时才会发生，但是仅仅是一次。

3.5　本章小结

经过本章的学习，完全可以使以前分散的线程对象进行通信与协作，线程任务不再是"单打独斗"，更具有"团结性"，因为它们之间可以互相通信，就像命令官与执行者一样，对任务的执行规划更加合理，不再具有随机性与盲目性。

第 4 章 Chapter 4

Lock 对象的使用

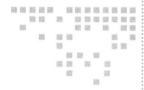

本章使用 Lock 对象实现同步的效果，Lock 对象在功能上比 synchronized 更加丰富，本章着重掌握如下两个知识点：

- ReentrantLock 类的使用；
- ReentrantReadWriteLock 类的使用。

4.1 使用 ReentrantLock 类

Java 多线程可以使用 synchronized 关键字来实现线程间同步，不过 JDK 1.5 新增加的 ReentrantLock 类也能达到同样的效果，并且在扩展功能上更加强大，如具有嗅探锁定、多路分支通知等功能。

4.1.1 使用 ReentrantLock 实现同步

既然 ReentrantLock 类在功能上相比 synchronized 更多，那么下面就以一个初步的程序示例来介绍一下 ReentrantLock 类的使用方法。

创建测试用的项目 ReentrantLockTest，创建类 MyService.java，代码如下：

```java
package service;

import java.util.concurrent.locks.Lock;
import java.util.concurrent.locks.ReentrantLock;

public class MyService {

    private Lock lock = new ReentrantLock();
```

```java
    public void testMethod() {
        lock.lock();
        for (int i = 0; i < 5; i++) {
            System.out.println("ThreadName=" + Thread.currentThread().getName()
                    + (" " + (i + 1)));
        }
        lock.unlock();
    }
}
```

调用 ReentrantLock 对象的 lock() 方法获取锁,调用 unlock() 方法释放锁,这两个方法成对使用。想要实现同步某些代码,把这些代码放在 lock() 和 unlock() 之间即可。

创建类 MyThread.java 代码如下:

```java
package extthread;

import service.MyService;

public class MyThread extends Thread {

    private MyService service;

    public MyThread(MyService service) {
        super();
        this.service = service;
    }

    @Override
    public void run() {
        service.testMethod();
    }
}
```

运行类 Run.java 代码如下:

```java
package test;

import service.MyService;
import extthread.MyThread;

public class Run {

    public static void main(String[] args) {

        MyService service = new MyService();

        MyThread a1 = new MyThread(service);
        MyThread a2 = new MyThread(service);
        MyThread a3 = new MyThread(service);
        MyThread a4 = new MyThread(service);
```

```
            MyThread a5 = new MyThread(service);

            a1.start();
            a2.start();
            a3.start();
            a4.start();
            a5.start();

    }

}
```

程序运行结果如图 4-1 所示。

从程序运行结果来看，只有当当前线程输出完毕之后将锁释放，其他线程才可以继续抢锁并输出，每个线程内输出的数据是有序的，从 1 到 5，因为当前线程已经持有锁，具有互斥排他性，但线程之间输出的顺序是随机的，即谁抢到锁，谁输出。

4.1.2 验证多代码块间的同步性

创建测试用的项目 ConditionTestMoreMethod，类 MyService.java 代码如下：

```
package service;

import java.util.concurrent.locks.Lock;
import java.util.concurrent.locks.ReentrantLock;

public class MyService {

    private Lock lock = new ReentrantLock();

    public void methodA() {
        try {
            lock.lock();
            System.out.println("methodA begin ThreadName="
                    + Thread.currentThread().getName()
                    + " time="
                    + System.currentTimeMillis());
            Thread.sleep(5000);
            System.out.println("methodA  end ThreadName="
                    + Thread.currentThread().getName() + " time="
                    + System.currentTimeMillis());
        } catch (InterruptedException e) {
            e.printStackTrace();
        } finally {
            lock.unlock();
        }
```

图 4-1 程序运行结果（同步）

```java
    }

    public void methodB() {
        try {
            lock.lock();
            System.out.println("methodB begin ThreadName="
                    + Thread.currentThread().getName() + " time="
                    + System.currentTimeMillis());
            Thread.sleep(5000);
            System.out.println("methodB  end ThreadName="
                    + Thread.currentThread().getName() + " time="
                    + System.currentTimeMillis());
        } catch (InterruptedException e) {
            e.printStackTrace();
        } finally {
            lock.unlock();
        }
    }

}
```

第一组线程类代码如图 4-2 所示。

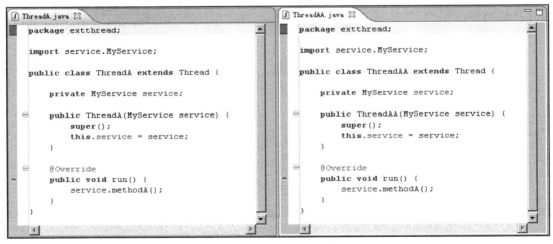

图 4-2　第一组线程类代码

第二组线程类代码如图 4-3 所示。

类 Run.java 代码如下：

```java
package test;

import service.MyService;
import extthread.ThreadA;
import extthread.ThreadAA;
```

```
ThreadB.java
package extthread;

import service.MyService;

public class ThreadB extends Thread {

    private MyService service;

    public ThreadB(MyService service) {
        super();
        this.service = service;
    }

    @Override
    public void run() {
        service.methodB();
    }
}
```

```
ThreadBB.java
package extthread;

import service.MyService;

public class ThreadBB extends Thread {

    private MyService service;

    public ThreadBB(MyService service) {
        super();
        this.service = service;
    }

    @Override
    public void run() {
        service.methodB();
    }
}
```

图 4-3　第二组线程类代码

```
import extthread.ThreadB;
import extthread.ThreadBB;

public class Run {

    public static void main(String[] args) throws InterruptedException {
        MyService service = new MyService();

        ThreadA a = new ThreadA(service);
        a.setName("A");
        a.start();
        ThreadAA aa = new ThreadAA(service);
        aa.setName("AA");
        aa.start();

        Thread.sleep(100);

        ThreadB b = new ThreadB(service);
        b.setName("B");
        b.start();

        ThreadBB bb = new ThreadBB(service);
        bb.setName("BB");
        bb.start();

    }

}
```

程序运行结果如图 4-4 所示。

```
methodA begin ThreadName=A time=1413945463859
methodA   end ThreadName=A time=1413945468859
methodA begin ThreadName=AA time=1413945468859
methodA   end ThreadName=AA time=1413945473859
methodB begin ThreadName=B time=1413945473859
methodB   end ThreadName=B time=1413945478859
methodB begin ThreadName=BB time=1413945478859
methodB   end ThreadName=BB time=1413945483859
```

图 4-4 线程全部同步运行了

此示例说明不管在一个方法还是多个方法的环境中,哪个线程持有锁,哪个线程就执行业务,其他线程只有等待锁被释放时再次争抢,抢到锁就开始执行业务,运行效果和使用 synchronized 关键字一样。

线程之间执行的顺序是随机的。

4.1.3 await() 方法的错误用法与更正

关键字 synchronized 与 wait()、notify()/notifyAll() 方法相结合可以实现 wait/notify 模式,ReentrantLock 类也可以实现同样的功能,但需要借助于 Condition 对象。Condition 类是 JDK 5 的技术,具有更好的灵活性,例如,可以实现多路通知功能,也就是在一个 Lock 对象中可以创建多个 Condition 实例,线程对象注册在指定的 Condition 中,从而可以有选择性地进行线程通知,在调度线程上更加灵活。

在使用 notify()/notifyAll() 方法进行通知时,被通知的线程由 JVM 进行选择,而方法 notifyAll() 会通知所有的 waiting 线程,没有选择权,会出现相当大的效率问题,但使用 ReentrantLock 结合 Condition 类可以实现"选择性通知",这个功能是 Condition 类默认提供的。

Condition 对象的作用是控制并处理线程的状态,它可以使线程呈 wait 状态,也可以让线程继续运行。

创建 Java 项目 UseConditionWaitNotifyError,类 MyService.java 代码如下:

```
package service;

import java.util.concurrent.locks.Condition;
import java.util.concurrent.locks.Lock;
import java.util.concurrent.locks.ReentrantLock;

public class MyService {

    private Lock lock = new ReentrantLock();
    private Condition condition = lock.newCondition();

    public void await() {
```

```
        try {
            condition.await();
        } catch (InterruptedException e) {
            e.printStackTrace();
        }
    }
}
```

await() 方法的作用使当前线程在接到通知或被中断之前一直处于等待 wait 状态。它和 wait() 方法的作用一样。

类 ThreadA.java 代码如下：

```
package extthread;

import service.MyService;

public class ThreadA extends Thread {

    private MyService service;

    public ThreadA(MyService service) {
        super();
        this.service = service;
    }

    @Override
    public void run() {
        service.await();
    }
}
```

类 Run.java 代码如下：

```
package test;

import service.MyService;
import extthread.ThreadA;

public class Run {

    public static void main(String[] args) {

        MyService service = new MyService();

        ThreadA a = new ThreadA(service);
        a.start();

    }

}
```

程序运行结果如图 4-5 所示。

```
<terminated> Run (1) [Java Application] C:\Program Files\Genuitec\Common\binary\com.sun.java.jdk.win32.x86_1.6.0.013\bin\javaw.exe
Exception in thread "Thread-0" java.lang.IllegalMonitorStateException
    at java.util.concurrent.locks.ReentrantLock$Sync.tryRelease(ReentrantLock.java:127)
    at java.util.concurrent.locks.AbstractQueuedSynchronizer.release(AbstractQueuedSynchronize
    at java.util.concurrent.locks.AbstractQueuedSynchronizer.fullyRelease(AbstractQueuedSynchr
    at java.util.concurrent.locks.AbstractQueuedSynchronizer$ConditionObject.await(AbstractQue
    at service.MyService.await(MyService.java:14)
    at extthread.ThreadA.run(ThreadA.java:16)
```

图 4-5 出现异常无监视器对象

报错的异常信息是监视器出错，解决这个问题的方法是必须在 condition.await() 方法调用之前调用 lock.lock() 代码获得锁。

创建名称为 z3_ok 的 Java 项目，类 MyService.java 代码如下：

```java
package service;

import java.util.concurrent.locks.Condition;
import java.util.concurrent.locks.ReentrantLock;

public class MyService {
    private ReentrantLock lock = new ReentrantLock();
    private Condition condition = lock.newCondition();

    public void waitMethod() {
        try {
            lock.lock();
            System.out.println("A");
            condition.await();
            System.out.println("B");
        } catch (InterruptedException e) {
            e.printStackTrace();
        } finally {
            lock.unlock();
            System.out.println("锁释放了！");
        }
    }
}
```

线程类代码如下：

```java
package extthread;

import service.MyService;

public class MyThreadA extends Thread {

    private MyService myService;
```

```java
    public MyThreadA(MyService myService) {
        super();
        this.myService = myService;
    }

    @Override
    public void run() {
        myService.waitMethod();
    }
}
```

运行类代码如下:

```java
package test;

import extthread.MyThreadA;
import service.MyService;

public class Run {

    public static void main(String[] args) {
        MyService myService = new MyService();
        MyThreadA a1 = new MyThreadA(myService);
        a1.start();
        MyThreadA a2 = new MyThreadA(myService);
        a2.start();
        MyThreadA a3 = new MyThreadA(myService);
        a3.start();
    }
}
```

程序运行结果如图 4-6 所示。

控制台输出 3 个字母 A，说明调用了 Condition 对象的 await() 方法将当前执行任务的线程转换成 wait 状态并释放锁。

4.1.4　使用 await() 和 signal() 实现 wait/notify 机制

创建项目 UseConditionWaitNotifyOK，类 MyService.java 代码如下:

图 4-6　打印 3 个字母 A

```java
package service;

import java.util.concurrent.locks.Condition;
import java.util.concurrent.locks.Lock;
import java.util.concurrent.locks.ReentrantLock;

public class MyService {

    private Lock lock = new ReentrantLock();
```

```java
    public Condition condition = lock.newCondition();

    public void await() {
        try {
            lock.lock();
            System.out.println(" await时间为" + System.currentTimeMillis());
            condition.await();
        } catch (InterruptedException e) {
            e.printStackTrace();
        } finally {
            lock.unlock();
        }
    }

    public void signal() {
        try {
            lock.lock();
            System.out.println("signal时间为" + System.currentTimeMillis());
            condition.signal();
        } finally {
            lock.unlock();
        }
    }
}
```

类 ThreadA.java 代码如下:

```java
package extthread;

import service.MyService;

public class ThreadA extends Thread {

    private MyService service;

    public ThreadA(MyService service) {
        super();
        this.service = service;
    }

    @Override
    public void run() {
        service.await();
    }
}
```

类 Run.java 代码如下:

```java
package test;

import service.MyService;
import extthread.ThreadA;
```

```
public class Run {

    public static void main(String[] args) throws InterruptedException {

        MyService service = new MyService();

        ThreadA a = new ThreadA(service);
        a.start();

        Thread.sleep(3000);

        service.signal();

    }
}
```

程序运行结果如图 4-7 所示。

此示例成功实现 wait/notify 模式。

Object 类中的 wait() 方法相当于 Condition 类中的 await() 方法。

Object 类中的 wait(long timeout) 方法相当于 Condition 类中的 await(long time, TimeUnit unit) 方法。

Object 类中的 notify() 方法相当于 Condition 类中的 signal() 方法。

图 4-7　正常运行

Object 类中的 notifyAll() 方法相当于 Condition 类中的 signalAll() 方法。

4.1.5　await() 方法暂停线程运行的原理

执行如下代码：

```
import java.util.concurrent.locks.Condition;
import java.util.concurrent.locks.Lock;
import java.util.concurrent.locks.ReentrantLock;

public class Test {
    public static void main(String[] args) throws InterruptedException {
        Lock lock = new ReentrantLock(true);
        lock.lock();
        Condition condition = lock.newCondition();
        System.out.println("await begin");
        condition.await();
        System.out.println("await   end");
    }
}
```

控制台只输出 await begin 信息，并没有输出 await end，说明线程执行 condition.await()

代码后出现了暂停的状态，不再向下继续运行，让执行 await() 方法的线程暂停运行是什么原理呢？其实并发包源代码内部执行了 Unsafe 类中的 public native void park(boolean isAbsolute, long time) 方法，让当前线程呈暂停状态，方法参数 isAbsolute 代表是否为绝对时间，方法参数 time 代表时间值。如果对参数 isAbsolute 传入 true，则第 2 个参数 time 时间单位为毫秒；如果传入 false，则第 2 个参数时间单位为纳秒。

下面创建 4 个类来测试一下 park() 方法的使用方法。

创建类代码如下：

```java
public class Test2 {
    public static void main(String[] args) throws InterruptedException, NoSuchFieldException, SecurityException,
            IllegalArgumentException, IllegalAccessException {
        Field f = Unsafe.class.getDeclaredField("theUnsafe");
        f.setAccessible(true);
        Unsafe unsafe = (Unsafe) f.get(null);
        System.out.println("begin " + System.currentTimeMillis());
        System.currentTimeMillis();
        // 如果传入true，则第2个参数时间单位为毫秒
        unsafe.park(true, System.currentTimeMillis() + 3000);
        System.out.println("  end " + System.currentTimeMillis());
    }
}
```

程序运行结果如下：

```
begin 1520513996506
    end 1520513999506
```

创建类代码如下：

```java
import java.lang.reflect.Field;

import sun.misc.Unsafe;

public class Test3 {
    public static void main(String[] args) throws InterruptedException, NoSuchFieldException, SecurityException,
            IllegalArgumentException, IllegalAccessException {
        Field f = Unsafe.class.getDeclaredField("theUnsafe");
        f.setAccessible(true);
        Unsafe unsafe = (Unsafe) f.get(null);
        System.out.println("begin " + System.currentTimeMillis());
        System.currentTimeMillis();
        // 3秒的纳秒值是3000000000
        // 3秒的微秒值是3000000
        // 3秒的毫秒值是3000
        // 3秒
        // 如果传入false，第2个参数时间单位为纳秒
        unsafe.park(false, 3000000000L);
```

```
            System.out.println("  end " + System.currentTimeMillis());
        }
    }
}
```

程序运行结果如下：

```
begin 1520514051547
    end 1520514054548
```

创建类代码如下：

```java
import java.lang.reflect.Field;

import sun.misc.Unsafe;

public class Test4 {
    public static void main(String[] args) throws InterruptedException, NoSuchField-
        Exception, SecurityException,
            IllegalArgumentException, IllegalAccessException {
        Field f = Unsafe.class.getDeclaredField("theUnsafe");
        f.setAccessible(true);
        Unsafe unsafe = (Unsafe) f.get(null);
        System.out.println("begin " + System.currentTimeMillis());
        System.currentTimeMillis();
        unsafe.park(true, 0L);
        System.out.println("  end " + System.currentTimeMillis());
    }
}
```

程序运行结果如下：

```
begin 1520514077801
    end 1520514077801
```

创建类代码如下：

```java
import java.lang.reflect.Field;

import sun.misc.Unsafe;

public class Test5 {
    public static void main(String[] args) throws InterruptedException, NoSuchField-
        Exception, SecurityException,
            IllegalArgumentException, IllegalAccessException {
        Field f = Unsafe.class.getDeclaredField("theUnsafe");
        f.setAccessible(true);
        Unsafe unsafe = (Unsafe) f.get(null);
        System.out.println("begin " + System.currentTimeMillis());
        System.currentTimeMillis();
        unsafe.park(false, 0L);
        System.out.println("  end " + System.currentTimeMillis());
```

        }
    }

程序运行结果如下:

```
begin 1520514131875
```

执行代码 unsafe.park(false, 0L) 后,当前线程呈暂停运行的状态,即实现了 wait 等待的效果,并发包源代码也执行了 unsafe.park(false, 0L) 这样的形式,效果如图 4-8 所示。

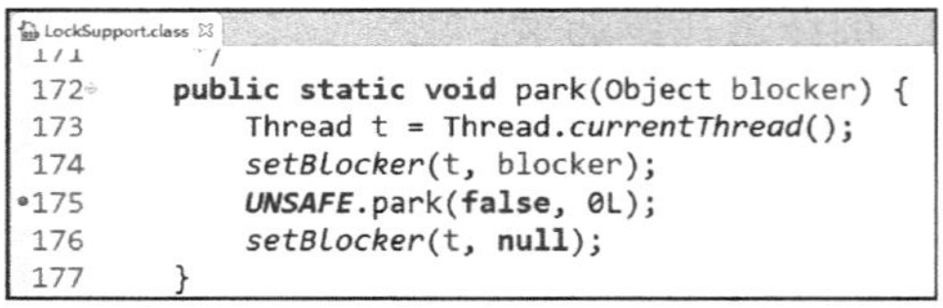

这就是 await() 方法实现暂停效果在源代码中的实现与原理。

图 4-8 执行了 park(false, 0L) 代码实现线程暂停

4.1.6 通知部分线程——错误用法

前面章节使用一个 Condition 对象来实现 wait/notify 模式,其实 Condition 对象也可以创建多个,那么一个 Condition 对象和多个 Condition 对象在使用上有什么区别呢?

创建 Java 项目 MustUseMoreCondition_Error,类 MyService.java 代码如下:

```java
package service;

import java.util.concurrent.locks.Condition;
import java.util.concurrent.locks.Lock;
import java.util.concurrent.locks.ReentrantLock;

public class MyService {

    private Lock lock = new ReentrantLock();
    public Condition condition = lock.newCondition();

    public void awaitA() {
        try {
            lock.lock();
            System.out.println("begin awaitA时间为" + System.currentTimeMillis()
                    + " ThreadName=" + Thread.currentThread().getName());
            condition.await();
            System.out.println("  end awaitA时间为" + System.currentTimeMillis()
                    + " ThreadName=" + Thread.currentThread().getName());
        } catch (InterruptedException e) {
            e.printStackTrace();
        } finally {
            lock.unlock();
        }
    }

    public void awaitB() {
```

```
        try {
            lock.lock();
            System.out.println("begin awaitB时间为" + System.currentTimeMillis()
                    + " ThreadName=" + Thread.currentThread().getName());
            condition.await();
            System.out.println("  end awaitB时间为" + System.currentTimeMillis()
                    + " ThreadName=" + Thread.currentThread().getName());
        } catch (InterruptedException e) {
            e.printStackTrace();
        } finally {
            lock.unlock();
        }
    }

    public void signalAll() {
        try {
            lock.lock();
            System.out.println("  signalAll时间为" + System.currentTimeMillis()
                    + " ThreadName=" + Thread.currentThread().getName());
            condition.signalAll();
        } finally {
            lock.unlock();
        }
    }
}
```

类 ThreadA.java 和 ThreadB.java 代码如图 4-9 所示。

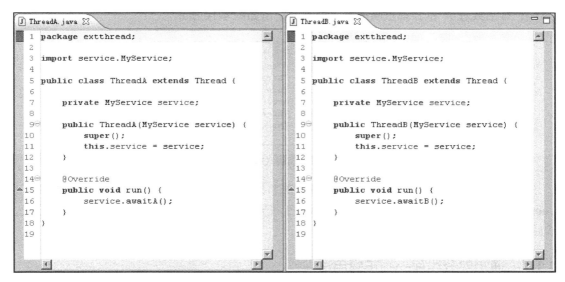

图 4-9　线程对象代码

类 Run.java 代码如下：

```
package test;

import service.MyService;
import extthread.ThreadA;
import extthread.ThreadB;

public class Run {

    public static void main(String[] args) throws InterruptedException {

        MyService service = new MyService();

        ThreadA a = new ThreadA(service);
        a.setName("A");
        a.start();

        ThreadB b = new ThreadB(service);
        b.setName("B");
        b.start();

        Thread.sleep(3000);

        service.signalAll();

    }

}
```

在 3s 后线程 A 和 B 都被唤醒了，控制台输出如图 4-10 所示。

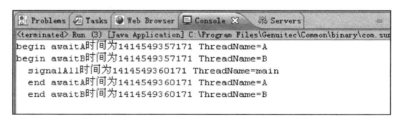

图 4-10　控制台输出（线程均被唤醒）

如果想单独唤醒部分线程，该怎么处理呢？这时就有必要使用多个 Condition 对象了，Condition 对象可以唤醒部分指定线程，有助于提升程序的运行效率，可以对线程进行分组，然后唤醒指定组中的线程。

4.1.7　通知部分线程——正确用法

创建 Java 项目 MustUseMoreCondition_OK，类 MyService.java 代码如下：

```
package service;
```

```java
import java.util.concurrent.locks.Condition;
import java.util.concurrent.locks.Lock;
import java.util.concurrent.locks.ReentrantLock;

public class MyService {

    private Lock lock = new ReentrantLock();
    public Condition conditionA = lock.newCondition();
    public Condition conditionB = lock.newCondition();

    public void awaitA() {
        try {
            lock.lock();
            System.out.println("begin awaitA时间为" + System.currentTimeMillis()
                    + " ThreadName=" + Thread.currentThread().getName());
            conditionA.await();
            System.out.println("  end awaitA时间为" + System.currentTimeMillis()
                    + " ThreadName=" + Thread.currentThread().getName());
        } catch (InterruptedException e) {
            e.printStackTrace();
        } finally {
            lock.unlock();
        }
    }

    public void awaitB() {
        try {
            lock.lock();
            System.out.println("begin awaitB时间为" + System.currentTimeMillis()
                    + " ThreadName=" + Thread.currentThread().getName());
            conditionB.await();
            System.out.println("  end awaitB时间为" + System.currentTimeMillis()
                    + " ThreadName=" + Thread.currentThread().getName());
        } catch (InterruptedException e) {
            e.printStackTrace();
        } finally {
            lock.unlock();
        }
    }

    public void signalAll_A() {
        try {
            lock.lock();
            System.out.println("  signalAll_A时间为" + System.currentTimeMillis()
                    + " ThreadName=" + Thread.currentThread().getName());
            conditionA.signalAll();
        } finally {
            lock.unlock();
        }
    }
```

```
    public void signalAll_B() {
        try {
            lock.lock();
            System.out.println(" signalAll_B时间为" + System.currentTimeMillis()
                    + " ThreadName=" + Thread.currentThread().getName());
            conditionB.signalAll();
        } finally {
            lock.unlock();
        }
    }
}
```

类 ThreadA.java 和 ThreadB.java 代码如图 4-11 所示。

```
package extthread;

import service.MyService;

public class ThreadA extends Thread {

    private MyService service;

    public ThreadA(MyService service) {
        super();
        this.service = service;
    }

    @Override
    public void run() {
        service.awaitA();
    }
}
```

```
package extthread;

import service.MyService;

public class ThreadB extends Thread {

    private MyService service;

    public ThreadB(MyService service) {
        super();
        this.service = service;
    }

    @Override
    public void run() {
        service.awaitB();
    }
}
```

图 4-11　线程代码

类 Run.java 代码如下：

```
package test;

import service.MyService;
import extthread.ThreadA;
import extthread.ThreadB;

public class Run {

    public static void main(String[] args) throws InterruptedException {

        MyService service = new MyService();

        ThreadA a = new ThreadA(service);
        a.setName("A");
        a.start();
```

```
            ThreadB b = new ThreadB(service);
            b.setName("B");
            b.start();

            Thread.sleep(3000);

            service.signalAll_A();
        }
    }
}
```

3s 后只有线程 A 被唤醒,控制台输出如图 4-12 所示。

图 4-12　线程 B 没有被唤醒

使用 Condition 对象可以唤醒指定种类的线程,这是控制部分线程行为的方便方式。

并发包 concurrent 是 Doug Lea 发布的基于 Java 语言的 Condition 并发工具包,提供了丰富的功能,使得 Java 语言在并发编程上较其他语言更有优势,弥补了 Java 语言在并发编程的空白。

4.1.8　实现生产者 / 消费者模式一对一交替输出

创建测试用的项目 ConditionTest,创建类 MyService.java,代码如下:

```
package service;

import java.util.concurrent.locks.Condition;
import java.util.concurrent.locks.ReentrantLock;

public class MyService {

    private ReentrantLock lock = new ReentrantLock();
    private Condition condition = lock.newCondition();
    private boolean hasValue = false;

    public void set() {
        try {
            lock.lock();
            if (hasValue == true) {
                condition.await();
```

```
            }
            System.out.println("打印★");
            hasValue = true;
            condition.signal();
        } catch (InterruptedException e) {
            e.printStackTrace();
        } finally {
            lock.unlock();
        }
    }

    public void get() {
        try {
            lock.lock();
            if (hasValue == false) {
                condition.await();
            }
            System.out.println("打印☆");
            hasValue = false;
            condition.signal();
        } catch (InterruptedException e) {
            e.printStackTrace();
        } finally {
            lock.unlock();
        }
    }

}
```

创建两个线程类代码，如图 4-13 所示。

```
package extthread;

import service.MyService;

public class MyThreadA extends Thread {

    private MyService myService;

    public MyThreadA(MyService myService) {
        super();
        this.myService = myService;
    }

    @Override
    public void run() {
        for (int i = 0; i < Integer.MAX_VALUE; i++) {
            myService.set();
        }
    }
}
```

```
package extthread;

import service.MyService;

public class MyThreadB extends Thread {

    private MyService myService;

    public MyThreadB(MyService myService) {
        super();
        this.myService = myService;
    }

    @Override
    public void run() {
        for (int i = 0; i < Integer.MAX_VALUE; i++) {
            myService.get();
        }
    }
}
```

图 4-13　两个线程类代码

运行类 Run.java 代码如下：

```
package test;

import service.MyService;
import extthread.MyThreadA;
import extthread.MyThreadB;

public class Run {

    public static void main(String[] args) throws InterruptedException {
        MyService myService = new MyService();

        MyThreadA a = new MyThreadA(myService);
        a.start();

        MyThreadB b = new MyThreadB(myService);
        b.start();

    }
}
```

程序运行结果如图 4-14 所示。

通过使用 Condition 对象，成功实现生产者与消费者交替运行的效果。

4.1.9 实现生产者/消费者模式多对多交替输出

创建新的项目 ConditionTestManyToMany，将 ConditionTest 项目中的所有源代码复制到新项目 ConditionTestManyToMany 中。

并更改 MyService.java 类代码如下：

图 4-14 交替运行

```
package service;

import java.util.concurrent.locks.Condition;
import java.util.concurrent.locks.ReentrantLock;

public class MyService {

    private ReentrantLock lock = new ReentrantLock();
    private Condition condition = lock.newCondition();
    private boolean hasValue = false;

    public void set() {
        try {
            lock.lock();
            while (hasValue == true) {
                System.out.println("有可能★★连续");
                condition.await();
```

```
            }
            System.out.println("打印★");
            hasValue = true;
            condition.signal();
        } catch (InterruptedException e) {
            e.printStackTrace();
        } finally {
            lock.unlock();
        }
    }

    public void get() {
        try {
            lock.lock();
            while (hasValue == false) {
                System.out.println("有可能☆☆连续");
                condition.await();
            }
            System.out.println("打印☆");
            hasValue = false;
            condition.signal();
        } catch (InterruptedException e) {
            e.printStackTrace();
        } finally {
            lock.unlock();
        }
    }

}
```

更改 Run.java 代码如下：

```
package test;

import service.MyService;
import extthread.MyThreadA;
import extthread.MyThreadB;

public class Run {

    public static void main(String[] args) throws InterruptedException {
        MyService service = new MyService();

        MyThreadA[] threadA = new MyThreadA[10];
        MyThreadB[] threadB = new MyThreadB[10];

        for (int i = 0; i < 10; i++) {
            threadA[i] = new MyThreadA(service);
            threadB[i] = new MyThreadB(service);
            threadA[i].start();
            threadB[i].start();
```

        }
    }
}

程序运行后出现假死现象,如图 4-15 所示。

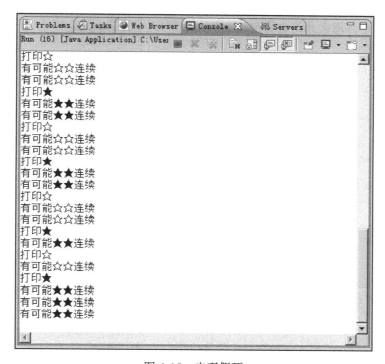

图 4-15　出现假死

根据第 3 章中的 notifyAll() 解决方案,可以使用 signalAll() 方法来解决假死的问题。将 MyService.java 类中的两处 signal() 代码改成 signalAll() 后,程序正确运行,如图 4-16 所示。

从控制台输出日志可以发现,程序运行后不再出现假死状态,假死问题得以解决。

控制台中"打印★"和"打印☆"是交替输出的,但是"有可能★★连续"和"有可能☆☆连续"却不是交替输出的,有时候出现连续输出的情况,原因是程序使用了一个 Condition 对象,再结合 signalAll() 方法来唤醒所有的线程,那么唤醒的线程就有可能是同类,所以就有连续输出"有可能★★连续"或"有可能☆☆连续"的情况了。

连续输出"有可能★★连续"或"有可能☆☆连续"就是唤醒同类最好的证明。

4.1.10　公平锁与非公平锁

公平锁:采用先到先得的策略,每次获取锁之前都会检查队列里面有没有排队等待的线程，没有才会尝试获取锁，如果有就将当前线程追加到队列中。

非公平锁：采用"有机会插队"的策略，一个线程获取锁之前要先去尝试获取锁而不是在队列中等待，如果获取锁成功，则说明线程虽然是后启动的，但先获得了锁，这就是"作弊插队"的效果。如果获取锁没有成功，那么才将自身追加到队列中进行等待。

创建 Java 项目 Fair_noFair_test，创建 MyService.java 类，代码如下：

```java
package test;

import java.util.concurrent.locks.Lock;
import java.util.concurrent.locks.ReentrantLock;

public class MyService {
    public Lock lock;

    public MyService(boolean fair) {
        lock = new ReentrantLock(fair);
    }

    public void testMethod() {
        try {
            lock.lock();
            System.out.println("testMethod " +
                Thread.currentThread().get-
                Name());
            // 此处的500ms是配合main方法中的500ms
            // 使"array2---"线程有机会在非公平情况
            //   下抢到锁
            Thread.sleep(500);
            lock.unlock();
        } catch (InterruptedException e) {
            e.printStackTrace();
        }
    }

}
```

图 4-16 假死的情况被解决

创建线程类代码如下：

```java
package test;

public class MyThread extends Thread {
    private MyService service;

    public MyThread(MyService service) {
```

```java
        super();
        this.service = service;
    }

    public void run() {
        service.testMethod();
    }
}
```

创建公平锁测试的运行类 Test1_1.java 代码如下:

```java
package test;

public class Test1_1 {
    public static void main(String[] args) throws InterruptedException {
        MyService service = new MyService(true);

        MyThread[] array1 = new MyThread[10];
        MyThread[] array2 = new MyThread[10];
        for (int i = 0; i < array1.length; i++) {
            array1[i] = new MyThread(service);
            array1[i].setName("array1+++" + (i + 1));
        }
        for (int i = 0; i < array1.length; i++) {
            array1[i].start();
        }

        for (int i = 0; i < array2.length; i++) {
            array2[i] = new MyThread(service);
            array2[i].setName("array2---" + (i + 1));
        }

        Thread.sleep(500);

        for (int i = 0; i < array2.length; i++) {
            array2[i].start();
        }

    }
}
```

程序运行结果如图 4-17 所示。

程序运行结果是 +++ 在前，--- 在后，说明 --- 没有任何机会抢到锁，这就是公平锁的特点。

创建非公平锁测试的运行类 Test1_2.java 代码如下:

```java
package test;

public class Test1_2 {
    public static void main(String[] args) throws
```

```
testMethod array1+++1
testMethod array1+++2
testMethod array1+++3
testMethod array1+++4
testMethod array1+++5
testMethod array1+++6
testMethod array1+++7
testMethod array1+++8
testMethod array1+++9
testMethod array1+++10
testMethod array2---1
testMethod array2---2
testMethod array2---3
testMethod array2---4
testMethod array2---5
testMethod array2---7
testMethod array2---6
testMethod array2---8
testMethod array2---9
testMethod array2---10
```

图 4-17　程序运行结果（公平锁）

```java
        InterruptedException {
    MyService service = new MyService(false);

    MyThread[] array1 = new MyThread[10];
    MyThread[] array2 = new MyThread[10];
    for (int i = 0; i < array1.length; i++) {
        array1[i] = new MyThread(service);
        array1[i].setName("array1+++" + (i + 1));
    }
    for (int i = 0; i < array1.length; i++) {
        array1[i].start();
    }

    for (int i = 0; i < array2.length; i++) {
        array2[i] = new MyThread(service);
        array2[i].setName("array2---" + (i + 1));
    }

    Thread.sleep(500);

    for (int i = 0; i < array2.length; i++) {
        array2[i].start();
    }

    }
}
```

多次运行程序，程序运行结果如图 4-18 所示。

程序多次运行后，使用非公平锁时有可能在第 2 次输出 ---，说明后启动的线程先抢到了锁，这就是非公平锁的特点。

A 线程持有锁后，B 线程不能执行的原理是在内部执行了 unsafe.park(false, 0L) 代码；A 线程释放锁后 B 线程可以运行的原理是当 A 线程执行 unlock() 方法时在内部执行了 unsafe.unpark(bThread)，B 线程得以继续运行。

4.1.11　public int getHoldCount() 方法的使用

public int getHoldCount() 方法的作用是查询"当前线程"保持此锁定的个数，即调用 lock() 方法的次数。

创建测试用的项目 lockMethodTest1。

创建名称为 test1 的 package，创建类 MyService.java，代码如下：

```
package test1;

import java.util.concurrent.locks.ReentrantLock;

public class MyService {
```

```
testMethod array1+++1
testMethod array2---1
testMethod array1+++2
testMethod array1+++3
testMethod array1+++4
testMethod array1+++5
testMethod array1+++6
testMethod array1+++7
testMethod array1+++8
testMethod array1+++9
testMethod array1+++10
testMethod array2---2
testMethod array2---4
testMethod array2---3
testMethod array2---5
testMethod array2---6
testMethod array2---7
testMethod array2---8
testMethod array2---9
testMethod array2---10
```

图 4-18　程序运行结果（非公平锁）

```java
    private ReentrantLock lock = new ReentrantLock(true);

    public void testMethod1() {
        System.out.println("A " + lock.getHoldCount());
        lock.lock();
        System.out.println("B " + lock.getHoldCount());
        testMethod2();
        System.out.println("F " + lock.getHoldCount());
        lock.unlock();
        System.out.println("G " + lock.getHoldCount());
    }

    public void testMethod2() {
        System.out.println("C " + lock.getHoldCount());
        lock.lock();
        System.out.println("D " + lock.getHoldCount());
        lock.unlock();
        System.out.println("E " + lock.getHoldCount());
    }
}
```

创建类 Test.java 代码如下：

```java
package test1;

public class Test {
    public static void main(String[] args) throws InterruptedException {
        MyService service = new MyService();
        service.testMethod1();
    }
}
```

程序运行结果如下：

```
A 0
B 1
C 1
D 2
E 1
F 1
G 0
```

执行 lock() 方法进行锁重入导致 count 计数呈加 1 的效果，执行 unlock() 方法会使 count 呈减 1 的效果。

4.1.12 public final int getQueueLength() 方法的使用

public final int getQueueLength() 方法的作用是返回正等待获取此锁的线程估计数，例如，这里有 5 个线程，其中 1 个线程长时间占有锁，那么调用 getQueueLength() 方法后，其返回值是 4，说明有 4 个线程同时在等待锁的释放。

创建名称为 test2 的 package,创建类 Service.java,代码如下:

```java
package test2;

import java.util.concurrent.locks.ReentrantLock;

public class Service {

    public ReentrantLock lock = new ReentrantLock();

    public void serviceMethod1() {
        try {
            lock.lock();
            System.out.println("ThreadName=" + Thread.currentThread().getName()
                    + "进入方法!");
            Thread.sleep(Integer.MAX_VALUE);
        } catch (InterruptedException e) {
            // TODO Auto-generated catch block
            e.printStackTrace();
        } finally {
            lock.unlock();
        }
    }

}
```

创建类 Run.java 代码如下:

```java
package test2;

public class Run {

    public static void main(String[] args) throws InterruptedException {
        final Service service = new Service();

        Runnable runnable = new Runnable() {
            @Override
            public void run() {
                service.serviceMethod1();
            }
        };

        Thread[] threadArray = new Thread[10];
        for (int i = 0; i < 10; i++) {
            threadArray[i] = new Thread(runnable);
        }
        for (int i = 0; i < 10; i++) {
            threadArray[i].start();
        }
        Thread.sleep(2000);
        System.out.println("有线程数:" + service.lock.getQueueLength() + "在等待获取锁!
```

");
    }
}
```

程序运行结果如图 4-19 所示。

## 4.1.13　public int getWaitQueueLength (Condition condition) 方法的使用

public int getWaitQueueLength(Condition condition) 方法的作用是返回等待与此锁相关的给定条件 Condition 的线程估计数。例如，这里有 5 个线程，每个线程都执行了同一个 Condition 对象的 await() 方法，则调用 getWaitQueueLength(Condition condition) 方法时，其返回的 int 值是 5。

图 4-19　getQueueLength() 方法运行示例

创建名称为 test3 的 package，创建类 Service.java，代码如下：

```
package test3;

import java.util.concurrent.locks.Condition;
import java.util.concurrent.locks.ReentrantLock;

public class Service {

    private ReentrantLock lock = new ReentrantLock();
    private Condition newCondition = lock.newCondition();

    public void waitMethod() {
        try {
            lock.lock();
            newCondition.await();
        } catch (InterruptedException e) {
            e.printStackTrace();
        } finally {
            lock.unlock();
        }
    }

    public void notifyMethod() {
        try {
            lock.lock();
            System.out.println("有" + lock.getWaitQueueLength(newCondition)
                    + "个线程正在等待newCondition");
            newCondition.signal();
        } finally {
            lock.unlock();
        }
    }
}
```
```

}

创建类 Run.java 代码如下：

```java
package test3;

public class Run {

    public static void main(String[] args) throws InterruptedException {
        final Service service = new Service();

        Runnable runnable = new Runnable() {
            @Override
            public void run() {
                service.waitMethod();
            }
        };

        Thread[] threadArray = new Thread[10];
        for (int i = 0; i < 10; i++) {
            threadArray[i] = new Thread(runnable);
        }
        for (int i = 0; i < 10; i++) {
            threadArray[i].start();
        }
        Thread.sleep(2000);
        service.notifyMethod();
    }
}
```

程序运行结果如图 4-20 所示。

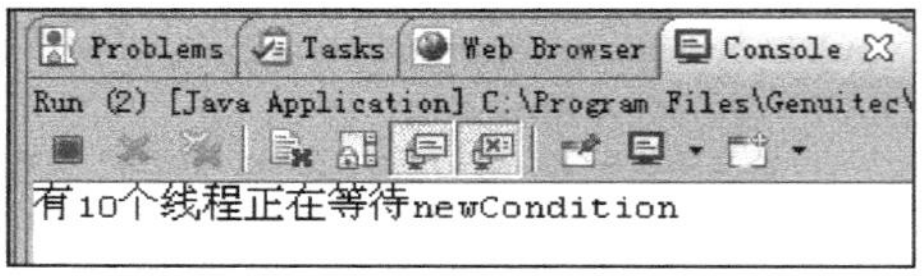

图 4-20 getWaitQueueLength(Condition condition) 方法运行示例

4.1.14 public final boolean hasQueuedThread(Thread thread) 方法的使用

public final boolean hasQueuedThread(Thread thread) 方法的作用是查询指定的线程是否正在等待获取此锁，也就是判断参数中的线程是否在等待队列中。

创建测试用的项目 lockMethodTest2。

创建名称为 test1 的 package，创建类 Service.java，代码如下：

```java
package test1;

import java.util.concurrent.locks.Condition;
import java.util.concurrent.locks.ReentrantLock;

public class Service {

    public ReentrantLock lock = new ReentrantLock();
    public Condition newCondition = lock.newCondition();

    public void waitMethod() {
```

```java
        try {
            lock.lock();
            Thread.sleep(Integer.MAX_VALUE);
        } catch (InterruptedException e) {
            e.printStackTrace();
        } finally {
            lock.unlock();
        }
    }
}
```

创建类 Run.java 代码如下：

```java
package test1;

public class Run {

    public static void main(String[] args) throws InterruptedException {
        final Service service = new Service();

        Runnable runnable = new Runnable() {
            @Override
            public void run() {
                service.waitMethod();
            }
        };

        Thread threadA = new Thread(runnable);
        threadA.start();

        Thread.sleep(500);

        Thread threadB = new Thread(runnable);
        threadB.start();

        Thread.sleep(500);
        System.out.println(service.lock.hasQueuedThread(threadA));
        System.out.println(service.lock.hasQueuedThread(threadB));
    }
}
```

程序运行结果如图 4-21 所示。

```
false
true
```

图 4-21 hasQueuedThread(Thread thread) 方法运行示例

4.1.15 public final boolean hasQueued-Threads() 方法的使用

public final boolean hasQueuedThreads() 方法的作用是查询是否有线程正在等待获取此锁，也就是等待队列中是否有等待的线程。

创建测试用的项目 lockMethodTest2。

创建名称为 test2 的 package，创建类 Service.java，代码如下：

```java
package test1;

import java.util.concurrent.locks.Condition;
import java.util.concurrent.locks.ReentrantLock;

public class Service {

    public ReentrantLock lock = new ReentrantLock();
    public Condition newCondition = lock.newCondition();

    public void waitMethod() {
        try {
            lock.lock();
            Thread.sleep(Integer.MAX_VALUE);
        } catch (InterruptedException e) {
            e.printStackTrace();
        } finally {
            lock.unlock();
        }
    }
}
```

创建类 Run.java 代码如下：

```java
package test2;

public class Run {

    public static void main(String[] args) throws InterruptedException {
        final Service service = new Service();

        Runnable runnable = new Runnable() {
            @Override
            public void run() {
                service.waitMethod();
            }
        };

        Thread threadA = new Thread(runnable);
        threadA.start();

        Thread.sleep(500);

        Thread threadB = new Thread(runnable);
        threadB.start();

        Thread.sleep(500);
        System.out.println(service.lock.hasQueuedThread(threadA));
        System.out.println(service.lock.hasQueuedThread(threadB));
        System.out.println(service.lock.hasQueuedThreads());
    }
}
```

4.1.16 public boolean hasWaiters(Condition condition) 方法的使用

public boolean hasWaiters(Condition condition) 方法的作用是查询是否有线程正在等待与此锁有关的 condition 条件，也就是是否有线程执行了 condition 对象中的 await() 方法而呈等待状态。而 public int getWaitQueueLength(Condition condition) 方法的作用是返回有多少个线程执行了 condition 对象中的 await() 方法而呈等待状态。

图 4-22　hasQueuedThreads() 方法运行示例

创建名称为 test3 的 package，创建类 Service.java，代码如下：

```java
package test3;

import java.util.concurrent.locks.Condition;
import java.util.concurrent.locks.ReentrantLock;

public class Service {

    private ReentrantLock lock = new ReentrantLock();
    private Condition newCondition = lock.newCondition();

    public void waitMethod() {
        try {
            lock.lock();
            newCondition.await();
        } catch (InterruptedException e) {
            e.printStackTrace();
        } finally {
            lock.unlock();
        }
    }

    public void notityMethod() {
        try {
            lock.lock();
            System.out.println("有没有线程正在等待newCondition? "
                    + lock.hasWaiters(newCondition) + " 线程数是多少? "
                    + lock.getWaitQueueLength(newCondition));
            newCondition.signal();
        } finally {
            lock.unlock();
        }
    }

}
```

创建类 Run.java 代码如下：

```
package test3;

public class Run {

    public static void main(String[] args) throws InterruptedException
        final Service service = new Service();

        Runnable runnable = new Runnable() {
            @Override
            public void run() {
                service.waitMethod();
            }
        };

        Thread[] threadArray = new Thread[10];
        for (int i = 0; i < 10; i++) {
            threadArray[i] = new Thread(runnable);
        }
        for (int i = 0; i < 10; i++) {
            threadArray[i].start();
        }
        Thread.sleep(2000);
        service.notityMethod();
    }
}
```

程序运行结果如图 4-23 所示。

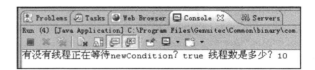

图 4-23 hasWaiters(Condition condition) 方法运行示例

4.1.17 public final boolean isFair() 方法的使用

public final boolean isFair() 方法的作用是判断是不是公平锁。

创建测试用的项目 lockMethodTest3。

创建名称为 test1 的 package，创建类 Run.java，代码如下：

```
package test1;

import java.util.concurrent.locks.ReentrantLock;

public class Run {
    public static void main(String[] args) throws InterruptedException {
        ReentrantLock lock1 = new ReentrantLock(true);
        System.out.println(lock1.isFair());
```

```
            ReentrantLock lock2 = new ReentrantLock(false);
            System.out.println(lock2.isFair());
            ReentrantLock lock3 = new ReentrantLock();
            System.out.println(lock3.isFair());
    }
}
```

程序运行结果如下：

```
true
false
false
```

在默认情况下，ReentrantLock 类使用的是非公平锁。

4.1.18　public boolean isHeldByCurrentThread() 方法的使用

public boolean isHeldByCurrentThread() 方法的作用是查询当前线程是否保持此锁。

创建名称为 test2 的 package，创建类 Service.java，代码如下：

```
package test2;

import java.util.concurrent.locks.ReentrantLock;

public class Service {

    private ReentrantLock lock = new ReentrantLock();

    public void serviceMethod() {
        try {
            System.out.println(lock.isHeldByCurrentThread());
            lock.lock();
            System.out.println(lock.isHeldByCurrentThread());
        } finally {
            lock.unlock();
        }
    }

}
```

创建类 Run.java 代码如下：

```
package test2;

public class Run {

    public static void main(String[] args) throws InterruptedException {
        final Service service1 = new Service();
        Runnable runnable = new Runnable() {
            @Override
            public void run() {
```

```
            service1.serviceMethod();
        }
    };
    Thread thread = new Thread(runnable);
    thread.start();
}
```

程序运行结果如图 4-24 所示。

4.1.19 public boolean isLocked() 方法的使用

public boolean isLocked() 方法的作用是查询此锁是否由任意线程保持,并没有释放。

创建名称为 test3 的 package,创建类 Service.java,代码如下:

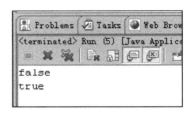

图 4-24 isHeldByCurrentThread() 方法运行示例

```
package test3;

import java.util.concurrent.locks.ReentrantLock;

public class Service {

    private ReentrantLock lock = new ReentrantLock();;

    public void serviceMethod() {
        try {
            System.out.println(lock.isLocked());
            lock.lock();
            System.out.println(lock.isLocked());
        } finally {
            lock.unlock();
        }
    }

}
```

创建类 Run.java 代码如下:

```
package test3;

public class Run {

    public static void main(String[] args) throws InterruptedException {
        final Service service1 = new Service();
        Runnable runnable = new Runnable() {
            @Override
            public void run() {
                service1.serviceMethod();
            }
```

```
            };
            Thread thread = new Thread(runnable);
            thread.start();
        }
    }
```

程序运行结果如图 4-25 所示。

4.1.20 public void lockInterruptibly() 方法的使用

图 4-25   isLocked() 方法运行示例

public void lockInterruptibly() 方法的作用是当某个线程尝试获得锁并且阻塞在 lockInterruptibly() 方法时，该线程可以被中断。

创建测试用的项目 lockInterruptiblyTest1，类 MyService.java 代码如下：

```
package test;

import java.util.concurrent.locks.ReentrantLock;

public class MyService {
    private ReentrantLock lock = new ReentrantLock();

    public void testMethod() {
        lock.lock();
        System.out.println("begin " + Thread.currentThread().getName() + " " +
            System.currentTimeMillis());
        for (int i = 0; i < Integer.MAX_VALUE / 10; i++) {
            String newString = new String();
            Math.random();
            // 为了不让currentThread()过高地占有CPU资源
            // 所以执行yield()方法
            Thread.currentThread().yield();
        }
        System.out.println("  end " + Thread.currentThread().getName() + " " +
            System.currentTimeMillis());
        lock.unlock();
    }
}
```

线程类代码如下：

```
package test;

public class ThreadA extends Thread {
    private MyService service;

    public ThreadA(MyService service) {
        super();
```

```java
        this.service = service;
    }

    @Override
    public void run() {
        service.testMethod();
    }
}
```

运行类 Test.java 代码如下：

```java
package test;

public class Test {
    public static void main(String[] args) throws InterruptedException {
        MyService service = new MyService();
        ThreadA a = new ThreadA(service);
        a.setName("a");
        a.start();

        Thread.sleep(500);

        ThreadA b = new ThreadA(service);
        b.setName("b");
        b.start();

        Thread.sleep(500);

        b.interrupt();

        System.out.println("main中断b，但并没有成功！");
    }
}
```

程序运行结果如图 4-26 所示。

前面示例使用的是 lock() 方法，说明 b 线程被中断了，那么执行 lock() 则不出现异常，而是正常执行。

```
begin a 1520817696774
main中断b，但并没有成功！
  end a 1520817714933
begin b 1520817714933
  end b 1520817732695
```

图 4-26　没有出现异常，a、b 线程正常结束按钮变灰

那如果使用 lockInterruptibly() 方法会是什么结果呢？

创建测试用的项目 lockInterruptiblyTest2，将项目 lockInterruptiblyTest1 中的所有源代码复制到 lockInterr-uptiblyTest2 项目中，更改类 MyService.java 中原有代码 " lock.lock ();" 为 " lock.lockInterr-uptibly();"。

程序运行结果如图 4-27 所示。

4.1.21　public boolean tryLock() 方法的使用

public boolean tryLock() 方法的作用是嗅探拿锁，如果当前线程发现锁被其他线程持有

了，则返回 false，程序继续执行后面的代码，而不是呈阻塞等待锁的状态。

```
begin a 1520818007369
main中断b，但并没有成功！
java.lang.InterruptedException
        at java.util.concurrent.locks.AbstractQueuedSynchronizer.doAcqu
        at java.util.concurrent.locks.AbstractQueuedSynchronizer.acquir
        at java.util.concurrent.locks.ReentrantLock.lockInterruptibly(R
        at test.MyService.testMethod(MyService.java:10)
        at test.ThreadA.run(ThreadA.java:13)
    end a 1520818056388
```

图 4-27　b 线程被中断后调用 lockInterruptibly() 方法报异常

创建测试用的项目 tryLockTest，类 MyService.java 代码如下：

```java
package service;

import java.util.concurrent.locks.ReentrantLock;

public class MyService {

    public ReentrantLock lock = new ReentrantLock();

    public void waitMethod() {
        if (lock.tryLock()) {
            System.out.println(Thread.currentThread().getName() + "获得锁");
        } else {
            System.out.println(Thread.currentThread().getName() + "没有获得锁");
        }
    }
}
```

运行类代码如下：

```java
package test;

import service.MyService;

public class Run {

    public static void main(String[] args) throws InterruptedException {
        final MyService service = new MyService();

        Runnable runnableRef = new Runnable() {
            @Override
            public void run() {
                service.waitMethod();
            }
        };

        Thread threadA = new Thread(runnableRef);
        threadA.setName("A");
        threadA.start();
```

```
            Thread threadB = new Thread(runnableRef);
            threadB.setName("B");
            threadB.start();
        }
    }
```

程序运行结果如图 4-28 所示。

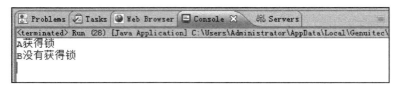

图 4-28　程序运行结果

4.1.22　public boolean tryLock(long timeout, TimeUnit unit) 方法的使用

public boolean tryLock(long timeout, TimeUnit unit) 方法的作用是嗅探拿锁，如果当前线程发现锁被其他线程持有了，则返回 false，程序继续执行后面的代码，而不是呈阻塞等待锁的状态。如果当前线程在指定的 timeout 内持有了锁，则返回值是 true，超过时间则返回 false。参数 timeout 代表当前线程抢锁的时间。

创建测试用的项目 tryLock_param，类 MyService.java 代码如下：

```
package service;

import java.util.concurrent.TimeUnit;
import java.util.concurrent.locks.ReentrantLock;

public class MyService {

    public ReentrantLock lock = new ReentrantLock();

    public void waitMethod() {
        try {
            if (lock.tryLock(3, TimeUnit.SECONDS)) {
                System.out.println("    " + Thread.currentThread().getName()
                        + "获得锁的时间: " + System.currentTimeMillis());
                Thread.sleep(10000);
            } else {
                System.out.println("    " + Thread.currentThread().getName()
                        + "没有获得锁");
            }
        } catch (InterruptedException e) {
            e.printStackTrace();
        } finally {
            if (lock.isHeldByCurrentThread()) {
                lock.unlock();
            }
```

            }
        }
    }

运行类 Run2.java 代码如下:

```java
package test;

import service.MyService;

public class Run {

    public static void main(String[] args) throws InterruptedException {
        final MyService service = new MyService();

        Runnable runnableRef = new Runnable() {
            @Override
            public void run() {
                System.out.println(Thread.currentThread().getName()
                        + "调用waitMethod时间: " + System.currentTimeMillis());
                service.waitMethod();
            }
        };

        Thread threadA = new Thread(runnableRef);
        threadA.setName("A");
        threadA.start();
        Thread threadB = new Thread(runnableRef);
        threadB.setName("B");
        threadB.start();
    }
}
```

程序运行结果如图 4-29 所示。

图 4-29　线程 B 超时未获得锁

4.1.23　public boolean await(long time, TimeUnit unit) 方法的使用

public boolean await(long time, TimeUnit unit) 方法的作用和 public final native void wait(long timeout) 方法一样，都具有自动唤醒线程的功能。

创建测试用的项目 awaitTest_method1。

创建类 MyService.java 代码如下:

```java
package test;

import java.util.concurrent.TimeUnit;
import java.util.concurrent.locks.Condition;
import java.util.concurrent.locks.Lock;
import java.util.concurrent.locks.ReentrantLock;

public class MyService {
    private Lock lock = new ReentrantLock();
    private Condition condition = lock.newCondition();

    public void testMethod() {
        try {
            lock.lock();
            System.out.println("await begin " + System.currentTimeMillis());
            condition.await(3, TimeUnit.SECONDS);
            System.out.println("await   end " + System.currentTimeMillis());
            lock.unlock();
        } catch (InterruptedException e) {
            e.printStackTrace();
        }
    }
}
```

创建类 ThreadA.java 代码如下:

```java
package test;

public class ThreadA extends Thread {
    private MyService service;

    public ThreadA(MyService service) {
        super();
        this.service = service;
    }

    @Override
    public void run() {
        service.testMethod();
    }
}
```

创建类 Test.java 代码如下:

```java
package test;

public class Test {
    public static void main(String[] args) throws InterruptedException {
```

```
        MyService service = new MyService();
        ThreadA a = new ThreadA(service);
        a.start();
    }
}
```

运行结果如下：

```
await begin 1520819682349
await   end 1520819685350
```

4.1.24　public long awaitNanos(long nanosTimeout) 方法的使用

public long awaitNanos(long nanosTimeout) 方法的作用和 public final native void wait(long timeout) 方法一样，都具有自动唤醒线程的功能，时间单位是纳秒（ns）。

1000ns 等于 1μs，1000μs 等于 1ms，1000ms 等于 1s。

创建测试用的项目 awaitTest_method2。

创建类 MyService.java 代码如下：

```java
package test;

import java.util.concurrent.locks.Condition;
import java.util.concurrent.locks.Lock;
import java.util.concurrent.locks.ReentrantLock;

public class MyService {
    private Lock lock = new ReentrantLock();
    private Condition condition = lock.newCondition();

    public void testMethod() {
        try {
            lock.lock();
            System.out.println("await begin " + System.currentTimeMillis());
            // 5000000000L====5s
            condition.awaitNanos(5000000000L);
            System.out.println("await   end " + System.currentTimeMillis());
            lock.unlock();
        } catch (InterruptedException e) {
            e.printStackTrace();
        }
    }

}
```

运行结果如下：

```
await begin 1520819794542
await   end 1520819799542
```

4.1.25 public boolean awaitUntil(Date deadline) 方法的使用

public boolean awaitUntil(Date deadline) 方法的作用是在指定的 Date 结束等待。创建测试用的项目 awaitUntilTest，两个线程类代码如图 4-30 所示。

```java
package extthread;

import service.Service;

public class MyThreadA extends Thread {

    private Service service;

    public MyThreadA(Service service) {
        super();
        this.service = service;
    }

    @Override
    public void run() {
        service.waitMethod();
    }
}
```

```java
package extthread;

import service.Service;

public class MyThreadB extends Thread {

    private Service service;

    public MyThreadB(Service service) {
        super();
        this.service = service;
    }

    @Override
    public void run() {
        service.notifyMethod();
    }
}
```

图 4-30　两个线程类代码

类 Service.java 代码如下：

```java
package service;

import java.util.Calendar;
import java.util.concurrent.locks.Condition;
import java.util.concurrent.locks.ReentrantLock;

public class Service {

    private ReentrantLock lock = new ReentrantLock();
    private Condition condition = lock.newCondition();

    public void waitMethod() {
        try {
            Calendar calendarRef = Calendar.getInstance();
            calendarRef.add(Calendar.SECOND, 10);
            lock.lock();
            System.out.println("wait begin timer=" + System.currentTimeMillis());
            condition.awaitUntil(calendarRef.getTime());
            System.out.println("wait  end timer=" + System.currentTimeMillis());
        } catch (InterruptedException e) {
            e.printStackTrace();
        } finally {
            lock.unlock();
        }
    }
```

```
    }
    public void notifyMethod() {
        try {
            Calendar calendarRef = Calendar.getInstance();
            calendarRef.add(Calendar.SECOND, 10);
            lock.lock();
            System.out.println("notify begin timer=" + System.currentTimeMillis());
            condition.signalAll();
            System.out.println("notify   end timer=" + System.currentTimeMillis());
        } finally {
            lock.unlock();
        }
    }
}
```

创建运行类 Run1.java 代码如下：

```
package test;

import service.Service;
import extthread.MyThreadA;
import extthread.MyThreadB;

public class Run1 {

    public static void main(String[] args) throws InterruptedException {
        Service service = new Service();
        MyThreadA myThreadA = new MyThreadA(service);
        myThreadA.start();
    }

}
```

程序运行结果如图 4-31 所示。

创建运行类 Run2.java 代码如下：

```
package test;

import service.Service;
import extthread.MyThreadA;
import extthread.MyThreadB;

public class Run2 {

    public static void main(String[] args)
        throws InterruptedException {
        Service service = new Service();
        MyThreadA myThreadA = new MyThreadA
            (service);
```

图 4-31 10s 后自动唤醒自己

```
            myThreadA.start();

            Thread.sleep(2000);

            MyThreadB myThreadB = new MyThreadB(service);
            myThreadB.start();
    }

}
```

程序运行结果如图 4-32 所示。

由程序运行结果可以看到，线程在等待时间到达前，可以被其他线程提前唤醒。

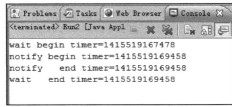

图 4-32　2s 后被其他线程唤醒

4.1.26　public void awaitUninterruptibly() 方法的使用

public void awaitUninterruptibly() 方法的作用是实现线程在等待的过程中，不允许被中断。

创建名称为 awaitUninterruptiblyTest_1 的项目，类 Service.java 代码如下：

```
package service;

import java.util.concurrent.locks.Condition;
import java.util.concurrent.locks.ReentrantLock;

public class Service {

    private ReentrantLock lock = new ReentrantLock();
    private Condition condition = lock.newCondition();

    public void testMethod() {
        try {
            lock.lock();
            System.out.println("wait begin");
            condition.await();
            System.out.println("wait   end");
        } catch (InterruptedException e) {
            e.printStackTrace();
            System.out.println("catch");
        } finally {
            lock.unlock();
        }

    }
}
```

线程类 MyThread.java 代码如下：

```
package extthread;

import service.Service;
```

```java
public class MyThread extends Thread {

    private Service service;

    public MyThread(Service service) {
        super();
        this.service = service;
    }

    @Override
    public void run() {
        service.testMethod();
    }

}
```

运行类 Run.java 代码如下：

```java
package test;

import service.Service;
import extthread.MyThread;

public class Run {

    public static void main(String[] args) {
        try {
            Service service = new Service();
            MyThread myThread = new MyThread(service);
            myThread.start();
            Thread.sleep(3000);
            myThread.interrupt();
        } catch (InterruptedException e) {
            e.printStackTrace();
        }
    }

}
```

程序运行后出现异常，是正常现象，如图 4-33 所示。

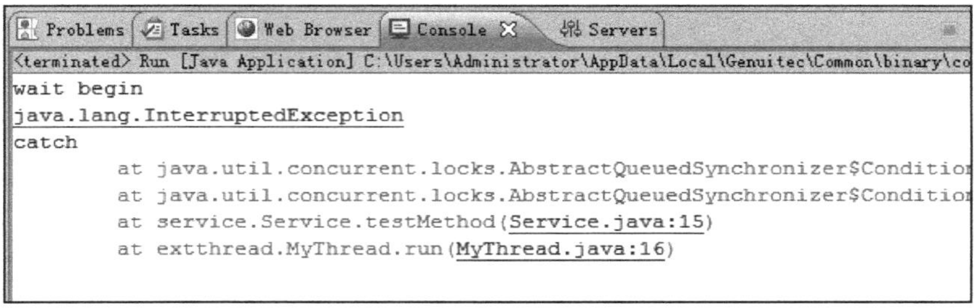

图 4-33　程序运行出现异常

由程序运行结果可以看出，await() 方法是可以被中断的。

创建测试用的项目 awaitUninterruptiblyTest_2，将 awaitUninterruptiblyTest_1 项目中的所有 Java 类复制到 awaitUninterruptiblyTest_2 中。

更改 Service.java 类代码如下：

```java
package service;

import java.util.concurrent.locks.Condition;
import java.util.concurrent.locks.ReentrantLock;

public class Service {

    private ReentrantLock lock = new ReentrantLock();
    private Condition condition = lock.newCondition();

    public void testMethod() {
        try {
            lock.lock();
            System.out.println("wait begin");
            condition.awaitUninterruptibly();
            System.out.println("wait   end");
        } finally {
            lock.unlock();
        }

    }
}
```

程序运行结果如图 4-34 所示。

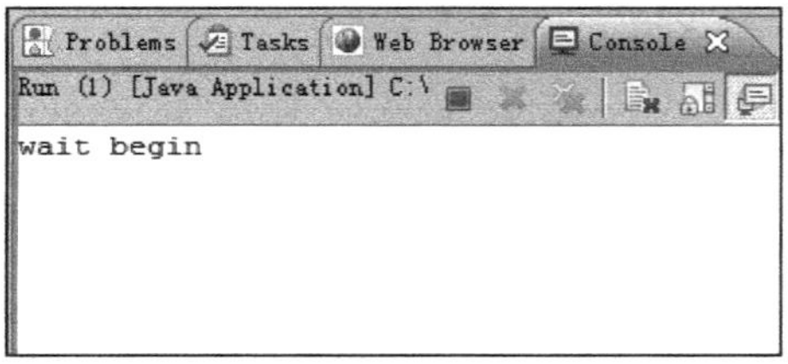

图 4-34　正常运行并没有异常发生

4.1.27　实现线程按顺序执行业务

创建测试用的项目 condition123。

创建类 MyService.java 代码如下：

```java
package test;

import java.util.concurrent.locks.Condition;
import java.util.concurrent.locks.ReentrantLock;

public class MyService {
    private ReentrantLock lock = new ReentrantLock();
    private Condition condition = lock.newCondition();

    volatile private int nextWhoPrint = 1;

    public void testMethod1() {
        try {
            lock.lock();
```

```java
            while (nextWhoPrint != 1) {
                condition.await();
            }
            System.out.println("AAA");
            nextWhoPrint = 2;
            condition.signalAll();
            lock.unlock();
        } catch (InterruptedException e) {
            e.printStackTrace();
        }
    }

    public void testMethod2() {
        try {
            lock.lock();
            while (nextWhoPrint != 2) {
                condition.await();
            }
            System.out.println("   BBB");
            nextWhoPrint = 3;
            condition.signalAll();
            lock.unlock();
        } catch (InterruptedException e) {
            e.printStackTrace();
        }
    }

    public void testMethod3() {
        try {
            lock.lock();
            while (nextWhoPrint != 3) {
                condition.await();
            }
            System.out.println("      CCC");
            nextWhoPrint = 1;
            condition.signalAll();
            lock.unlock();
        } catch (InterruptedException e) {
            e.printStackTrace();
        }
    }
}
```

创建 3 个线程类代码如下：

```java
package test;

public class ThreadA extends Thread {
    private MyService service;

    public ThreadA(MyService service) {
```

```java
        super();
        this.service = service;
    }

    @Override
    public void run() {
        service.testMethod1();
    }
}

package test;

public class ThreadB extends Thread {
    private MyService service;

    public ThreadB(MyService service) {
        super();
        this.service = service;
    }

    @Override
    public void run() {
        service.testMethod2();
    }
}

package test;

public class ThreadC extends Thread {
    private MyService service;

    public ThreadC(MyService service) {
        super();
        this.service = service;
    }

    @Override
    public void run() {
        service.testMethod3();
    }
}
```

创建运行类 Run.java 代码如下:

```java
package test;

public class Test {
    public static void main(String[] args) throws InterruptedException {
        MyService service = new MyService();
        for (int i = 0; i < 5; i++) {
            ThreadA a = new ThreadA(service);
            a.start();
```

```
            ThreadB b = new ThreadB(service);
            b.start();
            ThreadC c = new ThreadC(service);
            c.start();
        }
    }
}
```

程序运行结果如图 4-35 所示。

4.2 使用 ReentrantReadWriteLock 类

ReentrantLock 类具有完全互斥排他的效果,同一时间只有一个线程在执行 ReentrantLock.lock() 方法后面的任务,这样做虽然保证了同时写实例变量的线程安全性,但效率是非常低下的,所以 JDK 提供了一种读写锁——ReentrantReadWriteLock 类,使用它可以在进行读操作时不需要同步执行,提升运行速度,加快运行效率。

读写锁有两个锁:一个是读操作相关的锁,也称共享锁;另一个是写操作相关的锁,也称排他锁。

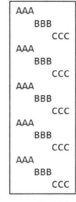

图 4-35 按顺序输出

读锁之间不互斥,读锁和写锁互斥,写锁与写锁互斥,因此只要出现写锁,就会出现互斥同步的效果。

读操作是指读取实例变量的值,写操作是指向实例变量写入值。

4.2.1 ReentrantLock 类的缺点

与 ReentrantReadWriteLock 类相比,ReentrantLock 类的主要缺点是使用 ReentrantLock 对象时,所有的操作都同步,哪怕只对实例变量进行读取操作,这样会耗费大量的时间,降低运行效率。

创建测试用的项目 ReentrantLock_end。

创建类 MyService.java 代码如下:

```
package test;

import java.util.concurrent.locks.ReentrantLock;

public class MyService {
    private ReentrantLock lock = new ReentrantLock();
    private String username = "abc";

    public void testMethod1() {
        try {
            lock.lock();
```

```java
            System.out.println("begin " + Thread.currentThread().getName() + " " +
                System.currentTimeMillis());
            System.out.println("print service " + username);
            Thread.sleep(4000);
            System.out.println("  end " + Thread.currentThread().getName() + " " +
                System.currentTimeMillis());
            lock.unlock();
        } catch (InterruptedException e) {
            e.printStackTrace();
        }
    }
}
```

创建类 ThreadA.java 代码如下：

```java
package test;

public class ThreadA extends Thread {
    private MyService service;

    public ThreadA(MyService service) {
        super();
        this.service = service;
    }

    @Override
    public void run() {
        service.testMethod1();
    }
}
```

创建类 Test.java 代码如下：

```java
package test;

public class Test {
    public static void main(String[] args) throws InterruptedException {
        MyService service = new MyService();
        ThreadA a = new ThreadA(service);
        a.start();
        ThreadA b = new ThreadA(service);
        b.start();
    }
}
```

运行结果如下：

```
begin Thread-0 1520824439727
print service abc
    end Thread-0 1520824443728
begin Thread-1 1520824443728
print service abc
    end Thread-1 1520824447728
```

从运行的总时间来看，两个线程读取实例变量共耗时 8s，每个线程占用 4s，非常浪费 CPU 资源，因为读取实例变量的操作是可以同时进行的，即读锁之间可以共享，有关此知识点的测试示例见 4.2.2 节。

4.2.2 ReentrantReadWriteLock 类的使用——读读共享

创建 Java 项目 ReadWriteLockBegin1，将 ReentrantLock_end 项目中的所有源代码复制到 ReadWriteLockBegin1 项目中，更改 MyService.java 代码如下：

```
package test;

import java.util.concurrent.locks.ReentrantReadWriteLock;

public class MyService {
    private ReentrantReadWriteLock lock = new ReentrantReadWriteLock();
    private String username = "abc";

    public void testMethod1() {
        try {
            lock.readLock().lock();
            System.out.println("begin " + Thread.currentThread().getName() + " " +
                System.currentTimeMillis());
            System.out.println("print service " + username);
            Thread.sleep(4000);
            System.out.println("  end " + Thread.currentThread().getName() + " " +
                System.currentTimeMillis());
            lock.readLock().unlock();
        } catch (InterruptedException e) {
            e.printStackTrace();
        }
    }
}
```

程序运行结果如图 4-36 所示。

从控制台输出的时间来看，两个线程几乎同时进入 lock() 方法后面的代码，共耗时 4s，说明在此使用 lock.readLock() 读锁可以提高程序运行效率，允许多个线程同时执行 lock() 方法后面的代码。

在此示例中，完全不使用锁，也可以实现异步运行的效果，那为什么要使用锁呢？这是因为有可能有第三个线程在执行写操作，在执行写操作时，这两个读操作就不能与写操作同时运行了，必须在写操作结束后，这两个读操作才可以同时运行，避免了出现非线程安全问题，而且提高了运行效率，此知识点会在后面的章节进行验证。

```
begin Thread-0 1520824602689
print service abc
print service abc
  end Thread-0 1520824606689
  end Thread-1 1520824606689
```

总结：读写互斥，写读互斥，写写互斥，读读异步。　　图 4-36　程序运行结果（读读共享）

4.2.3 ReentrantReadWriteLock 类的使用——写写互斥

创建 Java 项目 ReadWriteLockBegin2，将 ReadWriteLockBegin1 中的所有源代码复制到项目 ReadWriteLockBegin2 中。

更改类 Service.java 代码如下：

```java
package service;

import java.util.concurrent.locks.ReentrantReadWriteLock;

public class Service {

    private ReentrantReadWriteLock lock = new ReentrantReadWriteLock();

    public void write() {
        try {
            try {
                lock.writeLock().lock();
                System.out.println("获得写锁" + Thread.currentThread().getName()
                        + " " + System.currentTimeMillis());
                Thread.sleep(10000);
            } finally {
                lock.writeLock().unlock();
            }
        } catch (InterruptedException e) {
            e.printStackTrace();
        }
    }
}
```

程序运行结果如图 4-37 所示。

图 4-37 程序运行结果（写写互斥）

使用写锁代码 lock.writeLock() 的效果是同一时间只允许一个线程执行 lock() 方法后面的代码。

4.2.4 ReentrantReadWriteLock 类的使用——读写互斥

创建 Java 项目 ReadWriteLockBegin3，将 ReadWriteLockBegin2 中的所有源代码复制到项目 ReadWriteLockBegin3 中。

更改类 Service.java 代码如下：

```java
package service;

import java.util.concurrent.locks.ReentrantReadWriteLock;

public class Service {

    private ReentrantReadWriteLock lock = new ReentrantReadWriteLock();

    public void read() {
        try {
            try {
                lock.readLock().lock();
                System.out.println("获得读锁" + Thread.currentThread().getName()
                        + " " + System.currentTimeMillis());
                Thread.sleep(10000);
            } finally {
                lock.readLock().unlock();
            }
        } catch (InterruptedException e) {
            e.printStackTrace();
        }
    }

    public void write() {
        try {
            try {
                lock.writeLock().lock();
                System.out.println("获得写锁" + Thread.currentThread().getName()
                        + " " + System.currentTimeMillis());
                Thread.sleep(10000);
            } finally {
                lock.writeLock().unlock();
            }
        } catch (InterruptedException e) {
            e.printStackTrace();
        }
    }
}
```

运行类 Run.java 代码更改如下：

```java
package test;

import service.Service;
import extthread.ThreadA;
import extthread.ThreadB;

public class Run {
```

```
public static void main(String[] args) throws InterruptedException {

    Service service = new Service();

    ThreadA a = new ThreadA(service);
    a.setName("A");
    a.start();

    Thread.sleep(1000);

    ThreadB b = new ThreadB(service);
    b.setName("B");
    b.start();

}

}
```

程序运行结果如图 4-38 所示。

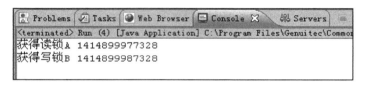

图 4-38　程序运行结果（读写互斥）

此示例说明"读写"操作是互斥的，而且下一个示例演示的"写读"操作也是互斥的。总结：只要出现"写操作"，就是互斥的。

4.2.5　ReentrantReadWriteLock 类的使用——写读互斥

创建 Java 项目 ReadWriteLockBegin4，将 ReadWriteLockBegin3 中的所有源代码复制到项目 ReadWriteLockBegin4 中。

更改运行类 Run.java 代码更改如下：

```
package test;

import service.Service;
import extthread.ThreadA;
import extthread.ThreadB;

public class Run {

    public static void main(String[] args) throws InterruptedException {

        Service service = new Service();
```

```
        ThreadB b = new ThreadB(service);
        b.setName("B");
        b.start();

        Thread.sleep(1000);

        ThreadA a = new ThreadA(service);
        a.setName("A");
        a.start();

    }

}
```

程序运行结果如图 4-39 所示。

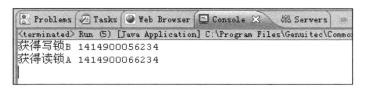

图 4-39　程序运行结果（写读互斥）

从控制台输出结果来看，写读也是互斥的。
"读写""写读""写写"都是互斥的，而"读读"是异步的，非互斥的。

4.3　本章小结

通过对本章的学习，完全可以使用 Lock 对象将 synchronized 关键字替换掉，而且其具有的独特功能也是 synchronized 所不具有的。在学习并发时，Lock 是 synchronized 关键字的进阶，掌握 Lock 有关知识有助于学习并发包中源代码的实现原理，在并发包中，大量类使用 Lock 接口作为同步的处理方式。

第 5 章
定时器 Timer

定时 / 计划功能在移动开发领域应用较多，如 Android 技术。定时 / 计划任务功能在 Java 中主要使用 Timer 对象实现，它在内部使用多线程的方式进行处理，所以它和线程技术有非常大的关联。本章着重掌握如下技术点：

- ❑ 如何实现按指定时间执行任务；
- ❑ 如何实现按指定周期执行任务。

5.1 定时器 Timer 的使用

在 JDK 库中，Timer 类的主要作用是设置计划任务，即在指定时间开始执行某一个任务。Timer 类的方法列表如图 5-1 所示。

TimerTask 类的主要作用是封装任务，该类的结构如图 5-2 所示。

执行计划任务的代码要放入 TimerTask 的子类中，因为 TimerTask 是一个抽象类。

本章将介绍与计划任务有关的方法。

5.1.1 schedule(TimerTask task, Date time) 方法的测试

该方法的作用是在指定日期执行一次某一任务。

1. 执行任务的时间晚于当前时间——在未来执行的效果

创建测试用的项目 timerTest1，创建类 MyTask.java 代码如下：

```
package mytask;

import java.util.TimerTask;
```

```java
public class MyTask extends TimerTask {
    @Override
    public void run() {
        System.out.println("任务执行了，时间为：" + System.currentTimeMillis());
    }
}
```

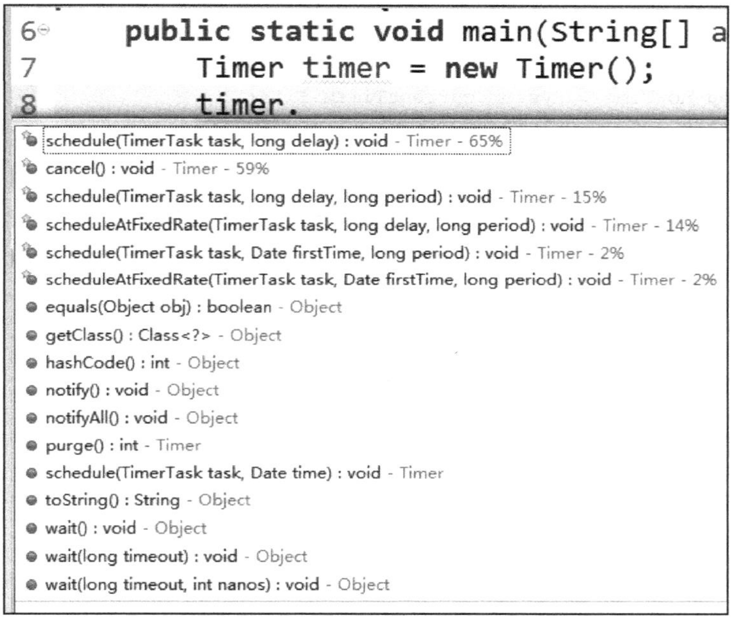

图 5-1　Timer 类的方法列表

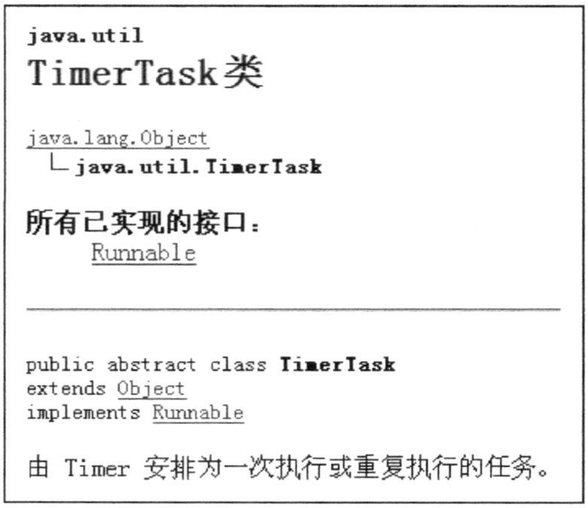
java.util
TimerTask类

java.lang.Object
  └ java.util.TimerTask

所有已实现的接口：
Runnable

public abstract class **TimerTask**
extends Object
implements Runnable

由 Timer 安排为一次执行或重复执行的任务。

图 5-2　TimerTask 类的结构

运行类 Test1.java 代码如下：

```java
package test;

import java.util.Date;
import java.util.Timer;

import mytask.MyTask;

public class Test1 {

    public static void main(String[] args) throws InterruptedException {
        long nowTime = System.currentTimeMillis();
        System.out.println("当前时间为: " + nowTime);

        long scheduleTime = (nowTime + 10000);
        System.out.println("计划时间为: " + scheduleTime);

        MyTask task = new MyTask();

        Timer timer = new Timer();
        Thread.sleep(1000);
        timer.schedule(task, new Date(scheduleTime));

        Thread.sleep(Integer.MAX_VALUE);
    }

}
```

程序运行结果如图 5-3 所示。

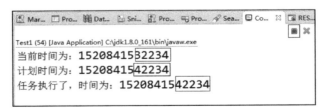

图 5-3　程序运行结果

10s 之后任务成功执行，任务虽然执行完了，但进程还未销毁，按钮呈红色显示，说明内部还有非守护线程正在执行，这一结果在 jvisualvm.exe 工具中也得到了证实，程序运行 10s 之后，Timer-0 线程还在运行，如图 5-4 所示。

为什么会出现这种情况？继续看下面的内容。

2.TimerThread 线程不销毁的原因

进程不销毁的原因是在创建 Timer 对象时启动了一个新的非守护线程，JDK 源代码如下：

```java
public Timer() {
    this("Timer-" + serialNumber());
}
```

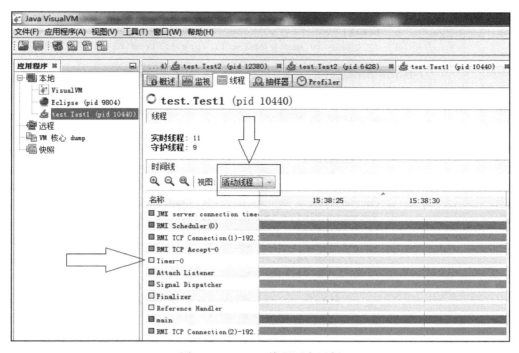

图 5-4　Timer-0 线程还在运行

此构造方法调用的构造方法如下：

```
private final TimerThread thread = new TimerThread(queue);
public Timer(String name) {
    thread.setName(name);
    thread.start();
}
```

查看构造方法可以得知，创建一个 Timer 对象时内部就启动了一个新的线程，可以用新启动的这个线程去执行计划任务。TimerThread 是线程类，其源代码如下：

```
class TimerThread extends Thread {
```

这个新启动的线程并不是守护线程，而且一直在运行，一直在运行的原因是新线程内部有一个死循环。TimerThread.java 类中的 mainLoop() 方法代码如下：

```
private void mainLoop() {
    while (true) {
        try {
            TimerTask task;
            boolean taskFired;
            synchronized(queue) {
                // Wait for queue to become non-empty
                while (queue.isEmpty() && newTasksMayBeScheduled)
                    queue.wait();
```

```
            if (queue.isEmpty())
                break;
                // Queue is empty and will forever remain; die

    // Queue nonempty; look at first evt and do the right thing
            long currentTime, executionTime;
            task = queue.getMin();
            synchronized(task.lock) {
                if (task.state == TimerTask.CANCELLED) {
                    queue.removeMin();
                    continue;
                    // No action required, poll queue again
                }
                currentTime = System.currentTimeMillis();
                executionTime = task.nextExecutionTime;
                if (taskFired = (executionTime<=currentTime)) {
                    if (task.period == 0) {
                    // Non-repeating, remove
                        queue.removeMin();
                        task.state = TimerTask.EXECUTED;
                    } else { // Repeating task, reschedule
                        queue.rescheduleMin(
                            task.period<0 ? currentTime - task.period : execution-
                               Time + task.period);
                    }
                }
            }
            if (!taskFired) // Task hasn't yet fired; wait
                queue.wait(executionTime - currentTime);
        }
        if (taskFired)  // Task fired; run it, holding no locks
            task.run();
    } catch(InterruptedException e) {
    }
  }
}
```

private void mainLoop() 方法内部使用 while (true) 死循环一直执行计划任务，并不退出 while (true) 死循环，但是根据源代码的执行流程，只有满足 if (queue.isEmpty()) 条件，才执行 break 退出 while(true) 死循环。退出逻辑的核心源代码如下：

```
while (queue.isEmpty() && newTasksMayBeScheduled)
    queue.wait();
if (queue.isEmpty())
    break; // Queue is empty and will forever remain; die
```

该源代码的逻辑如下：

1）使用 while 循环对 queue.isEmpty() && newTasksMayBeScheduled 条件进行判断。

2）当 && 两端运算结果都为 true 时，执行 wait() 方法使当前线程暂停运行，等待被唤醒。

3）唤醒线程的时机是执行了 public void schedule(TimerTask task, Date time) 方法。

4）唤醒线程后 while 继续判断 queue.isEmpty() && newTasksMayBeScheduled 条件。如果 queue.isEmpty() 结果为 true，则说明队列中并没有任务，而且布尔变量 newTasksMayBeScheduled 的值由 true 变成 false，继续执行下面的 if 语句。

5）if (queue.isEmpty()) 中的 queue.isEmpty() 结果为 true，说明队列为空，那么就执行 break 语句退出 while(true) 死循环。

6）执行 public void cancel() 方法会使布尔变量 newTasksMayBeScheduled 的值由 true 变成 false。

7）不执行 public void cancel()，则变量 newTasksMayBeScheduled 的值就不会是 false，进程一直呈死循环的状态，进程不销毁就是这个原因。public void cancel() 方法的源代码如下：

```
public void cancel() {
    synchronized(queue) {
        thread.newTasksMayBeScheduled = false;
        queue.clear();
        queue.notify();  // In case queue was already empty.
    }
}
```

上面 7 步就是进程不销毁的原因，以及退出死循环 while(true) 的逻辑。

3. 使用 public void cancel() 方法实现线程 TimerThread 销毁

Timer 类中 public void cancel() 方法的作用是终止此计时器，丢弃所有当前已安排的任务。这不会干扰当前正在执行的任务（如果存在）。一旦终止了计时器，那么它的执行线程也会终止，并且无法根据它安排更多的任务。注意，在此计时器调用的计时器任务的 run() 方法内调用此方法，可以确保正在执行的任务是此计时器所执行的最后一个任务。可以重复调用此方法，但是第二次和后续调用无效。

根据对上面源代码的分析可知，当队列为空，并且 newTasksMayBeScheduled 值是 false 时，退出 while(true) 死循环，导致 TimerThread 线程结束运行并销毁。

创建类 Test2.java 代码如下：

```
package test;

import java.util.Date;
import java.util.Timer;

import mytask.MyTask;

public class Test2 {

    public static void main(String[] args) throws InterruptedException {
```

```
            long nowTime = System.currentTimeMillis();
            System.out.println("当前时间为: " + nowTime);

            long scheduleTime = (nowTime + 15000);
            System.out.println("计划时间为: " + scheduleTime);

            MyTask task = new MyTask();

            Timer timer = new Timer();
            timer.schedule(task, new Date(scheduleTime));

            Thread.sleep(18000);
            timer.cancel();
            Thread.sleep(Integer.MAX_VALUE);
        }
}
```

程序运行 18s 之后，在 jvisualvm.exe 工具中可以看到 TimerThread 线程销毁了，如图 5-5 所示。

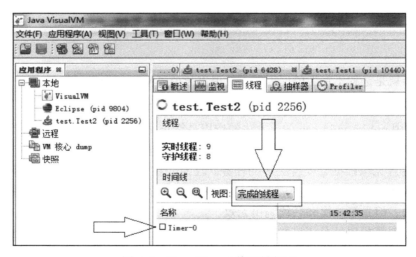

图 5-5 TimerThread 线程销毁了

TimerThread 线程销毁了，但进程还是处于执行状态，因为 main 线程一直在执行 Thread.sleep(Integer.MAX_VALUE) 代码，控制台输出结果如下：

```
当前时间为: 1520841820588
计划时间为: 1520841835588
任务执行了，时间为: 1520841835589
```

4. 计划时间早于当前时间——立即运行的效果

如果执行任务的时间早于当前时间，则立即执行 task 任务。

创建名称为 timerTest2 的 Java 项目，创建类 MyTask.java 代码如下：

```java
package mytask;

import java.util.TimerTask;

public class MyTask extends TimerTask {
    @Override
    public void run() {
        System.out.println("任务执行了，时间为：" + System.currentTimeMillis());
    }
}
```

示例代码 Test1.java 如下：

```java
package test;

import java.util.Date;
import java.util.Timer;

import mytask.MyTask;

public class Test1 {

    public static void main(String[] args) {
        long nowTime = System.currentTimeMillis();
        System.out.println("当前时间为：" + nowTime);

        long scheduleTime = (nowTime - 5000);
        System.out.println("计划时间为：" + scheduleTime);

        MyTask task = new MyTask();

        Timer timer = new Timer();
        timer.schedule(task, new Date(scheduleTime));

    }

}
```

程序运行结果如图 5-6 所示。

5. 在 Timer 中执行多个 TimerTask 任务

在 Timer 中可以执行多个 TimerTask 任务。
创建类 Test2.java，代码如下：

```java
package test;

import java.util.Date;
import java.util.Timer;
```

```
当前时间为：1520841108806
计划时间为：1520841103806
任务执行了，时间为：1520841108808
```

图 5-6　立即执行 task 任务

```
import mytask.MyTask;

public class Test2 {

    public static void main(String[] args) {
        long nowTime = System.currentTimeMillis();
        System.out.println("当前时间为: " + nowTime);

        long scheduleTime1 = (nowTime + 5000);
        long scheduleTime2 = (nowTime + 8000);
        System.out.println("计划时间1为: " + scheduleTime1);
        System.out.println("计划时间2为: " + scheduleTime2);

        MyTask task1 = new MyTask();
        MyTask task2 = new MyTask();

        Timer timer = new Timer();
        timer.schedule(task1, new Date(scheduleTime1));
        timer.schedule(task2, new Date(scheduleTime2));
    }
}
```

程序运行结果如图 5-7 所示。

```
当前时间为: 1520842020683
计划时间1为: 1520842025683
计划时间2为: 1520842028683
任务执行了,时间为: 1520842025684
任务执行了,时间为: 1520842028683
```

图 5-7 在一个 Timer 中可以运行多个 TimerTask 任务

6. 延时执行 TimerTask 的测试

TimerTask 以队列的方式逐一按顺序执行,所以执行的时间有可能和预期的时间不一致,因为前面任务有可能消耗的时间较长,而后面任务的运行时间也可能会被延后,请看下面的示例。

创建测试用的 Java 项目 taskLater,类 MyTaskA.java 代码如下:

```
package mytask;

import java.util.TimerTask;

public class MyTaskA extends TimerTask {
    @Override
    public void run() {
        try {
            System.out.println("A begin timer=" + System.currentTimeMillis());
            Thread.sleep(20000);
            System.out.println("A   end timer=" + System.currentTimeMillis());
        } catch (InterruptedException e) {
            e.printStackTrace();
        }
    }
}
```

类 MyTaskB.java 代码如下:

```java
package mytask;

import java.util.TimerTask;

public class MyTaskB extends TimerTask {
    @Override
    public void run() {
        System.out.println("B begin timer=" + System.currentTimeMillis());
        System.out.println("B   end timer=" + System.currentTimeMillis());
    }
}
```

类 Test.java,代码如下:

```java
package test;

import java.util.Date;
import java.util.Timer;

import mytask.MyTaskA;
import mytask.MyTaskB;

public class Test {
    public static void main(String[] args) {
        long nowTime = System.currentTimeMillis();
        System.out.println("当前时间为: " + nowTime);

        long scheduleTime1 = nowTime;
        long scheduleTime2 = nowTime + 5000;
        System.out.println("计划时间1为: " + scheduleTime1);
        System.out.println("计划时间2为: " + scheduleTime2);

        MyTaskA task1 = new MyTaskA();
        MyTaskB task2 = new MyTaskB();

        Timer timer = new Timer();
        timer.schedule(task1, new Date(scheduleTime1));
        timer.schedule(task2, new Date(scheduleTime2));
    }
}
```

程序运行结果如图 5-8 所示。

在代码中,long scheduleTime2 = nowTime + 5000 原计划设置任务 1 和任务 2 的运行间隔时间为 5s,但是由于 task1 需要用时 20s 执行完任务,而 task1 的结束时间就是

```
计划时间1为: 1520842270473
计划时间2为: 1520842275473
A begin timer=1520842270474
A   end timer=1520842290474
B begin timer=1520842290474
B   end timer=1520842290474
```

图 5-8 任务 2 被延时

task2 的开始时间，所以 task2 不再以 5s 作为参考，而是以 20s 作为参考，这是因为创建了一个 Timer 对象，即创建一个 TimerThread 线程，一个 TimerThread 线程管理一个队列，队列按顺序运行 Task 任务。

5.1.2　schedule(TimerTask task, Date firstTime, long period) 方法的测试

该方法的作用是在指定日期之后按指定的间隔周期无限循环地执行某一任务。

1. 计划时间晚于当前时间——在未来执行的效果

创建测试用的项目 timerTest2_period，类 MyTask.java 代码如下：

```java
package mytask;

import java.util.TimerTask;

public class MyTask extends TimerTask {
    @Override
    public void run() {
        System.out.println("任务执行了，时间为：" + System.currentTimeMillis());
    }
}
```

创建类 Test1.java 代码如下：

```java
package test;

import java.util.Date;
import java.util.Timer;

import mytask.MyTask;

public class Test1 {

    public static void main(String[] args) {
        long nowTime = System.currentTimeMillis();
        System.out.println("当前时间为：" + nowTime);

        long scheduleTime = (nowTime + 10000);
        System.out.println("计划时间为：" + scheduleTime);

        MyTask task = new MyTask();

        Timer timer = new Timer();
        timer.schedule(task, new Date(scheduleTime), 4000);

    }

}
```

程序运行结果如图 5-9 所示。

从程序运行结果来看,每隔 4s 运行 TimerTask 任务,并且无限期重复执行 task 任务。

2. 计划时间早于当前时间——立即运行的效果

如果计划时间早于当前时间,则立即执行 task 任务。
创建类 Test2.java 代码如下:

```
package test;

import java.util.Date;
import java.util.Timer;

import mytask.MyTask;

public class Test2 {

    public static void main(String[] args) {
        long nowTime = System.currentTimeMillis();
        System.out.println("当前时间为: " + nowTime);

        long scheduleTime = (nowTime - 10000);
        System.out.println("计划时间为: " + scheduleTime);
        MyTask task = new MyTask();

        Timer timer = new Timer();
        timer.schedule(task, new Date(scheduleTime), 4000);
    }

}
```

```
当前时间为: 1520843078248
计划时间为: 1520843088248
任务执行了,时间为: 1520843088248
任务执行了,时间为: 1520843092249
任务执行了,时间为: 1520843096249
任务执行了,时间为: 1520843100249
任务执行了,时间为: 1520843104249
任务执行了,时间为: 1520843108250
```

图 5-9 程序运行结果

程序运行结果如图 5-10 所示。
控制台输出的结果是程序运行后立即执行 task 任务,并且每隔 4s 一打印。

```
当前时间为: 1520843444381
计划时间为: 1520843434381
任务执行了,时间为: 1520843444382
任务执行了,时间为: 1520843448382
任务执行了,时间为: 1520843452382
任务执行了,时间为: 1520843456383
任务执行了,时间为: 1520843460383
```

图 5-10 立即执行 task 任务

3. 延时执行 TimerTask 的测试

创建 Java 项目,名称为 timerTest2_periodLater,创建类 MyTaskA.java 代码如下:

```
package mytask;

import java.util.TimerTask;

public class MyTaskA extends TimerTask {

    @Override
    public void run() {
        try {
            System.out.println("A begin timer=" + System.currentTimeMillis());
            Thread.sleep(5000);
            System.out.println("A   end timer=" + System.currentTimeMillis());
```

```java
        } catch (InterruptedException e) {
            e.printStackTrace();
        }
    }
}
```

运行类 Test1.java 代码如下：

```java
package test;

import java.util.Date;
import java.util.Timer;

import mytask.MyTaskA;

public class Test1 {

    public static void main(String[] args) {
        long nowTime = System.currentTimeMillis();
        System.out.println("当前时间为: " + nowTime);

        System.out.println("计划时间为: " + nowTime);

        MyTaskA task = new MyTaskA();

        Timer timer = new Timer();
        timer.schedule(task, new Date(nowTime), 3000);

    }

}
```

代码 timer.schedule(task, new Date(nowTime), 3000) 计划让每个任务执行的间隔时间是 3s，但是结果是 5s，因为任务执行了 Thread.sleep(5000) 代码。程序运行结果如图 5-11 所示。

```
当前时间为: 1520843735638
计划时间为: 1520843735638
A begin timer=1520843735639
A   end timer=1520843740639
A begin timer=1520843740639
A   end timer=1520843745639
A begin timer=1520843745639
A   end timer=1520843750639
A begin timer=1520843750639
A   end timer=1520843755640
```

图 5-11　任务被延时成 5s 而不是 3s

4. TimerTask 类中的 cancel() 方法

TimerTask 类中的 cancel() 方法的作用是将自身从任务队列中清除。该方法的源代码如下：

```java
public boolean cancel() {
    synchronized(lock) {
        boolean result = (state == SCHEDULED);
        state = CANCELLED;
        return result;
    }
}
```

从方法的代码可以分析出，执行 TimerTask 类中的 cancel() 方法时，当前 TimerTask 任务的 state 状态改成 CANCELLED（取消）。

创建名称为 TimerTaskCancelMethod 的项目，类 MyTaskA.java 代码如下：

```java
package mytask;

import java.util.TimerTask;

public class MyTaskA extends TimerTask {
    @Override
    public void run() {
        System.out.println("A run timer=" + System.currentTimeMillis());
        this.cancel();
        System.out.println("A任务自己移除自己");
    }
}
```

类 MyTaskB.java 代码如下：

```java
package mytask;

import java.util.TimerTask;

public class MyTaskB extends TimerTask {
    @Override
    public void run() {
        System.out.println("B run timer=" + System.currentTimeMillis());
    }
}
```

创建 Test.java 文件，代码如下：

```java
package test;

import java.util.Date;
import java.util.Timer;

import mytask.MyTaskA;
import mytask.MyTaskB;

public class Test {
    public static void main(String[] args) {
        long nowTime = System.currentTimeMillis();
        System.out.println("当前时间为: " + nowTime);

        System.out.println("计划时间为: " + nowTime);

        MyTaskA task1 = new MyTaskA();
        MyTaskB task2 = new MyTaskB();
```

```
        Timer timer = new Timer();
        timer.schedule(task1, new Date(nowTime), 4000);
        timer.schedule(task2, new Date(nowTime), 4000);
    }
}
```

程序运行结果如图 5-12 所示。

TimerTask 类的 cancel() 方法是将自身从任务队列中消除，其他任务不受影响。

5. Timer 类中的 cancel() 方法

和 TimerTask 类中的 cancel() 方法清除自身不同，Timer 类中的 cancel() 方法的作用是将任务队列中的全部任务清空，源代码如下：

```
当前时间为: 1520844509939
计划时间为: 1520844509939
A run timer=1520844509940
A任务自己移除自己
B run timer=1520844509940
B run timer=1520844513940
B run timer=1520844517940
B run timer=1520844521940
B run timer=1520844525941
B run timer=1520844529941
B run timer=1520844533941
B run timer=1520844537941
B run timer=1520844541942
B run timer=1520844545942
```

图 5-12 TimerTaskA 仅运行一次后被取消了

```java
public void cancel() {
    synchronized(queue) {
        thread.newTasksMayBeScheduled = false;
        queue.clear();
        queue.notify();    // In case queue was already empty.
    }
}
```

创建名称为 TimerCancelMethod 的项目，类 MyTaskA.java 代码如下：

```java
package mytask;

import java.util.TimerTask;

public class MyTaskA extends TimerTask {
    @Override
    public void run() {
        System.out.println("A run timer=" + System.currentTimeMillis());
    }
}
```

类 MyTaskB.java 代码如下：

```java
package mytask;

import java.util.TimerTask;

public class MyTaskB extends TimerTask {
    @Override
    public void run() {
        System.out.println("B run timer=" + System.currentTimeMillis());
    }
}
```

类 Test.java，代码如下：

```java
package test;

import java.util.Date;
import java.util.Timer;

import mytask.MyTaskA;
import mytask.MyTaskB;

public class Test {

    public static void main(String[] args) throws InterruptedException {
        long nowTime = System.currentTimeMillis();
        System.out.println("当前时间为：" + nowTime);

        System.out.println("计划时间为：" + nowTime);

        MyTaskA task1 = new MyTaskA();
        MyTaskB task2 = new MyTaskB();

        Timer timer = new Timer();
        timer.schedule(task1, new Date(nowTime), 2000);
        timer.schedule(task2, new Date(nowTime), 2000);

        Thread.sleep(10000);

        timer.cancel();//全部任务都取消

    }

}
```

程序运行 10s 后，进程销毁，如图 5-13 所示。

全部任务都被清除，并且进程被销毁，按钮由红色变成灰色。

6. 间隔执行 Task 任务的算法

当队列中有 3 个任务 ABC 时，这 3 个任务执行顺序的算法是每次将最后一个任务放入队列头，再执行队列头中 Task 任务的 run() 方法，算法效果如下：

1）ABC；
2）CAB（将 C 放入 AB 之前）；
3）BCA（将 B 放入 CA 之前）。

创建新的任务类代码如下：

```
package mytask;
```

```
当前时间为：1520844720890
计划时间为：1520844720890
A run timer=1520844720891
B run timer=1520844720891
B run timer=1520844722891
A run timer=1520844722891
A run timer=1520844724891
B run timer=1520844724891
B run timer=1520844726891
A run timer=1520844726891
A run timer=1520844728891
B run timer=1520844728891
B run timer=1520844730891
```

图 5-13　进程销毁

```java
import java.util.TimerTask;

public class MyTaskC extends TimerTask {
    @Override
    public void run() {
        System.out.println("C run timer=" + System.currentTimeMillis());
    }
}
```

运行类代码如下:

```java
package test;

import java.util.Date;
import java.util.Timer;

import mytask.MyTaskA;
import mytask.MyTaskB;
import mytask.MyTaskC;

public class Test2 {

    public static void main(String[] args) throws InterruptedException {
        long nowTime = System.currentTimeMillis();
        System.out.println("当前时间为: " + nowTime);

        System.out.println("计划时间为: " + nowTime);

        MyTaskA task1 = new MyTaskA();
        MyTaskB task2 = new MyTaskB();
        MyTaskC task3 = new MyTaskC();

        Timer timer = new Timer();
        timer.schedule(task1, new Date(nowTime), 2000);
        timer.schedule(task2, new Date(nowTime), 2000);
        timer.schedule(task3, new Date(nowTime), 2000);

        Thread.sleep(Integer.MAX_VALUE);
    }

}
```

程序运行结果如下:

```
当前时间为: 1520846683801
计划时间为: 1520846683801
A run timer=1520846683803
B run timer=1520846683803
C run timer=1520846683803
C run timer=1520846685803
A run timer=1520846685803
B run timer=1520846685803
```

```
B run timer=1520846687803
C run timer=1520846687803
A run timer=1520846687803
```

7. Timer 类中的 cancel() 方法的使用注意事项

调用 Timer 类中的 cancel() 方法有时并不一定会停止计划任务，即计划任务正常执行。

创建名称为 TimerCancelError 的项目，类 **MyTaskA.java** 代码如下：

```java
package mytask;

import java.util.TimerTask;

public class MyTaskA extends TimerTask {

    private int i;

    public MyTaskA(int i) {
        super();
        this.i = i;
    }

    @Override
    public void run() {
        System.out.println("第" + i + "次没有被cancel取消");
    }

}
```

创建 Test.java 类，代码如下：

```java
package test;

import java.util.Date;
import java.util.Timer;

import mytask.MyTaskA;

public class Test {

    public static void main(String[] args) throws InterruptedException {
        int i = 0;

        long nowTime = System.currentTimeMillis();
        System.out.println("当前时间为: " + nowTime);
        System.out.println("计划时间为: " + nowTime);

        while (true) {
            i++;
            Timer timer = new Timer();
            MyTaskA task1 = new MyTaskA(i);
            timer.schedule(task1, new Date(nowTime));
```

```
            timer.cancel();
        }
    }
}
```

程序运行结果如图 5-14 所示。

由程序运行结果可以看出，任务并没有停止，原因就是 Timer 类中的 cancel() 方法有时并没有争抢到 queue 锁，所以 TimerTask 类中的任务正常执行。

> 当前时间为：1520845565499
> 计划时间为：1520845565499
> 第1次没有被cancel取消
> 第512次没有被cancel取消
> 第726594次没有被cancel取消

图 5-14　任务并没有停止

5.1.3　schedule(TimerTask task, long delay) 方法的测试

该方法的作用是以执行 schedule(TimerTask task, long delay) 方法当前的时间为参考时间，在此时间基础上延迟指定的毫秒数后执行一次 TimerTask 任务。

创建测试用的项目 timerTest3，创建类 Run.java 代码如下：

```java
package test;

import java.util.Timer;
import java.util.TimerTask;

public class Run {
    static public class MyTask extends TimerTask {
        @Override
        public void run() {
            System.out.println("运行了! 时间为: " + System.currentTimeMillis());
        }
    }

    public static void main(String[] args) {
        MyTask task = new MyTask();
        Timer timer = new Timer();
        System.out.println("当前时间: " + System.currentTimeMillis());
        timer.schedule(task, 7000);
    }
}
```

程序运行结果如图 5-15 所示。

由程序运行结果可以看出，任务 task 被延迟 7s 执行。

> 当前时间：1520846850725
> 运行了! 时间为：1520846857725

图 5-15　程序运行结果

5.1.4　schedule(TimerTask task, long delay, long period) 方法的测试

该方法的作用是以执行 schedule(TimerTask task, long delay, long period) 方法当前的时间为参考时间，在此时间基础上延迟指定的毫秒数再以某一间隔时间无限次数地执行某一任务。

创建测试用的项目 timerTest4，创建类 Run.java 代码如下：

```java
package test;

import java.util.Timer;
import java.util.TimerTask;

public class Run {
    static public class MyTask extends TimerTask {
        @Override
        public void run() {
            System.out.println("运行了！时间为: " + System.currentTimeMillis());
        }
    }

    public static void main(String[] args) {
        MyTask task = new MyTask();
        Timer timer = new Timer();
        System.out.println("当前时间: " + System.currentTimeMillis());
        timer.schedule(task, 3000, 5000);
    }
}
```

程序运行结果如图 5-16 所示。

凡是带有 period 参数的方法，均无限循环执行 TimerTask 中的任务。

```
当前时间: 1520847026272
运行了！时间为: 1520847029272
运行了！时间为: 1520847034272
运行了！时间为: 1520847039272
运行了！时间为: 1520847044273
运行了！时间为: 1520847049273
```

图 5-16　程序运行结果

5.1.5　scheduleAtFixedRate(TimerTask task, Date firstTime, long period) 方法的测试

schedule() 方法和 scheduleAtFixedRate() 方法的主要区别在于有没有追赶特性。

1. 测试 schedule() 方法任务不延时——Date 类型

创建项目 timerTest5，创建 Java 类 Test1.java，代码如下：

```java
package test;

import java.util.Date;
import java.util.Timer;
import java.util.TimerTask;

public class Test1 {
    static class MyTask extends TimerTask {
        public void run() {
            try {
                System.out.println("begin timer=" + System.currentTimeMillis());
                Thread.sleep(1000);
                System.out.println("  end timer=" + System.currentTimeMillis());
            } catch (InterruptedException e) {
                e.printStackTrace();
            }
        }
```

```
        }
    }

    public static void main(String[] args) {
        MyTask task = new MyTask();

        long nowTime = System.currentTimeMillis();

        Timer timer = new Timer();
        timer.schedule(task, new Date(nowTime), 3000);
    }
}
```

程序运行结果如图 5-17 所示。

程序运行结果证明，在不延时的情况下，如果任务没有被延时执行，则下一次执行任务的开始时间是上一次任务的开始时间加上 period 时间。

"不延时"是指执行任务的时间小于 period 间隔时间。

```
begin timer=1520903502972
  end timer=1520903503973
begin timer=1520903505972
  end timer=1520903506972
begin timer=1520903508972
  end timer=1520903509972
begin timer=1520903511972
  end timer=1520903512972
```

图 5-17　任务没有延时的运行结果（Date 类型）

2. 测试 schedule() 方法任务不延时——long 类型

创建 Java 类 Test2.java，代码如下：

```java
package test;

import java.util.Timer;
import java.util.TimerTask;

public class Test2 {

    static class MyTask extends TimerTask {
        public void run() {
            try {
                System.out.println("begin timer=" + System.currentTimeMillis());
                Thread.sleep(1000);
                System.out.println("  end timer=" + System.currentTimeMillis());
            } catch (InterruptedException e) {
                e.printStackTrace();
            }
        }
    }

    public static void main(String[] args) {
        MyTask task = new MyTask();
        System.out.println("当前时间: " + System.currentTimeMillis());
        Timer timer = new Timer();
        timer.schedule(task, 3000, 4000);
```

            }
        }

程序运行结果如图 5-18 所示。

程序运行结果证明，在不延时的情况下，如果任务没有被延时执行，则第一次执行任务的时间是任务开始时间加上 delay 时间，接下来执行任务的时间是上一次任务的开始时间加上 period 时间。

```
当前时间：1438241834750
begin timer=1438241837750
  end timer=1438241838750
begin timer=1438241841750
  end timer=1438241842750
begin timer=1438241845750
  end timer=1438241846750
begin timer=1438241849750
  end timer=1438241850750
begin timer=1438241853750
```

图 5-18　任务没有延时的运行结果（long 类型）

3. 测试 s.chedule() 方法任务延时——Date 类型

创建 Test3.java 类，代码如下：

```java
package test;

import java.util.Date;
import java.util.Timer;
import java.util.TimerTask;

public class Test3 {
    static class MyTask extends TimerTask {
        public void run() {
            try {
                System.out.println("begin timer=" + System.currentTimeMillis());
                Thread.sleep(5000);
                System.out.println("end timer=" + System.currentTimeMillis());
            } catch (InterruptedException e) {
                e.printStackTrace();
            }
        }
    }

    public static void main(String[] args) {
        MyTask task = new MyTask();

        long nowTime = System.currentTimeMillis();

        Timer timer = new Timer();
        timer.schedule(task, new Date(nowTime),
            2000);
    }
}
```

```
begin timer=1520903958857
  end timer=1520903963857
begin timer=1520903963857
  end timer=1520903968857
begin timer=1520903968857
  end timer=1520903973858
```

图 5-19　任务延时的结果（Date 类型）

程序运行结果如图 5-19 所示。

从控制台打印的结果来看，在延时的情况下，如果任务被延时执行，那么下一次任务的执行时间参考的是上一次任务"结束"时的时间来开始的。

4. 测试 schedule() 方法任务延时——long 类型

创建 Test4.java 类，代码如下：

```java
package test;

import java.util.Timer;
import java.util.TimerTask;

public class Test4 {

    static class MyTask extends TimerTask {
        public void run() {
            try {
                System.out.println("begin timer=" + System.currentTimeMillis());
                Thread.sleep(5000);
                System.out.println("  end timer=" + System.currentTimeMillis());
            } catch (InterruptedException e) {
                e.printStackTrace();
            }
        }
    }

    public static void main(String[] args) {
        MyTask task = new MyTask();
        System.out.println("当前时间: " + System.currentTimeMillis());
        Timer timer = new Timer();
        timer.schedule(task, 3000, 2000);

    }

}
```

程序运行结果如图 5-20 所示。

从控制台打印的结果来看，在延时的情况下，如果任务被延时执行，那么下一次任务的执行时间参考的是上一次任务"结束"时的时间来开始的。

5. 测试 scheduleAtFixedRate() 方法任务不延时——Date 类型

创建 Test5.java 文件，本示例使用 scheduleAtFixed-Rate() 方法进行测试。

完整代码如下：

```java
package test;

import java.util.Date;
import java.util.Timer;
```

```
当前时间: 1438242120812
begin timer=1438242123812
  end timer=1438242128812
begin timer=1438242128812
  end timer=1438242133812
begin timer=1438242133812
  end timer=1438242138812
begin timer=1438242138812
  end timer=1438242143812
begin timer=1438242143812
```

图 5-20　任务延时的结果（long 类型）

```java
import java.util.TimerTask;

public class Test5 {

    static class MyTask extends TimerTask {
        public void run() {
            try {
                System.out.println("begin timer=" + System.currentTimeMillis());
                Thread.sleep(1000);
                System.out.println("  end timer=" + System.currentTimeMillis());
            } catch (InterruptedException e) {
                e.printStackTrace();
            }
        }
    }

    public static void main(String[] args) {
        MyTask task = new MyTask();

        long nowTime = System.currentTimeMillis();

        Timer timer = new Timer();
        timer.scheduleAtFixedRate(task, new Date(nowTime), 3000);
    }
}
```

程序运行结果如图 5-21 所示。

程序运行结果证明，在不延时的情况下，如果任务没有被延时执行，则下一次执行任务的时间是上一次任务的开始时间加上 period 时间。

```
begin timer=1520904128657
  end timer=1520904129659
begin timer=1520904131656
  end timer=1520904132656
begin timer=1520904134656
  end timer=1520904135656
begin timer=1520904137656
  end timer=1520904138656
```

图 5-21　任务没有被延时的运行结果（Date 类型）

6. 测试 scheduleAtFixedRate() 方法任务不延时——long 类型

创建 Test6.java 文件，本示例使用 scheduleAtFixedRate() 方法进行测试。

完整代码如下：

```java
package test;

import java.util.Timer;
import java.util.TimerTask;

public class Test6 {

    static class MyTask extends TimerTask {
        public void run() {
            try {
```

```
            System.out.println("begin timer=" + System.currentTimeMillis());
            Thread.sleep(1000);
            System.out.println("end timer=" + System.currentTimeMillis());
        } catch (InterruptedException e) {
            e.printStackTrace();
        }
    }
}

public static void main(String[] args) {
    MyTask task = new MyTask();
    System.out.println("当前时间: " + System.currentTimeMillis());
    Timer timer = new Timer();
    timer.scheduleAtFixedRate(task, 3000, 4000);
}

}
```

程序运行结果如图 5-22 所示。

程序运行结果证明,在不延时的情况下,如果任务没有被延时执行,则第一次执行任务的时间是任务开始时间加上 delay 时间,接下来执行任务的时间是上一次任务的开始时间加上 period 时间。

```
当前时间: 1438242357218
begin timer=1438242360218
  end timer=1438242361218
begin timer=1438242364218
  end timer=1438242365218
begin timer=1438242368218
  end timer=1438242369218
begin timer=1438242372218
  end timer=1438242373218
```

图 5-22 任务没有被延时的运行结果(long 类型)

7. 测试 scheduleAtFixedRate() 方法任务延时——Date 类型

创建 Test7.java 文件,代码如下:

```java
package test;

import java.util.Date;
import java.util.Timer;
import java.util.TimerTask;

public class Test7 {

    static class MyTask extends TimerTask {
        public void run() {
            try {
                System.out.println("begin timer=" + System.currentTimeMillis());
                Thread.sleep(5000);
                System.out.println("end timer=" + System.currentTimeMillis());
            } catch (InterruptedException e) {
                e.printStackTrace();
            }
        }
    }
```

```java
    public static void main(String[] args) {
        MyTask task = new MyTask();

        long nowTime = System.currentTimeMillis();

        Timer timer = new Timer();
        timer.scheduleAtFixedRate(task, new Date(nowTime), 2000);
    }

}
```

程序运行结果如图 5-23 所示。

从程序运行结果来看，在延时的情况下，如果任务被延时执行，那么下一次任务的执行时间是参考上一次任务"结束"时的时间来计算的。

```
begin timer=1520904377023
  end timer=1520904382023
begin timer=1520904382023
  end timer=1520904387023
begin timer=1520904387023
  end timer=1520904392024
```

图 5-23　任务延时的运行结果（Date 类型）

8. 测试 scheduleAtFixedRate() 方法任务延时——long 类型

创建 Test8.java 文件，代码如下：

```java
package test;

import java.util.Timer;
import java.util.TimerTask;

public class Test8 {

    static class MyTask extends TimerTask {
        public void run() {
            try {
                System.out.println("begin timer=" + System.currentTimeMillis());
                Thread.sleep(5000);
                System.out.println("  end timer=" + System.currentTimeMillis());
            } catch (InterruptedException e) {
                e.printStackTrace();
            }
        }
    }

    public static void main(String[] args) {
        MyTask task = new MyTask();
        System.out.println("当前时间：" + System.
            currentTimeMillis());
        Timer timer = new Timer();
        timer.scheduleAtFixedRate(task, 3000,
            2000);
    }

}
```

程序运行结果如图 5-24 所示。

```
当前时间：1438242567640
begin timer=1438242570640
  end timer=1438242575640
begin timer=1438242575640
  end timer=1438242580640
begin timer=1438242580640
  end timer=1438242585640
begin timer=1438242585640
```

图 5-24　任务延时的运行结果（long 类型）

从程序运行结果来看，在延时的情况下，如果任务被延时执行，那么下一次任务的执行时间参考的是上一次任务"结束"时的时间来计算。

从上面代码的运行结果来看，schedule() 方法和 scheduleAtFixedRate() 方法的运行效果并没有什么区别，那它们之间到底有什么区别呢？那就是是否具有追赶执行性。

9. 验证 schedule() 方法不具有追赶执行性

创建 Java 类 Test9.java，代码如下：

```java
package test;

import java.util.Date;
import java.util.Timer;
import java.util.TimerTask;

public class Test9 {
    static class MyTask extends TimerTask {
        public void run() {
            System.out.println("begin timer=" + System.currentTimeMillis());
            System.out.println("  end timer=" + System.currentTimeMillis());
        }
    }

    public static void main(String[] args) {
        MyTask task = new MyTask();
        long nowTime = System.currentTimeMillis();
        System.out.println("现在执行时间：" + nowTime);
        long runTime = nowTime - 20000;
        System.out.println("计划执行时间：" + runTime);
        Timer timer = new Timer();
        timer.schedule(task, new Date(runTime), 2000);
    }
}
```

程序运行结果如图 5-25 所示。

1520904922596 到 1520904902596 之间的时间所对应的 Task 任务被取消，不被执行，这说明 Task 任务不具有追赶执行性。

10. 验证 scheduleAtFixedRate() 方法具有追赶执行性

创建 Java 类 Test10.java，代码如下：

```java
package test;

import java.util.Date;
import java.util.Timer;
import java.util.TimerTask;
```

```
现在执行时间：1520904922596
计划执行时间：1520904902596
begin timer=1520904922598
  end timer=1520904922598
begin timer=1520904924598
  end timer=1520904924598
begin timer=1520904926598
  end timer=1520904926598
begin timer=1520904928620
  end timer=1520904928622
begin timer=1520904930620
  end timer=1520904930620
begin timer=1520904932620
  end timer=1520904932620
begin timer=1520904934621
  end timer=1520904934621
```

图 5-25　不具有追赶执行性

```java
public class Test10 {
    static class MyTask extends TimerTask {
        public void run() {
            System.out.println("begin timer=" + System.currentTimeMillis());
            System.out.println("  end timer=" + System.currentTimeMillis());
        }
    }

    public static void main(String[] args) {
        MyTask task = new MyTask();
        long nowTime = System.currentTimeMillis();
        System.out.println("现在执行时间: " + nowTime);
        long runTime = nowTime - 20000;
        System.out.println("计划执行时间: " + runTime);
        Timer timer = new Timer();
        timer.scheduleAtFixedRate(task, new Date(runTime), 2000);
    }
}
```

程序运行结果如下：

现在执行时间：1520905227591
计划执行时间：1520905207591
begin timer=1520905227593
  end timer=1520905227593
begin timer=1520905227593
  end timer=1520905227593
begin timer=1520905227593
  end timer=1520905227593
begin timer=1520905227593
  end timer=1520905227593
begin timer=1520905227593
  end timer=1520905227593
begin timer=1520905227593
  end timer=1520905227593
begin timer=1520905227593
  end timer=1520905227593
begin timer=1520905227593
  end timer=1520905227593
begin timer=1520905227593
  end timer=1520905227593
begin timer=1520905227593
  end timer=1520905227593
begin timer=1520905227593
  end timer=1520905227593
begin timer=1520905229591
  end timer=1520905229591
begin timer=1520905231592
  end timer=1520905231592
begin timer=1520905233591
  end timer=1520905233591

```
begin timer=1520905235591
  end timer=1520905235591
```

输出时间 1520905227593 都是曾经流逝时间的任务追赶，也就是将之前没有执行的任务追加执行，将 20s 之内执行任务的次数输出完，再每间隔 2s 执行一次任务。

将两个时间段内的时间所对应的 Task 任务被"弥补"地执行，也就是在指定时间段内的运行次数必须运行完整，这就是 Task 任务的追赶特性。

5.2 本章小结

通过对本章的学习，应该掌握如何在 Java 中使用定时任务功能，并且可以对这些定时任务使用指定的 API 进行处理。这些示例代码完全可以应用在 Android 技术中，实现类似于轮询、动画等常见的主要功能。

第 6 章 Chapter 6

单例模式与多线程

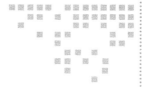

本章的知识点非常重要，通过将单例模式与多线程技术相结合，在这个过程中能发现很多以前从未考虑过的情况，这些不良的程序设计写法应用在商业项目中会使其遇到非常大的麻烦，本章的案例也充分说明，将线程与某些技术相结合时要考虑的事情会更多。在学习本章时只需要考虑一件事情，那就是如何使单例模式遇到多线程是安全、正确的。

在标准的 23 个设计模式中，单例模式是比较常见的，但在常规的教学资料中，多数并没有结合多线程技术作为参考，这就造成在使用结合多线程的单例模式时会出现一些意想不到的意外，这样的代码应用在生产环境中会使其出现异常，有可能造成灾难性后果。本章将介绍单例模式结合多线程技术在使用时的相关知识。

6.1 立即加载 / 饿汉模式

立即加载是指使用类的时候已经将对象创建完毕，常见的实现办法是直接用 new 实例化。从中文的语境看，立即加载有"着急""急迫"的意味，所以也称为"饿汉模式"。

在立即加载 / 饿汉模式中，调用方法前，实例已经被工厂创建了，下面来看一下实现代码。

创建测试用的项目，名称为 singleton_0，创建类 MyObject.java 代码如下：

```
package test;

public class MyObject {
    // 立即加载方式/饿汉模式
    private static MyObject myObject = new MyObject();
```

```java
    private MyObject() {
    }

    public static MyObject getInstance() {
        return myObject;
    }
}
```

创建线程类 MyThread.java 代码如下：

```java
package extthread;

import test.MyObject;

public class MyThread extends Thread {

    @Override
    public void run() {
        System.out.println(MyObject.getInstance().hashCode());
    }

}
```

创建运行类 Run.java 代码如下：

```java
package test.run;

import extthread.MyThread;

public class Run {

    public static void main(String[] args) {
        MyThread t1 = new MyThread();
        MyThread t2 = new MyThread();
        MyThread t3 = new MyThread();

        t1.start();
        t2.start();
        t3.start();

    }
}
```

程序运行结果如图 6-1 所示。

控制台输出的 hashCode 是同一个值，说明对象都是同一个，也就实现了立即加载型单例模式。

此代码版本为立即加载型，但此版本代码的缺点是不能有其他实例变量，因为 getInstance() 方法没有同步，所以有可能出现非线程安全问题，例如，出现如下代码：

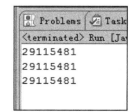

图 6-1　饿汉模式的运行结果

```java
public class MyObject {
    // 立即加载/饿汉模式
    private static MyObject myObject = new MyObject();

    private MyObject() {
    }

    private static String username;
    private static String password;

    public static MyObject getInstance() {
        username = "从不同的服务器取出值有可能不一样，并赋值";
        password = "从不同的服务器取出值有可能不一样，并赋值";
        // 上面的赋值并没有被同步syn，所以极易出现非线程安全问题，变量值被覆盖
        return myObject;
    }
}
```

要解决这个问题，请看后面章节中的相关内容。

6.2 延迟加载 / 懒汉模式

延迟加载是指调用 get() 方法时实例才被工厂创建，常见的实现办法是在 get() 方法中进行 new 实例化。从中文的语境来看，延迟加载有"缓慢""不急迫"的意味，所以也称为"懒汉模式"。

6.2.1 延迟加载 / 懒汉模式解析

在延迟加载 / 懒汉模式中，调用方法时实例才被工厂创建，下面来看一下实现代码。
创建测试用的项目，名称为 singleton_1，创建类 MyObject.java 代码如下：

```java
package test;

public class MyObject {

    private static MyObject myObject;

    private MyObject() {
    }

    public static MyObject getInstance() {
        // 延迟加载
        if (myObject != null) {
        } else {
            myObject = new MyObject();
        }
        return myObject;
    }
```

创建线程类 MyThread.java 代码如下：

```java
package extthread;

import test.MyObject;

public class MyThread extends Thread {

    @Override
    public void run() {
        System.out.println(MyObject.getInstance().hashCode());
    }

}
```

创建运行类 Run.java 代码如下：

```java
package test.run;

import extthread.MyThread;

public class Run {

    public static void main(String[] args) {
        MyThread t1 = new MyThread();
        t1.start();
    }

}
```

程序运行结果如图 6-2 所示。

此示例虽然取得一个对象的实例，但如果在多线程环境中则会出现取出多个实例的情况，这与单例模式的初衷是背离的。

图 6-2　懒汉模式成功取出一个实例

6.2.2　延迟加载 / 懒汉模式的缺点

前面两个示例虽然使用"立即加载"和"延迟加载"实现了单例模式，但在多线程环境中，前面"延迟加载"示例中的代码就是错误的，不能实现保持单例的状态。下面来看一下如何在多线程环境中结合"错误的单例模式"创建出"多例"的情况。

创建测试用的项目，名称为 singleton_2，创建类 MyObject.java 代码如下：

```java
package test;

public class MyObject {
```

```java
    private static MyObject myObject;

    private MyObject() {
    }

    public static MyObject getInstance() {
        try {
            if (myObject != null) {
            } else {
                // 模拟在创建对象之前做一些准备性的工作
                Thread.sleep(3000);
                myObject = new MyObject();
            }
        } catch (InterruptedException e) {
            e.printStackTrace();
        }
        return myObject;
    }

}
```

创建线程类 MyThread.java 代码如下:

```java
package extthread;

import test.MyObject;

public class MyThread extends Thread {

    @Override
    public void run() {
        System.out.println(MyObject.getInstance().hashCode());
    }

}
```

创建运行类 Run.java 代码如下:

```java
package test.run;

import extthread.MyThread;

public class Run {

    public static void main(String[] args) {
        MyThread t1 = new MyThread();
        MyThread t2 = new MyThread();
        MyThread t3 = new MyThread();

        t1.start();
        t2.start();
```

```
            t3.start();
        }
    }
}
```

程序运行结果如图 6-3 所示。

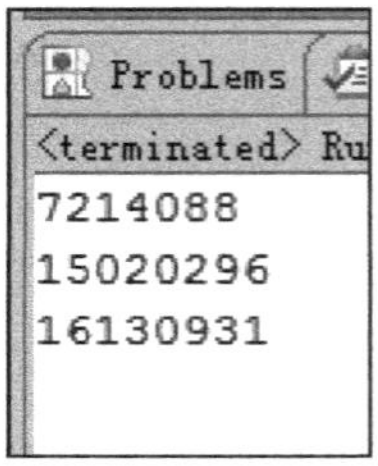

控制台输出了 3 种 hashCode，说明创建了 3 个对象，并不是单例的，这就是"错误的单例模式"，如何解决呢？

6.2.3 延迟加载 / 懒汉模式的解决方案

1. 声明 synchronized 关键字

既然多个线程可以同时进入 getInstance() 方法，那么只需要对 getInstance() 方法声明 synchronized 关键字即可。

图 6-3 非单例模式

创建测试用的项目，名称为 singleton_2_1，创建类 MyObject.java 代码如下：

```java
package test;

public class MyObject {

    private static MyObject myObject;

    private MyObject() {
    }

    // 设置同步方法效率太低
    // 整个方法被上锁
    synchronized public static MyObject getInstance() {
        try {
            if (myObject != null) {
            } else {
                // 模拟在创建对象之前做一些准备性的工作
                Thread.sleep(3000);
                myObject = new MyObject();
            }
        } catch (InterruptedException e) {
            e.printStackTrace();
        }
        return myObject;
    }

}
```

创建线程类 MyThread.java 代码如下：

```java
package extthread;

import test.MyObject;
```

```
public class MyThread extends Thread {

    @Override
    public void run() {
        System.out.println(MyObject.getInstance().hashCode());
    }

}
```

创建运行类 Run.java 代码如下：

```
package test.run;

import extthread.MyThread;

public class Run {

    public static void main(String[] args) {
        MyThread t1 = new MyThread();
        MyThread t2 = new MyThread();
        MyThread t3 = new MyThread();

        t1.start();
        t2.start();
        t3.start();

    }

}
```

程序运行结果如图 6-4 所示。

此方法加入同步 synchronized 关键字得到相同实例的对象，但这种方法的运行效率非常低，是同步运行的。下一个线程想要取得对象，则上一个线程必须释放完锁之后，下一个线程才可以继续执行。那么换成同步代码块可以解决吗？

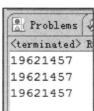

图 6-4　声明 synchronized 关键字的运行结果

2. 尝试同步代码块

同步方法用于对方法的整体进行同步，这其运行效率是比较低的，那么改成同步代码块能改进吗？

创建测试用的项目，名称为 singleton_2_2，创建类 MyObject.java 代码如下：

```
package test;

public class MyObject {

    private static MyObject myObject;

    private MyObject() {
```

```java
    }

    public static MyObject getInstance() {
        try {
            // 此种写法等同于:
            // synchronized public static MyObject getInstance()
            // 的写法,效率一样很低,全部代码被同步
            synchronized (MyObject.class) {
                if (myObject != null) {
                } else {
                    // 模拟在创建对象之前做一些准备性的工作
                    Thread.sleep(3000);

                    myObject = new MyObject();
                }
            }
        } catch (InterruptedException e) {
            e.printStackTrace();
        }
        return myObject;
    }

}
```

创建线程类 MyThread.java 代码如下:

```java
package extthread;

import test.MyObject;

public class MyThread extends Thread {

    @Override
    public void run() {
        System.out.println(MyObject.getInstance().hashCode());
    }

}
```

创建运行类 Run.java 代码如下:

```java
package test.run;

import test.MyObject;
import extthread.MyThread;

public class Run {

    public static void main(String[] args) {
        MyThread t1 = new MyThread();
        MyThread t2 = new MyThread();
```

```
        MyThread t3 = new MyThread();

        t1.start();
        t2.start();
        t3.start();

        // 此版本代码虽然是正确的
        // 但public static MyObject getInstance()方法
        // 中的全部代码都是同步的，这样做也有损效率
    }
}
```

程序运行结果如图 6-5 所示。

此方法加入同步 synchronized 语句块得到相同实例的对象，但这种方法的运行效率也非常低，和 synchronized 同步方法一样是同步运行的。继续更改代码尝试解决这个缺点。

图 6-5 使用同步代码块的运行结果

3. 针对某些重要代码进行单独同步

同步代码块可以针对某些重要代码进行单独同步，而其他代码则不需要同步，运行效率完全可以得到大幅提升。

创建测试用的项目，名称为 singleton_3，创建类 MyObject.java 代码如下：

```java
package test;

public class MyObject {

    private static MyObject myObject;

    private MyObject() {
    }

    public static MyObject getInstance() {
        try {
            if (myObject != null) {
            } else {
                // 模拟在创建对象之前做一些准备性的工作
                Thread.sleep(3000);
                // 使用synchronized (MyObject.class)
                // 虽然部分代码被上锁
                // 但还是有非线程安全问题
                // 多次创建MyObject类的对象，并不是单例的效果
                synchronized (MyObject.class) {
                    myObject = new MyObject();
                }
            }
        } catch (InterruptedException e) {
            e.printStackTrace();
```

            }
            return myObject;
    }
}
```

创建线程类 MyThread.java 代码如下：

```
package extthread;

import test.MyObject;

public class MyThread extends Thread {

    @Override
    public void run() {
        System.out.println(MyObject.getInstance().hashCode());
    }

}
```

创建运行类 Run.java 代码如下：

```
package test.run;

import extthread.MyThread;

public class Run {

    public static void main(String[] args) {
        MyThread t1 = new MyThread();
        MyThread t2 = new MyThread();
        MyThread t3 = new MyThread();

        t1.start();
        t2.start();
        t3.start();

    }
}
```

程序运行结果如图 6-6 所示。

此方法使同步 synchronized 语句块只对实例化对象的关键代码进行同步，从语句的结构来讲，运行效率的确得到了提升，但在多线程的情况下还是无法解决得到同一个实例对象的结果。那到底如何解决"懒汉模式"遇到多线程的情况呢？

图 6-6　对重要代码单独同步的运行结果

4. 使用 DCL 机制

可以使用 DCL（Double-Check Locking，双检查锁）机制来实现
```

多线程环境中的延迟加载单例模式。

创建测试用的项目，名称为 singleton_5，创建类 MyObject.java 代码如下：

```java
package test;

public class MyObject {
    private volatile static MyObject myObject;

    private MyObject() {
    }

    public static MyObject getInstance() {
        try {
            if (myObject != null) {
            } else {
                // 模拟在创建对象之前做一些准备性的工作
                Thread.sleep(3000);
                synchronized (MyObject.class) {
                    if (myObject == null) {
                        myObject = new MyObject();
                    }
                }
            }
        } catch (InterruptedException e) {
            e.printStackTrace();
        }
        return myObject;
    }
    // 此版本的代码称为：
    // 双重检查Double-Check Locking
}
```

使用 volatile 修改变量 myObject 使该变量在多个线程间达到可见性，另外也禁止了 myObject = new MyObject() 代码重排序，因为 myObject = new MyObject() 代码在内部分为 3 个步骤：

```
1) memory = allocate();        //分配对象的内存空间
2) ctorInstance(memory);       //初始化对象
3) myObject = memory;          //设置instance指向刚分配的内存地址
```

JIT 编译器有可能将这 3 个步骤重排序成：

```
1) memory = allocate();        //分配对象的内存空间
2) myObject = memory;          //设置instance指向刚分配的内存地址
3) ctorInstance(memory);       //初始化对象
```

这时就会出现以下情况：虽然构造方法还没有执行，但 myObject 对象具有了内存地址，值不是 null，当访问 myObject 对象中的实量时还是数据类型的默认值，此知识点在后面的章节会进行详细介绍。

创建线程类 MyThread.java 代码如下：

```java
package extthread;

import test.MyObject;

public class MyThread extends Thread {

    @Override
    public void run() {
        System.out.println(MyObject.getInstance().hashCode());
    }

}
```

创建运行类 Run.java 代码如下：

```java
package test.run;

import extthread.MyThread;

public class Run {

    public static void main(String[] args) {

        MyThread t1 = new MyThread();
        MyThread t2 = new MyThread();
        MyThread t3 = new MyThread();

        t1.start();
        t2.start();
        t3.start();

    }

}
```

程序运行结果如图 6-7 所示。

使用 DCL 机制成功解决了懒汉模式遇到多线程的问题。DCL 也是大多数多线程结合单例模式使用的解决方案。

5. DCL 使用 volatile 的必要性

前面已经介绍，myObject = new MyObject() 代码中的 3 个步骤会发生重排序的情况，这导致取得实例变量的值不是构造方法初始化后的值，下面对此进行验证。

创建测试用的项目 dcl_and_volatile。

创建产生单例对象的业务类，代码如下：

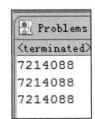

图 6-7 使用 DCL 机制的运行结果

```java
package test1;

import java.util.Random;

public class OneInstanceService {

    public int i_am_has_state = 0;

    private static OneInstanceService test;

    private OneInstanceService() {
        i_am_has_state = new Random().nextInt(200) + 1;
    }

    public static OneInstanceService getTest1() {
        if (test == null) {
            synchronized (OneInstanceService.class) {
                if (test == null) {
                    test = new OneInstanceService();
                }
            }
        }
        return test;
    }

    public static void reset() {
        test = null;
    }
}
```

创建运行类代码如下:

```java
package test1;

import java.util.concurrent.CountDownLatch;

public class Test1 {
    public static void main(String[] args) throws InterruptedException {
        for (;;) {
            CountDownLatch latch = new CountDownLatch(1);
            CountDownLatch end = new CountDownLatch(100);
            for (int i = 0; i < 100; i++) {
                Thread t1 = new Thread() {
                    @Override
                    public void run() {
                        try {
                            latch.await();
                            OneInstanceService one = OneInstanceService.getTest1();
                            if (one.i_am_has_state == 0) {
                                System.out.println("one.i_am_has_state == 0 进程结束");
                                System.exit(0);
```

```
                    }
                    end.countDown();
                } catch (InterruptedException e) {
                    e.printStackTrace();
                }
            }
        };
        t1.start();
    }
    latch.countDown();
    end.await();
    OneInstanceService.reset();
    }
  }
}
```

程序在运行时添加 VM 参数 -server 会更容易获得预期的结果，运行后控制台会出现如下信息：

```
one.i_am_has_state == 0 进程结束
```

说明 myObject = new MyObject() 的确发生了重排序而出现错误。

更改代码如下：

```
package test2;

import java.util.Random;

public class OneInstanceService {

    public int i_am_has_state = 0;

    // 添加volatile关键字禁止重排序
    volatile private static OneInstanceService test;

    private OneInstanceService() {
        i_am_has_state = new Random().nextInt(200) + 1;
    }

    public static OneInstanceService getTest1() {
        if (test == null) {
            synchronized (OneInstanceService.class) {
                if (test == null) {
                    test = new OneInstanceService();
                }
            }
        }
        return test;
    }

    public static void reset() {
```

```
            test = null;
    }
}
```

程序运行后不再输出任何信息，说明禁止了重排序后，实例变量 i_am_has_state 永远不是 0 了。也就是步骤（A）开辟空间（B）执行构造方法（C）赋值，在赋值代码插入屏障，防止 B 跑到 C 的后面，执行顺序永远是 A、B、C，而且使用 volatile 可保持变量的值在多个线程间的可见性。

6.3 使用静态内置类实现单例模式

DCL 可以解决多线程单例模式的非线程安全问题，当然，还可以使用其他办法达到同样的效果。

创建测试用的项目，名称为 singleton_7，创建类 MyObject.java 代码如下：

```java
package test;

public class MyObject {

    // 内部类方式
    private static class MyObjectHandler {
        private static MyObject myObject = new MyObject();
    }

    private MyObject() {
    }

    public static MyObject getInstance() {
        return MyObjectHandler.myObject;
    }
}
```

创建线程类 MyThread.java 代码如下：

```java
package extthread;

import test.MyObject;

public class MyThread extends Thread {

    @Override
    public void run() {
        System.out.println(MyObject.getInstance().hashCode());
    }
}
```

创建运行类 Run.java 代码如下：

```java
package run;

import extthread.MyThread;

public class Run {

    public static void main(String[] args) {
        MyThread t1 = new MyThread();
        MyThread t2 = new MyThread();
        MyThread t3 = new MyThread();

        t1.start();
        t2.start();
        t3.start();

    }
}
```

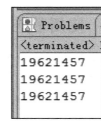

图 6-8　用静态内置类实现单例模式

程序运行结果如图 6-8 所示。

6.4　序列化与反序列化的单例模式实现

当将单例的对象进行序列化时，使用默认的反序列行为取出的对象是多例的。

创建测试用的项目，名称为 singleton_7_1，

创建实体类代码如下：

```java
package entity;

public class Userinfo {

}
```

创建类 MyObject.java 代码如下：

```java
package test1;

import java.io.Serializable;

import entity.Userinfo;

public class MyObject implements Serializable {

    private static final long serialVersionUID = 888L;

    public static Userinfo userinfo = new Userinfo();
```

```java
    private static MyObject myObject = new MyObject();

    private MyObject() {
    }

    public static MyObject getInstance() {
        return myObject;
    }

    // protected Object readResolve() throws ObjectStreamException {
    // System.out.println("调用了readResolve方法！");
    // return MyObject.myObject;
    // }

}
```

protected Object readResolve() 方法的作用是反序列化时不创建新的 MyObject 对象，而是复用原有的 MyObject 对象。

创建业务类 SaveAndRead.java 代码如下：

```java
package test1;

import java.io.File;
import java.io.FileInputStream;
import java.io.FileNotFoundException;
import java.io.FileOutputStream;
import java.io.IOException;
import java.io.ObjectInputStream;
import java.io.ObjectOutputStream;

public class SaveAndRead {
    public static void main(String[] args) {
        try {
            MyObject myObject = MyObject.getInstance();
            System.out.println("序列化-myObject=" + myObject.hashCode() + "user-" +
                "info=" + myObject.userinfo.hashCode());
            FileOutputStream fosRef = new FileOutputStream(new File("myObject-" +
                "File.txt"));
            ObjectOutputStream oosRef = new ObjectOutputStream(fosRef);
            oosRef.writeObject(myObject);
            oosRef.close();
            fosRef.close();
        } catch (FileNotFoundException e) {
            e.printStackTrace();
        } catch (IOException e) {
            e.printStackTrace();
        }

        try {
            FileInputStream fisRef = new FileInputStream(new File("myObjectFile.txt"));
```

```
            ObjectInputStream iosRef = new ObjectInputStream(fisRef);
            MyObject myObject = (MyObject) iosRef.readObject();
            iosRef.close();
            fisRef.close();
            System.out.println("序列化-myObject=" + myObject.hashCode() + "user-
                info=" + myObject.userinfo.hashCode());
        } catch (FileNotFoundException e) {
            e.printStackTrace();
        } catch (IOException e) {
            e.printStackTrace();
        } catch (ClassNotFoundException e) {
            e.printStackTrace();
        }

    }

}
```

程序运行结果如图 6-9 所示。

```
序列化-myObject=29115481 userinfo=19621457
序列化-myObject=4121220 userinfo=19621457
```

图 6-9 不是同一个 MyObject 对象

从程序运行结果可以分析，在反序列化时创建了新的 MyObject 对象，内存中产生了两个 MyObject 对象，MyObject 并不是单例的，但 Userinfo 对象得到复用，因为 hashcode 是同一个 19621457，为了实现 MyObject 在内存中一直呈单例的效果，解决办法就是在反序列化时使用 readResolve() 方法，对原有的 MyObject 对象进行复用，去掉如下代码注释：

```
protected Object readResolve() throws ObjectStreamException {
    System.out.println("调用了readResolve方法！");
    return MyObject.myObject;
}
```

程序运行结果如图 6-10 所示。

```
序列化-myObject=29115481 userinfo=19621457
调用了readResolve方法！
序列化-myObject=29115481 userinfo=19621457
```

图 6-10 是同一个对象

protected Object readResolve() 方法的作用是在反序列化时不创建新的 MyObject 对象，而是复用 JVM 内存中原有的 MyObject 单例对象，Userinfo 对象被复用，也就实现了对 MyObject 序列化与反序列化时保持单例性的效果。

> 注意 如果将序列化和反序列化操作分别放入两个 class 中，则反序列化时会产生新的 MyObject 对象，放在两个 class 类中分别执行其实相当于创建了两个 JVM 虚拟机，每个虚拟机里面的确只有一个 MyObject 对象，我们想要实现的是在一个 JVM 虚拟机中进行序列化与反序列化时保持 MyObject 单例性的效果，而不是创建两个 JVM 虚拟机。

6.5 使用 static 代码块实现单例模式

静态代码块中的代码在使用类的时候就已经执行，所以可以应用静态代码块的这个特性来实现单例模式。

创建测试用的项目,名称为 singleton_8,创建类 MyObject.java 代码如下:

```java
package test;

public class MyObject {

    private static MyObject instance = null;

    private MyObject() {
    }

    static {
        instance = new MyObject();
    }

    public static MyObject getInstance() {
        return instance;
    }

}
```

创建线程类 MyThread.java 代码如下:

```java
package extthread;

import test.MyObject;

public class MyThread extends Thread {

    @Override
    public void run() {
        for (int i = 0; i < 5; i++) {
            System.out.println(MyObject.getInstance().hashCode());
        }
    }
}
```

创建运行类 Run.java 代码如下:

```java
package test.run;

import extthread.MyThread;

public class Run {

    public static void main(String[] args) {
        MyThread t1 = new MyThread();
        MyThread t2 = new MyThread();
        MyThread t3 = new MyThread();
```

```
        t1.start();
        t2.start();
        t3.start();

    }

}
```

程序运行结果如图 6-11 所示。

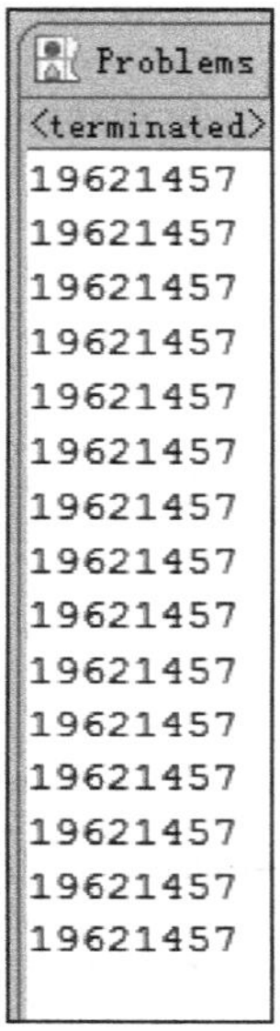

6.6 使用 enum 枚举数据类型实现单例模式

enum 枚举数据类型的特性和静态代码块的特性相似，在使用枚举类时，构造方法会被自动调用，可以应用这个特性实现单例模式。

创建测试用的项目，名称为 singleton_9，创建类 MyObject.java 代码如下：

图 6-11　用 static 代码块实现单例模式

```java
package test;

import java.sql.Connection;
import java.sql.DriverManager;
import java.sql.SQLException;

public enum MyObject {
    connectionFactory;

    private Connection connection;

    private MyObject() {
        try {
            System.out.println("调用了MyObject的构造");
            String url = "jdbc:sqlserver://localhost:1079;databaseName=ghydb";
            String username = "sa";
            String password = "";
            String driverName = "com.microsoft.sqlserver.jdbc.SQLServerDriver";
            Class.forName(driverName);
            connection = DriverManager.getConnection(url, username, password);
        } catch (ClassNotFoundException e) {
            e.printStackTrace();
        } catch (SQLException e) {
            e.printStackTrace();
        }
    }

    public Connection getConnection() {
        return connection;
    }
}
```

创建线程类 MyThread.java 代码如下：

```java
package extthread;

import test.MyObject;

public class MyThread extends Thread {

    @Override
    public void run() {
        for (int i = 0; i < 5; i++) {
            System.out.println(MyObject.connectionFactory.getConnection()
                    .hashCode());
        }
    }
}
```

创建运行类 Run.java 代码如下：

```java
package test.run;

import test.MyObject;
import extthread.MyThread;

public class Run {

    public static void main(String[] args) {
        MyThread t1 = new MyThread();
        MyThread t2 = new MyThread();
        MyThread t3 = new MyThread();

        t1.start();
        t2.start();
        t3.start();

    }
}
```

程序运行结果如图 6-12 所示。

6.7 完善使用 enum 枚举数据类型实现单例模式

将项目 singleton_9 中的所有源代码复制到 singleton_10 项目中，更改类 MyObject.java 代码如下：

```java
package test;

import java.sql.Connection;
import java.sql.DriverManager;
```

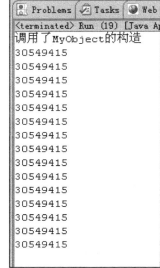

图 6-12 用 enum 枚举数据类型实现单例模式

```java
import java.sql.SQLException;

public class MyObject {

    public enum MyEnumSingleton {
        connectionFactory;

        private Connection connection;

        private MyEnumSingleton() {
            try {
                System.out.println("创建MyObject对象");
                String url = "jdbc:sqlserver://localhost:1079;databaseName=y2";
                String username = "sa";
                String password = "";
                String driverName = "com.microsoft.sqlserver.jdbc.SQLServerDriver";
                Class.forName(driverName);
                connection = DriverManager.getConnection(url, username,
                        password);
            } catch (ClassNotFoundException e) {
                e.printStackTrace();
            } catch (SQLException e) {
                e.printStackTrace();
            }
        }

        public Connection getConnection() {
            return connection;
        }
    }

    public static Connection getConnection() {
        return MyEnumSingleton.connectionFactory.getConnection();
    }

}
```

更改 MyThread.java 类代码如下:

```java
package extthread;

import test.MyObject;

public class MyThread extends Thread {

    @Override
    public void run() {
        for (int i = 0; i < 5; i++) {
            System.out.println(MyObject.getConnection().hashCode());
        }
    }
```

}

程序运行结果如图 6-13 所示。

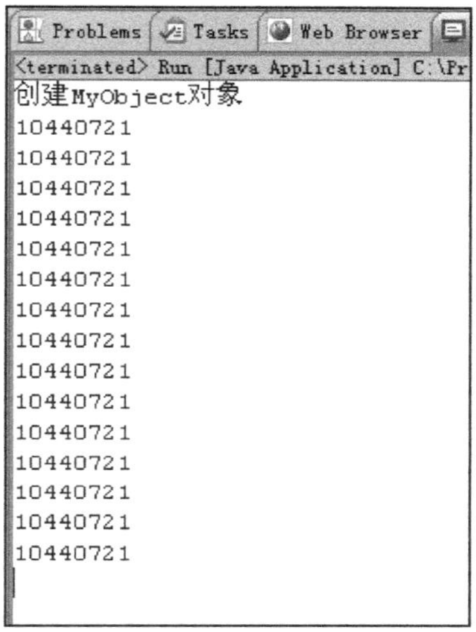

图 6-13　成功实现单例模式

6.8　本章小结

本章通过若干 Demo 案例阐述了单例模式与多线程结合时遇到的情况与解决方法，并复习了不同单例模式的使用方法。通过对本章的学习，我们以后遇到单例模式时就能从容面对多线程环境的情况了。

第 7 章 拾遗增补

本章是本书的最后一章，本章将对前面若干章节遗漏的知识点进行补充，丰富多线程案例的完整性。在开发此类应用时这些案例能起到推波助澜的作用，至少遇到某些情况时你会回想起这些案例的初衷与解决办法。

在本章中，读者应该掌握如下知识点：
- 线程组的使用；
- 线程状态是如何切换的；
- SimpleDataFormat 类与多线程的解决办法；
- 如何对线程的异常进行处理。

7.1 线程的状态

线程对象在不同的运行时期存在不同的状态，状态信息存在于 State 枚举类型中，如图 7-1 所示。有关枚举类型的解释如图 7-2 所示。

调用与线程有关的方法是造成线程状态改变的主要原因，方法与状态示意图如图 7-3 所示。

由图 7-3 可以得知，在调用与线程有关的方法后，线程会进入不同的状态。在这些状态之间，有些是双向切换的，例如，WAITING 和 RUNNING 状态之间可以循环地进行切换；而有些是单向切换的，例如，线程销毁后并不能自动进入 RUNNING 状态。

图 7-1 枚举类型 State 信息

```
public static enum Thread.State
extends Enum<Thread.State>
```

线程状态。线程可以处于下列状态之一：

- NEW
  至今尚未启动的线程处于这种状态。
- RUNNABLE
  正在 Java 虚拟机中执行的线程处于这种状态。
- BLOCKED
  受阻塞并等待某个监视器锁的线程处于这种状态。
- WAITING
  无限期地等待另一个线程来执行某一特定操作的线程处于这种状态。
- TIMED_WAITING
  等待另一个线程来执行取决于指定等待时间的操作的线程处于这种状态。
- TERMINATED
  已退出的线程处于这种状态。

图 7-2　有关枚举类型的解释

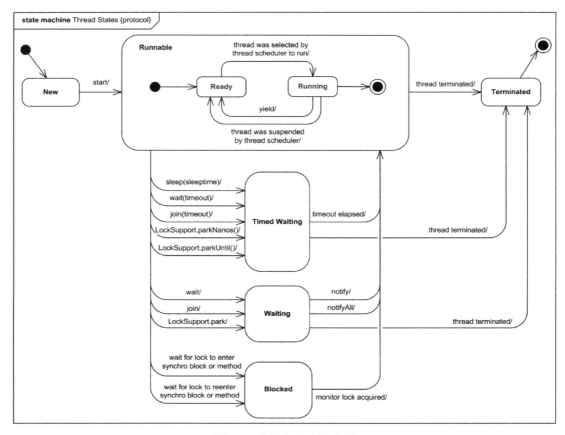

图 7-3　方法与状态示意图

下面将对这 6 种线程状态用程序代码的方式进行验证。

7.1.1 验证 NEW、RUNNABLE 和 TERMINATED

下面使用代码的方式验证线程的所有状态值。了解线程的状态有助于程序员监控线程对象所处的情况，如哪些线程从未启动、哪些线程正在执行、哪些线程正在阻塞、哪些线程正在等待、哪些线程已经销毁了等与线程生命周期相关的信息。

首先验证的是 NEW、RUNNABLE 及 TERMINATED 状态。NEW 状态下，线程实例化后还从未执行 start() 方法；RUNNABLE 状态下，线程进入运行状态；TERMINATED 状态下，线程被销毁了。

创建名称为 stateTest1 的项目，类 MyThread.java 代码如下：

```java
package extthread;

public class MyThread extends Thread {
    public MyThread() {
        System.out.println("构造方法中的状态 Thread.currentThread().getState()=" +
            Thread.currentThread().getState());
        System.out.println("构造方法中的状态 this.getState()=" + this.getState());
    }
    @Override
    public void run() {
        System.out.println("run方法中的状态：" + Thread.currentThread().getState());
    }
}
```

类 Run.java 代码如下：

```java
package test;

import extthread.MyThread;

public class Run {

    // NEW,
    // RUNNABLE,
    // TERMINATED,

    // BLOCKED,
    // WAITING,
    // TIMED_WAITING,

    public static void main(String[] args) {
        try {
            MyThread t = new MyThread();
            System.out.println("main方法中的状态1: " + t.getState());
            Thread.sleep(1000);
            t.start();
            Thread.sleep(1000);
            System.out.println("main方法中的状态2: " + t.getState());
```

```
            } catch (InterruptedException e) {
                // TODO Auto-generated catch block
                e.printStackTrace();
            }
        }
    }
}
```

程序运行结果如图 7-4 所示。

```
构造方法中的状态 Thread.currentThread().getState()=RUNNABLE
构造方法中的状态 this.getState()=NEW
main方法中的状态1: NEW
run方法中的状态: RUNNABLE
main方法中的状态2: TERMINATED
```

图 7-4 NEW、RUNNABLE 和 TERMINATED 验证结果

 由构造方法的输出日志可以看出，main 主线程的状态为 RUNNABLE。

7.1.2 验证 TIMED_WAITING

线程状态 TIMED_WAITING 代表线程执行了 Thread.sleep() 方法，呈等待状态，等待时间到达，然后继续向下运行。

创建名称为 stateTest2 的项目，类 MyThread.java 代码如下：

```
package extthread;

public class MyThread extends Thread {

    @Override
    public void run() {
        try {
            System.out.println("begin sleep");
            Thread.sleep(10000);
            System.out.println("  end sleep");
        } catch (InterruptedException e) {
            e.printStackTrace();
        }
    }
}
```

类 Run.java 代码如下：

```
package test;

import extthread.MyThread;
```

```
public class Run {

    // NEW,
    // RUNNABLE,
    // TERMINATED,

    // BLOCKED,
    // WAITING,
    // TIMED_WAITING,

    public static void main(String[] args) {
        try {
            MyThread t = new MyThread();
            t.start();
            Thread.sleep(1000);
            System.out.println("main方法中的状态: " + t.getState());
        } catch (InterruptedException e) {
            // TODO Auto-generated catch block
            e.printStackTrace();
        }

    }

}
```

程序运行结果如图 7-5 所示。

执行 sleep() 方法后线程的状态枚举值就是 TIMED_WAITING。

图 7-5　TIMED_WAITING 验证结果

7.1.3　验证 BLOCKED

BLOCKED 状态出现在某个线程在等待锁的时候。

创建名称为 stateTest3 的项目，创建业务对象 MyService.java，代码如下：

```
package service;

public class MyService {

    synchronized static public void serviceMethod() {
        try {
            System.out.println(Thread.currentThread().getName() + "进入了业务方法！");
            Thread.sleep(10000);
        } catch (InterruptedException e) {
            e.printStackTrace();
        }
    }

}
```

类 MyThread1.java 代码如下：

```java
package extthread;

import service.MyService;

public class MyThread1 extends Thread {

    @Override
    public void run() {
        MyService.serviceMethod();
    }

}
```

类 MyThread2.java 代码如下：

```java
package extthread;

import service.MyService;

public class MyThread2 extends Thread {

    @Override
    public void run() {
        MyService.serviceMethod();
    }

}
```

类 Run.java 代码如下：

```java
package test;

import extthread.MyThread1;
import extthread.MyThread2;

public class Run {

    // NEW,
    // RUNNABLE,
    // TERMINATED,

    // BLOCKED,
    // WAITING,
    // TIMED_WAITING,

    public static void main(String[] args) throws InterruptedException{
        MyThread1 t1 = new MyThread1();
        t1.setName("a");
        t1.start();
        Thread.sleep(1000);
```

```
            MyThread2 t2 = new MyThread2();
            t2.setName("b");
            t2.start();
            Thread.sleep(1000);
            System.out.println("main方法中的t2状态: " + t2.getState());

        }
    }
```

程序运行结果如图 7-6 所示。

从程序运行结果来看，t2 线程一直在等待 t1 对锁进行释放，所以 t2 当时处于 BLOCKED 状态。

```
a进入了业务方法!
main方法中的t2状态: BLOCKED
```

图 7-6 BLOCKED 验证结果

7.1.4 验证 WAITING

WAITING 状态是线程执行了 Object.wait() 方法后所处的状态。

创建名称为 stateTest4 的项目，类 Lock.java 代码如下：

```java
package service;

public class Lock {

    public static final Byte lock = new Byte("0");

}
```

类 MyThread.java 代码如下：

```java
package extthread;

import service.Lock;

public class MyThread extends Thread {

    @Override
    public void run() {
        try {
            synchronized (Lock.lock) {
                Lock.lock.wait();
            }
        } catch (InterruptedException e) {
            e.printStackTrace();
        }
    }

}
```

类 Run.java 代码如下：

```java
package test;
```

```
import extthread.MyThread;

public class Run {
    // NEW,
    // RUNNABLE,
    // TERMINATED,
    // BLOCKED,
    // WAITING,
    // TIMED_WAITING,

    public static void main(String[] args) {
        try {
            MyThread t = new MyThread();
            t.start();
            Thread.sleep(1000);
            System.out.println("main方法中的t状态: " + t.getState());
        } catch (InterruptedException e) {
            // TODO Auto-generated catch block
            e.printStackTrace();
        }
    }
}
```

程序运行结果如图 7-7 所示。

执行 wait() 方法后线程的状态枚举值就是 WAITING。

图 7-7　WAITING 验证结果

7.2　线程组

为了便于对某些具有相同功能的线程进行管理，可以把线程归属到某一个线程组中，线程组中既可以有线程对象，也可以有线程组，组中也可以有线程，这样的组织结构类似于树的形式，如图 7-8 所示。

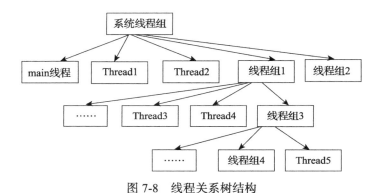

图 7-8　线程关系树结构

线程组可以批量地管理线程或线程组对象，有效地对线程或线程组对象进行组织。

7.2.1 线程对象关联线程组：一级关联

一级关联就是父对象中有子对象，但并不创建子孙对象，这种情况经常出现在开发中，例如，创建一些线程时为了有效地对这些线程进行组织管理，通常的情况是创建一个线程组，然后将部分线程归属到该组中，这样处理可以对零散的线程对象进行有效的组织与规划。

创建名称为 groupAddThread 的项目。创建两个线程类，如图 7-9 所示。

```java
package extthread;

public class ThreadA extends Thread {
    @Override
    public void run() {
        try {
            while (!Thread.currentThread().isInterrupted()) {
                System.out
                    .println("ThreadName=" +
                        Thread.currentThread().
                        getName());
                Thread.sleep(3000);
            }
        } catch (InterruptedException e) {
            // TODO Auto-generated catch block
            e.printStackTrace();
        }
    }
}
```

```java
package extthread;

public class ThreadB extends Thread {
    @Override
    public void run() {
        try {
            while (!Thread.currentThread().isInterrupted()) {
                System.out
                    .println("ThreadName=" +
                        Thread.currentThread().
                        getName());
                Thread.sleep(3000);
            }
        } catch (InterruptedException e) {
            // TODO Auto-generated catch block
            e.printStackTrace();
        }
    }
}
```

图 7-9　两个线程类代码

创建运行类 Run.java 代码如下：

```java
package test;

import extthread.ThreadA;
import extthread.ThreadB;

public class Run {

    public static void main(String[] args) {
        ThreadA aRunnable = new ThreadA();
        ThreadB bRunnable = new ThreadB();

        ThreadGroup group = new ThreadGroup("高洪岩的线程组");

        Thread aThread = new Thread(group, aRunnable);
        Thread bThread = new Thread(group, bRunnable);
        aThread.start();
        bThread.start();

        System.out.println("活动的线程数为：" + group.activeCount());
        System.out.println("线程组的名称为：" + group.getName());
```

            }
        }

在代码中将 aThread 和 bThread 对象关联线程组 group，然后对 aThread 和 bThread 对象执行 start() 方法，执行 ThreadA 和 ThreadB 中的 run() 方法。

程序运行结果如图 7-10 所示。

图 7-10　一级关联程序运行结果

控制台输出的信息是线程组中有两个线程，并且输出了线程组的名称，另外两个线程一直无限并且每隔 3s 输出日志。

7.2.2　线程对象关联线程组：多级关联

多级关联就是父对象中有子对象，子对象中再创建子对象，即子孙对象，但是这种写法在开发中不太常见。设计非常复杂的线程树结构反而不利于线程对象的管理，但 JDK 提供了支持多级关联的线程树结构，可以实现多级关联的关键代码是 ThreadGroup 类的构造方法：

public ThreadGroup(ThreadGroup parent, String name)

创建一个组时可以指定一个父组。

创建名称为 groupAddThreadMoreLevel 的项目。

创建运行类 Run.java 代码如下：

```
package test.run;

public class Run {

    public static void main(String[] args) {
        // 在main组中添加一个线程组A，然后在这个A组中添加线程对象Z
        // 方法activeGroupCount()和activeCount()的值不是固定的
        // 是系统中环境的一个快照
        ThreadGroup mainGroup = Thread.currentThread().getThreadGroup();
        // 取得main主线程所在的线程组
```

```
            ThreadGroup group = new ThreadGroup(mainGroup, "A");
            Runnable runnable = new Runnable() {
                @Override
                public void run() {
                    try {
                        System.out.println("runMethod!");
                        Thread.sleep(10000);// 线程必须在运行状态才可以受组管理
                    } catch (InterruptedException e) {
                        // TODO Auto-generated catch block
                        e.printStackTrace();
                    }
                }
            };

            Thread newThread = new Thread(group, runnable);
            newThread.setName("Z");
            newThread.start();
            // 线程必须启动然后才归到组A中，因为在调用start()方法时会调用
            // group.add(this);
            // ///
            ThreadGroup[] listGroup = new ThreadGroup[Thread.currentThread()
                    .getThreadGroup().activeGroupCount()];
            Thread.currentThread().getThreadGroup().enumerate(listGroup);
            System.out.println("main线程中有多少个子线程组: " + listGroup.length + " 名字为: "
                    + listGroup[0].getName());
            Thread[] listThread = new Thread[listGroup[0].activeCount()];
            listGroup[0].enumerate(listThread);
            System.out.println(listThread[0].getName());

        }
    }
```

程序运行结果如图 7-11 所示。

本程序代码的结构就是 main 组创建一个新组，然后在该新组中添加线程，并取得相关信息。

图 7-11 多级关联程序运行结果

7.2.3 线程组自动归属特性

自动归属就是自动归到当前线程组中。

创建名称为 autoAddGroup 的项目。

创建运行类 Run.java 代码如下：

```
package test.run;

public class Run {
    public static void main(String[] args) {
        // 方法activeGroupCount()取得当前线程组对象中的子线程组数量
        // 方法enumerate()的作用是将线程组中的子线程组以复制的形式
```

```
                // 复制到ThreadGroup[]数组对象中
                System.out.println("A处线程: " + Thread.currentThread().getName()
                        + " 所属的线程组名为: "
                        + Thread.currentThread().getThreadGroup().getName() + " "
                        + " 中有线程组数量: "
                        + Thread.currentThread().getThreadGroup().activeGroupCount());
                ThreadGroup group = new ThreadGroup("新的组");// 自动加到main组中
                System.out.println("B处线程: " + Thread.currentThread().getName()
                        + " 所属的线程组名为: "
                        + Thread.currentThread().getThreadGroup().getName() + " "
                        + " 中有线程组数量: "
                        + Thread.currentThread().getThreadGroup().activeGroupCount());
                ThreadGroup[] threadGroup = new ThreadGroup[Thread.currentThread()
                        .getThreadGroup().activeGroupCount()];
                Thread.currentThread().getThreadGroup().enumerate(threadGroup);
                for (int i = 0; i < threadGroup.length; i++) {
                    System.out.println("第一个线程组名称为: " + threadGroup[i].getName());
                }
            }
        }
```

程序运行结果如图 7-12 所示。

图 7-12 线程自动归属程序运行结果

本示例要证明的是，在实例化一个 ThreadGroup 线程组 x 时，如果不指定所属的线程组，则 x 线程组自动归到当前线程对象所属的线程组中，也就是以隐式的方式在一个线程组中添加了一个子线程组，所以在控制台输出的线程组数量值由 0 变成 1。

7.2.4 获取根线程组

创建名称为 getGroupParent 的项目。
创建运行类 Run.java 代码如下：

```java
package test.run;

public class Run {
    public static void main(String[] args) {
        System.out.println("线程: " + Thread.currentThread().getName()
                + " 所在的线程组名为: "
                + Thread.currentThread().getThreadGroup().getName());
        System.out
```

```
                    .println("main线程所在的线程组的父线程组的名称是："
                + Thread.currentThread().getThreadGroup().getParent()
                    .getName());
        System.out.println("main线程所在的线程组的父线程组的父线程组的名称是："
                + Thread.currentThread().getThreadGroup().getParent()
                    .getParent().getName());
    }
}
```

程序运行结果如图 7-13 所示。

图 7-13　获取根线程组程序运行结果

程序运行结果说明 JVM 的根线程组是 system，再取父线程组则出现空异常。

7.2.5　线程组中加线程组

创建名称为 mainGroup 的项目，创建类 Run.java 代码如下：

```
package test.run;

public class Run {

    public static void main(String[] args) {

        System.out.println("线程组名称："
                + Thread.currentThread().getThreadGroup().getName());
        System.out.println("线程组中活动的线程数量："
                + Thread.currentThread().getThreadGroup().activeCount());
        System.out.println("线程组中线程组的数量-加之前："
                + Thread.currentThread().getThreadGroup().activeGroupCount());
        ThreadGroup newGroup = new ThreadGroup(Thread.currentThread()
                .getThreadGroup(), "newGroup");
        System.out.println("线程组中线程组的数量-加之后："
                + Thread.currentThread().getThreadGroup().activeGroupCount());
        System.out
                .println("父线程组名称："
                        + Thread.currentThread().getThreadGroup().getParent()
                            .getName());
    }

}
```

程序运行结果如图 7-14 所示。

本示例用显式的方式在一个线程组中添加了一个子线程组。

```
线程组名称: main
线程组中活动的线程数量: 1
线程组中线程组的数量-加之前: 0
线程组中线程组的数量-加之后: 1
父线程组名称: system
```

图 7-14 线程组中加线程组
程序运行结果

7.2.6 组内的线程批量停止

前面介绍了线程组 ThreadGroup，并没有发现使用与不使用线程组 ThreadGroup 的区别。使用线程组 ThreadGroup 的优点是可以批量处理本组内的线程对象，例如，可以批量中断组中的线程。

创建名称为 groupInnerStop 的项目。

创建 MyThread.java 类代码如下：

```java
package mythread;

public class MyThread extends Thread {

    public MyThread(ThreadGroup group, String name) {
        super(group, name);
    }

    @Override
    public void run() {
        System.out.println("ThreadName=" + Thread.currentThread().getName()
                + "准备开始死循环了: )");
        while (!this.isInterrupted()) {
        }
        System.out.println("ThreadName=" + Thread.currentThread().getName()
                + "结束了: )");
    }

}
```

创建类 Run.java 代码如下：

```java
package test.run;

import mythread.MyThread;

public class Run {

    public static void main(String[] args) {
        try {
            ThreadGroup group = new ThreadGroup("我的线程组");

            for (int i = 0; i < 5; i++) {
                MyThread thread = new MyThread(group, "线程" + (i + 1));
                thread.start();
            }
            Thread.sleep(5000);
```

```
                group.interrupt();
                System.out.println("调用了interrupt()方法");
            } catch (InterruptedException e) {
                System.out.println("停了停了! ");
                e.printStackTrace();
            }
        }
    }
}
```

程序运行结果如图 7-15 所示。

通过将线程归属到线程组中，当调用线程组 ThreadGroup 的 interrupt() 方法时可以中断该组中所有正在运行的线程。

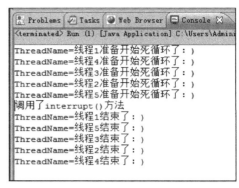

图 7-15　组内线程批量停止程序运行结果

7.2.7　递归取得与非递归取得组内对象

创建名称为 groupRecurseTest 的项目。

创建类 Run.java 代码如下：

```java
package test.run;

public class Run {

    public static void main(String[] args) {

        ThreadGroup mainGroup = Thread.currentThread().getThreadGroup();
        ThreadGroup groupA = new ThreadGroup(mainGroup, "A");
        Runnable runnable = new Runnable() {
            @Override
            public void run() {
                try {
                    System.out.println("runMethod!");
                    Thread.sleep(10000);
                } catch (InterruptedException e) {
                    e.printStackTrace();
                }
            }
        };
        ThreadGroup groupB = new ThreadGroup(groupA, "B");

        // 分配空间，但不一定全部用完
        ThreadGroup[] listGroup1 = new ThreadGroup[Thread.currentThread()
                .getThreadGroup().activeGroupCount()];
        // 传入true是递归取得子组及子孙组
        Thread.currentThread().getThreadGroup().enumerate(listGroup1, true);
        for (int i = 0; i < listGroup1.length; i++) {
            if (listGroup1[i] != null) {
```

```
                System.out.println(listGroup1[i].getName());
            }
        }
        ThreadGroup[] listGroup2 = new ThreadGroup[Thread.currentThread()
                .getThreadGroup().activeGroupCount()];
        Thread.currentThread().getThreadGroup().enumerate(listGroup2, false);
        for (int i = 0; i < listGroup2.length; i++) {
            if (listGroup2[i] != null) {
                System.out.println(listGroup2[i].getName());
            }
        }
    }
}
```

类 ThreadGroup 中的 activeGroupCount() 方法取得子孙组的数量。

程序运行结果如图 7-16 所示。

图 7-16　递归与非递归取得组内对象程序运行结果

7.3　Thread.activeCount() 方法的使用

Thread.activeCount() 方法的作用是返回当前线程的线程组中活动线程的数目。

创建测试用的代码如下：

```
package test5;

public class Test1 {
    public static void main(String[] args) throws InterruptedException {
        System.out.println(Thread.activeCount());
    }
}
```

程序运行结果如下：

1

程序输出 1 的原因是当前线程所在的线程组中只有一个活动线程。

7.4　Thread.enumerate(Thread tarray[]) 方法的使用

Thread.enumerate(Thread tarray[]) 方法的作用是将当前线程的线程组及其子组中的每一个活动线程复制到指定的数组中。该方法只调用当前线程的线程组的 enumerate() 方法，且带有数组参数。

创建测试用的代码如下：

```
package test7;

public class Test1 {
    public static void main(String[] args) {
        Thread[] threadArray = new Thread[Thread.currentThread().activeCount()];
        Thread.enumerate(threadArray);
        for (int i = 0; i < threadArray.length; i++) {
            System.out.println(threadArray[i].getName());
        }
    }
}
```

程序运行结果如下:

main

程序输出 main 的原因是当前线程所在的线程组中只有一个活动线程 main。

7.5 再次实现线程执行有序性

创建名称为 threadRunSyn 的 Java 项目。

创建 MyService.java 业务类，代码如下:

```
package test;

public class MyService {
    private ThreadLocal<Integer> printCountLocal = new ThreadLocal<>();
    private static int currentPrintPosition = 0;
    private static int finalPrintPosition = 0;

    synchronized public void printMethod(String eachThreadPrintChar, Integer
            eachThreadPrintPosition) {
        printCountLocal.set(0);
        while (printCountLocal.get() < 3) {
            if (currentPrintPosition == 3) {
                currentPrintPosition = 0;
            }
            while (eachThreadPrintPosition - 1 % 3 != currentPrintPosition) {
                try {
                    this.wait();
                } catch (InterruptedException e) {
                    e.printStackTrace();
                }
            }
            finalPrintPosition++;
            System.out.println(Thread.currentThread().getName() + " " + each-
                ThreadPrintChar + " "
                    + "currentPrintPosition=" + currentPrintPosition + " print-
                        CountLocal.get()="
```

```
                    + (printCountLocal.get() + 1) + " finalPrintPosition=" + (final-
                        PrintPosition));
                currentPrintPosition++;
                printCountLocal.set(printCountLocal.get() + 1);
                this.notifyAll();
            }
        }
    }
```

创建 MyThread.java 线程类，代码如下：

```
package test;

public class MyThread extends Thread {

    private MyService service;
    private String eachThreadPrintChar;
    private Integer eachThreadPrintPosition;

    public MyThread(MyService service, String eachThreadPrintChar, Integer each-
        ThreadPrintPosition) {
        super();
        this.service = service;
        this.eachThreadPrintChar = eachThreadPrintChar;
        this.eachThreadPrintPosition = eachThreadPrintPosition;
    }

    @Override
    public void run() {
        service.printMethod(eachThreadPrintChar, eachThreadPrintPosition);
    }
}
```

创建 Test.java 运行类，代码如下：

```
package test;

public class Test {
    public static void main(String[] args) {
        MyService service = new MyService();

        MyThread a = new MyThread(service, "A", 1);
        a.setName("线程1");
        a.start();
        MyThread b = new MyThread(service, "B", 2);
        b.setName("线程2");
        b.start();
        MyThread c = new MyThread(service, "C", 3);
        c.setName("线程3");
        c.start();
```

        }
   }
}

程序运行结果如图 7-17 所示。

```
线程1 A currentPrintPosition=0 printCountLocal.get()=1 finalPrintPosition=1
线程2 B currentPrintPosition=1 printCountLocal.get()=1 finalPrintPosition=2
线程3 C currentPrintPosition=2 printCountLocal.get()=1 finalPrintPosition=3
线程1 A currentPrintPosition=0 printCountLocal.get()=2 finalPrintPosition=4
线程2 B currentPrintPosition=1 printCountLocal.get()=2 finalPrintPosition=5
线程3 C currentPrintPosition=2 printCountLocal.get()=2 finalPrintPosition=6
线程1 A currentPrintPosition=0 printCountLocal.get()=3 finalPrintPosition=7
线程2 B currentPrintPosition=1 printCountLocal.get()=3 finalPrintPosition=8
线程3 C currentPrintPosition=2 printCountLocal.get()=3 finalPrintPosition=9
```

图 7-17 输出 3 批 ABC

7.6 SimpleDateFormat 非线程安全

SimpleDateFormat 类的作用是对日期进行解析与格式化，但在使用时如果不想使用 0 进行填充，例如，将 2000-01-02 转换成 2000-1-2，是需要在代码上进行处理的，示例代码如下：

```java
public static void main(String[] args) throws InterruptedException, ParseException {
    String dateString1 = "2000-1-1";
    String dateString2 = "2000-11-18";
    SimpleDateFormat format1 = new SimpleDateFormat("yyyy-M-d");
    SimpleDateFormat format2 = new SimpleDateFormat("yyyy-MM-dd");
    System.out.println(format1.format(format1.parse(dateString1)));
    System.out.println(format2.format(format2.parse(dateString1)));
    System.out.println();
    System.out.println(format1.format(format1.parse(dateString2)));
    System.out.println(format2.format(format2.parse(dateString2)));
}
```

程序运行结果如下：

```
2000-1-1
2000-01-01

2000-11-18
2000-11-18
```

使用 SimpleDateFormat 类在多线程环境中容易造成数据转换及处理的不准确，因为 SimpleDateFormat 类并不是线程安全的。

7.6.1 出现异常

本示例将演示使用 SimpleDateFormat 类在多线程环境下处理日期得出的结果却是错误

的情况，这也是在多线程环境开发中容易遇到的问题。

创建项目，名称为 formatError，类 MyThread.java 代码如下：

```java
package extthread;

import java.text.ParseException;
import java.text.SimpleDateFormat;
import java.util.Date;

public class MyThread extends Thread {

    private SimpleDateFormat sdf;
    private String dateString;

    public MyThread(SimpleDateFormat sdf, String dateString) {
        super();
        this.sdf = sdf;
        this.dateString = dateString;
    }

    @Override
    public void run() {
        try {
            Date dateRef = sdf.parse(dateString);
            String newDateString = sdf.format(dateRef).toString();
            if (!newDateString.equals(dateString)) {
                System.out.println("ThreadName=" + this.getName()
                        + "报错了  日期字符串：" + dateString + "  转换成的日期为："+
                        newDateString);
            }
        } catch (ParseException e) {
            e.printStackTrace();
        }

    }

}
```

运行类 Test.java 代码如下：

```java
package test.run;

import java.text.SimpleDateFormat;

import extthread.MyThread;

public class Test {

    public static void main(String[] args) {

        SimpleDateFormat sdf = new SimpleDateFormat("yyyy-MM-dd");
```

```
        String[] dateStringArray = new String[] { "2000-01-01", "2000-01-02",
                "2000-01-03", "2000-01-04", "2000-01-05", "2000-01-06",
                "2000-01-07", "2000-01-08", "2000-01-09", "2000-01-10" };

        MyThread[] threadArray = new MyThread[10];
        for (int i = 0; i < 10; i++) {
            threadArray[i] = new MyThread(sdf, dateStringArray[i]);
        }
        for (int i = 0; i < 10; i++) {
            threadArray[i].start();
        }

    }
}
```

程序运行结果如图 7-18 所示。

图 7-18 出现异常

从控制台输出的结果来看，使用单例的 SimpleDateFormat 类在多线程环境中处理日期极易出现日期转换错误的情况。

7.6.2 解决异常的方法 1

创建项目，名称为 formatOK1，类 MyThread.java 代码如下：

```java
package extthread;

import java.text.ParseException;
import java.text.SimpleDateFormat;
import java.util.Date;

import tools.DateTools;

public class MyThread extends Thread {

    private SimpleDateFormat sdf;
    private String dateString;

    public MyThread(SimpleDateFormat sdf, String dateString) {
        super();
        this.sdf = sdf;
```

```java
            this.dateString = dateString;
        }

        @Override
        public void run() {
            try {
                Date dateRef = DateTools.parse("yyyy-MM-dd", dateString);
                String newDateString = DateTools.format("yyyy-MM-dd", dateRef)
                        .toString();
                if (!newDateString.equals(dateString)) {
                    System.out.println("ThreadName=" + this.getName()
                            + "报错了  日期字符串: " + dateString + "  转换成的日期为: "+
                            newDateString);
                }
            } catch (ParseException e) {
                e.printStackTrace();
            }
        }

    }
```

类 DateTools.java 代码如下:

```java
package tools;

import java.text.ParseException;
import java.text.SimpleDateFormat;
import java.util.Date;

public class DateTools {

    public static Date parse(String formatPattern, String dateString)
            throws ParseException {
        return new SimpleDateFormat(formatPattern).parse(dateString);
    }
    public static String format(String formatPattern, Date date) {
        return new SimpleDateFormat(formatPattern).format(date).toString();
    }

}
```

运行类 Test.java 代码与前面章节保持一致。

程序运行结果如图 7-19 所示。

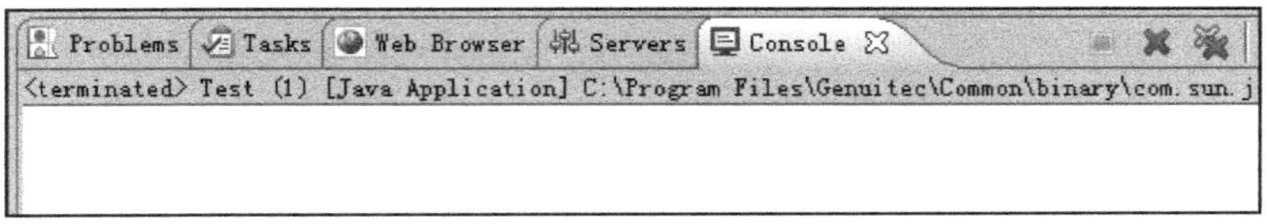

图 7-19　解决异常方法 1 的运行结果

控制台没有异常输出，处理错误的原理其实就是创建多个 SimpleDateFormat 类的实例。

7.6.3 解决异常的方法 2

前面介绍过 ThreadLocal 类能使线程绑定到指定对象，使用该类也可以解决多线程环境下 SimpleDateFormat 类处理错误的情况。

创建项目，名称为 formatOK2，类 MyThread.java 代码如下：

```java
package extthread;

import java.text.ParseException;
import java.text.SimpleDateFormat;
import java.util.Date;

import tools.DateTools;

public class MyThread extends Thread {

    private SimpleDateFormat sdf;
    private String dateString;

    public MyThread(SimpleDateFormat sdf, String dateString) {
        super();
        this.sdf = sdf;
        this.dateString = dateString;
    }

    @Override
    public void run() {
        try {
            Date dateRef = DateTools.getSimpleDateFormat("yyyy-MM-dd").parse(
                    dateString);
            String newDateString = DateTools.getSimpleDateFormat("yyyy-MM-dd")
                    .format(dateRef).toString();
            if (!newDateString.equals(dateString)) {
                System.out.println("ThreadName=" + this.getName()
                        + "报错了  日期字符串:" + dateString + "  转换成的日期为:"+
                        newDateString);
            }
        } catch (ParseException e) {
            e.printStackTrace();
        }
    }

}
```

类 DateTools.java 代码如下：

```java
package tools;
```

```
import java.text.SimpleDateFormat;

public class DateTools {

    private static ThreadLocal<SimpleDateFormat> tl = new ThreadLocal<SimpleDateFormat>();

    public static SimpleDateFormat getSimpleDateFormat(String datePattern) {
        SimpleDateFormat sdf = null;
        sdf = tl.get();
        if (sdf == null) {
            sdf = new SimpleDateFormat(datePattern);
            tl.set(sdf);
        }
        return sdf;
    }

}
```

运行类 Test.java 代码与前面章节保持不变。

程序运行结果如图 7-20 所示。

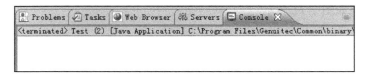

图 7-20　解决异常方法 2 的运行结果

控制台没有信息输出，由此可以看出程序运行结果是正确的，说明 ThreadLocal 结合 SimpleDateFormat 解决了 SimpleDateFormat 非线程安全的问题。

7.7　线程中出现异常的处理

当在线程中出现异常时，只能用 run() 方法中的 catch 语句进行处理，当有多个线程需要处理异常时，需要用每一个线程的 run() 方法中的 catch 语句分别进行处理，这样易造成代码严重冗余，所以可以使用 setDefaultUncaughtExceptionHandler() 和 setUncaught-ExceptionHandler() 方法来集中处理线程的异常。

7.7.1　线程出现异常的默认行为

创建项目 threadCreateException，创建线程类，MyThread.java 代码如下：

```
package extthread;

public class MyThread extends Thread {
    @Override
```

```java
    public void run() {
        String username = null;
        System.out.println(username.hashCode());
    }
}
```

创建 Main1.java 文件代码如下：

```java
package controller;

import extthread.MyThread;

public class Main1 {

    public static void main(String[] args) {
        MyThread t = new MyThread();
        t.start();
    }

}
```

程序运行结果如图 7-21 所示。

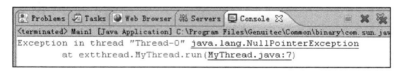

图 7-21　出现异常

程序运行后在控制台输出空指针异常，在 Java 多线程技术中，可以对多线程中的异常进行捕捉，使用的是 UncaughtExceptionHandler 接口，从而可以对发生的异常进行有效处理。

UncaughtExceptionHandler 接口的主要作用是当线程出现异常而终止时，JVM 捕获到此情况，自动调用 UncaughtExceptionHandler 接口中的 void uncaughtException(Thread t, Throwable e) 方法来处理异常。

7.7.2　使用 setUncaughtExceptionHandler() 方法进行异常处理

创建 Main2.java 文件代码如下：

```java
package controller;

import java.lang.Thread.UncaughtExceptionHandler;

import extthread.MyThread;
```

```java
public class Main2 {

    public static void main(String[] args) {
        MyThread t1 = new MyThread();
        t1.setName("线程t1");
        t1.setUncaughtExceptionHandler(new UncaughtExceptionHandler() {
            @Override
            public void uncaughtException(Thread t, Throwable e) {
                System.out.println("线程:" + t.getName() + " 出现了异常: ");
                e.printStackTrace();
            }
        });
        t1.start();

        MyThread t2 = new MyThread();
        t2.setName("线程t2");
        t2.start();
    }
}
```

setUncaughtExceptionHandler() 方法的作用是对指定的线程对象设置默认的异常处理器。程序运行结果如图 7-22 所示。

图 7-22　用 setUncaughtExceptionHandler() 方法处理异常

setUncaughtExceptionHandler() 方法用于给指定线程对象设置异常处理器，在 Thread 类中还可以使用 setDefaultUncaughtExceptionHandler() 方法对所有线程对象设置异常处理器。

7.7.3　使用 setDefaultUncaughtExceptionHandler() 方法进行异常处理

创建 Main3.java 文件代码如下：

```java
package controller;

import java.lang.Thread.UncaughtExceptionHandler;

import extthread.MyThread;

public class Main3 {

    public static void main(String[] args) {
        MyThread.setDefaultUncaughtExceptionHandler(new UncaughtExceptionHandler() {
```

```
            @Override
            public void uncaughtException(Thread t, Throwable e) {
                System.out.println("线程:" + t.getName() + " 出现了异常:");
                e.printStackTrace();

            }
        });

        MyThread t1 = new MyThread();
        t1.setName("线程t1");
        t1.start();

        MyThread t2 = new MyThread();
        t2.setName("线程t2");
        t2.start();
    }
}
```

setDefaultUncaughtExceptionHandler() 方法的作用是对指定线程类的所有线程对象设置默认的异常处理器。

程序运行结果如图 7-23 所示。

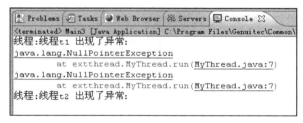

图 7-23　用 setDefaultUncaughtExceptionHandler() 方法处理异常

7.8　线程组内处理异常

创建项目 threadGroup_1，类 MyThread.java 代码如下：

```
package extthread;

public class MyThread extends Thread {

    private String num;

    public MyThread(ThreadGroup group, String name, String num) {
        super(group, name);
        this.num = num;
    }

    @Override
```

```java
    public void run() {
        int numInt = Integer.parseInt(num);
        while (true) {
            System.out.println("死循环中: " + Thread.currentThread().getName());
        }
    }
}
```

类 Run.java 代码如下：

```java
package test.run;

import extthread.MyThread;

public class Run {
    public static void main(String[] args) {
        ThreadGroup group = new ThreadGroup("我的线程组");
        MyThread[] myThread = new MyThread[10];
        for (int i = 0; i < myThread.length; i++) {
            myThread[i] = new MyThread(group, "线程" + (i + 1), "1");
            myThread[i].start();
        }
        MyThread newT = new MyThread(group, "报错线程", "a");
        newT.start();
    }
}
```

程序运行后，其中一个线程出现了异常，而其他线程一直以死循环的方式持续输出，如图 7-24 所示。

```
死循环中: 线程10
死循环中: 线程10
死循环中: 线程10
死循环中: 线程10Exception in thread "报错线程" java.lang.NumberFormatException: For input string: "a"
        at java.lang.NumberFormatException.forInputString(NumberFormatException.java:65)
        at java.lang.Integer.parseInt(Integer.java:580)
        at java.lang.Integer.parseInt(Integer.java:615)
        at extthread.MyThread.run(MyThread.java:14)

死循环中: 线程10
死循环中: 线程10
```

图 7-24 出现异常

此实验需要手动结束进程，因为 while(true) 死循环是无限输出的。

从程序运行结果来看，在默认情况下，线程组中的一个线程出现异常后不会影响其他线程的运行。

如果想实现线程组内一个线程出现异常后全部线程都停止，则该如何实现呢？

创建项目 threadGroup_2，创建新的线程组 MyThreadGroup.java 类，代码如下：

```java
package extthreadgroup;

public class MyThreadGroup extends ThreadGroup {

    public MyThreadGroup(String name) {
        super(name);
    }

    @Override
    public void uncaughtException(Thread t, Throwable e) {
        super.uncaughtException(t, e);
        this.interrupt();
    }

}
```

注意，要使用 this 关键字停止线程。this 代表的是线程组。

public void uncaughtException(Thread t, Throwable e)方法的t参数是出现异常的线程对象

类 MyThread.java 代码如下：

```java
package extthread;

public class MyThread extends Thread {

    private String num;

    public MyThread(ThreadGroup group, String name, String num) {
        super(group, name);
        this.num = num;
    }

    @Override
    public void run() {
        int numInt = Integer.parseInt(num);
        while (this.isInterrupted() == false) {
            System.out.println("死循环中：" + Thread.currentThread().getName());
        }
    }

}
```

需要注意的是，使用自定义 java.lang.ThreadGroup 线程组，并且重写 uncaughtException() 方法处理组内线程中断行为时，每个线程对象中的 run() 方法内部不要有异常 catch 语句，如果有 catch 语句，则 public void uncaughtException(Thread t, Throwable e) 方法不执行。

类 Run.java 代码如下：

```
package test.run;

import extthread.MyThread;
import extthreadgroup.MyThreadGroup;

public class Run {

    public static void main(String[] args) {
        MyThreadGroup group = new MyThreadGroup("我的线程组");
        MyThread[] myThread = new MyThread[10];
        for (int i = 0; i < myThread.length; i++) {
            myThread[i] = new MyThread(group, "线程" + (i + 1), "1");
            myThread[i].start();
        }
        MyThread newT = new MyThread(group, "报错线程", "a");
        newT.start();
    }

}
```

程序运行后，其中一个线程出现异常，其他线程全部停止了，如图 7-25 所示。

图 7-25　程序运行结果

7.9　线程异常处理的优先性

前面介绍了若干个线程异常处理的方式，如果将这些处理方式放在一起运行，会出现什么运行结果呢？

创建测试用的项目 threadExceptionMove，类 MyThread.java 代码如下：

```
package extthread;

public class MyThread extends Thread {

    private String num = "a";

    public MyThread() {
```

```
        super();
    }

    public MyThread(ThreadGroup group, String name) {
        super(group, name);
    }

    @Override
    public void run() {
        int numInt = Integer.parseInt(num);
        System.out.println("在线程中打印: " + (numInt + 1));
    }

}
```

类 MyThreadGroup.java 代码如下:

```
package extthreadgroup;

public class MyThreadGroup extends ThreadGroup {

    public MyThreadGroup(String name) {
        super(name);
    }

    @Override
    public void uncaughtException(Thread t, Throwable e) {
        super.uncaughtException(t, e);
        System.out.println("线程组的异常处理");
        e.printStackTrace();
    }

}
```

类 ObjectUncaughtExceptionHandler.java 代码如下:

```
package test.extUncaughtExceptionHandler;

import java.lang.Thread.UncaughtExceptionHandler;

public class ObjectUncaughtExceptionHandler implements UncaughtExceptionHandler {

    @Override
    public void uncaughtException(Thread t, Throwable e) {
        System.out.println("对象的异常处理");
        e.printStackTrace();
    }

}
```

类 StateUncaughtExceptionHandler.java 代码如下:

```java
package test.extUncaughtExceptionHandler;

import java.lang.Thread.UncaughtExceptionHandler;

public class StateUncaughtExceptionHandler implements UncaughtExceptionHandler {

    @Override
    public void uncaughtException(Thread t, Throwable e) {
        System.out.println("静态的异常处理");
        e.printStackTrace();
    }

}
```

创建运行类 Run1.java，代码如下：

```java
package test;

import test.extUncaughtExceptionHandler.ObjectUncaughtExceptionHandler;
import test.extUncaughtExceptionHandler.StateUncaughtExceptionHandler;
import extthread.MyThread;

public class Run1 {

    public static void main(String[] args) {
        MyThread myThread = new MyThread();
        // 对象
        myThread.setUncaughtExceptionHandler(new ObjectUncaughtExceptionHandler());
        // 类
        MyThread.setDefaultUncaughtExceptionHandler(new StateUncaughtException-
            Handler());
        myThread.start();

    }
}
```

程序运行结果如图 7-26 所示。

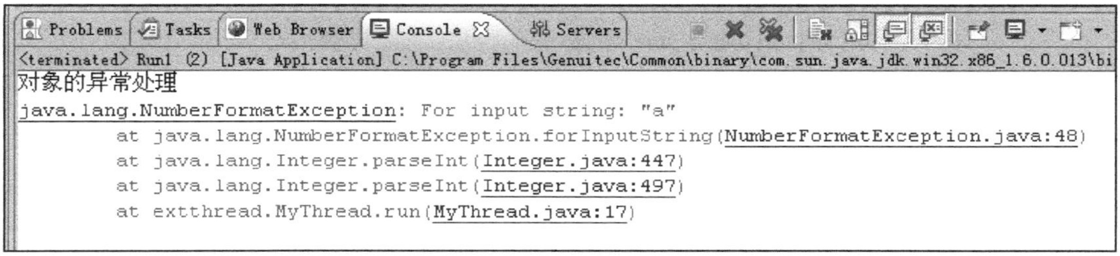

图 7-26　对象异常处理被运行

更改 Run1.java 代码如下：

```java
public class Run1 {

    public static void main(String[] args) {
        MyThread myThread = new MyThread();
        // 对象
        // smyThread.setUncaughtExceptionHandler(new ObjectUncaughtException-
            Handler());
        // 类
        MyThread.setDefaultUncaughtExceptionHandler(new StateUncaughtException-
            Handler());
        myThread.start();
    }
}
```

程序运行结果如图 7-27 所示。

```
静态的异常处理
java.lang.NumberFormatException: For input string: "a"
        at java.lang.NumberFormatException.forInputString(NumberFormatException.java:48)
        at java.lang.Integer.parseInt(Integer.java:447)
        at java.lang.Integer.parseInt(Integer.java:497)
        at extthread.MyThread.run(MyThread.java:17)
```

图 7-27 Run1.java 运行结果

创建类 Run2.java，代码如下：

```java
package test;

import test.extUncaughtExceptionHandler.ObjectUncaughtExceptionHandler;
import test.extUncaughtExceptionHandler.StateUncaughtExceptionHandler;
import extthread.MyThread;
import extthreadgroup.MyThreadGroup;

public class Run2 {

    public static void main(String[] args) {
        MyThreadGroup group = new MyThreadGroup("我的线程组");
        MyThread myThread = new MyThread(group, "我的线程");
        // 对象
        myThread.setUncaughtExceptionHandler(new ObjectUncaughtExceptionHandler());
        // 类
        MyThread.setDefaultUncaughtExceptionHandler(new StateUncaughtException-
            Handler());
        myThread.start();

    }
}
```

程序运行结果如图 7-28 所示。

```
对象的异常处理
java.lang.NumberFormatException: For input string: "a"
        at java.lang.NumberFormatException.forInputString(NumberFormatException.java:48)
        at java.lang.Integer.parseInt(Integer.java:447)
        at java.lang.Integer.parseInt(Integer.java:497)
        at extthread.MyThread.run(MyThread.java:17)
```

图 7-28　Run2.java 运行结果

更改 Run2.java 代码如下：

```java
package test;

import test.extUncaughtExceptionHandler.ObjectUncaughtExceptionHandler;
import test.extUncaughtExceptionHandler.StateUncaughtExceptionHandler;
import extthread.MyThread;
import extthreadgroup.MyThreadGroup;

public class Run2 {

    public static void main(String[] args) {
        MyThreadGroup group = new MyThreadGroup("我的线程组");
        MyThread myThread = new MyThread(group, "我的线程");
        // 对象
        // myThread.setUncaughtExceptionHandler(new ObjectUncaughtExceptionHandler());
        // 类
        MyThread.setDefaultUncaughtExceptionHandler(new StateUncaughtException-
            Handler());
        myThread.start();

    }
}
```

程序运行结果如图 7-29 所示。

```
java.lang.NumberFormatException: For input string: "a"静态的异常处理
        at java.lang.NumberFormatException.forInputString(NumberFormatException.java:48)
        at java.lang.Integer.parseInt(Integer.java:447)
        at java.lang.Integer.parseInt(Integer.java:497)
        at extthread.MyThread.run(MyThread.java:17)
java.lang.NumberFormatException: For input string: "a"
        at java.lang.NumberFormatException.forInputString(NumberFormatException.java:48)
        at java.lang.Integer.parseInt(Integer.java:447)
        at java.lang.Integer.parseInt(Integer.java:497)
        at extthread.MyThread.run(MyThread.java:17)线程组的异常处理
```

图 7-29　更改 Run2.java 后的运行结果

本示例想要输出"静态的异常处理"信息，则必须在 public void uncaughtException (Thread t, Throwable e) 方法中加上"super.uncaughtException(t, e);"代码。

继续更改 Run2.java 代码如下：

```java
public class Run2 {

    public static void main(String[] args) {
        MyThreadGroup group = new MyThreadGroup("我的线程组");
        MyThread myThread = new MyThread(group, "我的线程");
        // 对象
        // myThread.setUncaughtExceptionHandler(new ObjectUncaughtExceptionHandler());
        // 类
        // MyThread.setDefaultUncaughtExceptionHandler(new
        // StateUncaughtExceptionHandler());
        myThread.start();

    }
}
```

程序运行结果如图 7-30 所示。

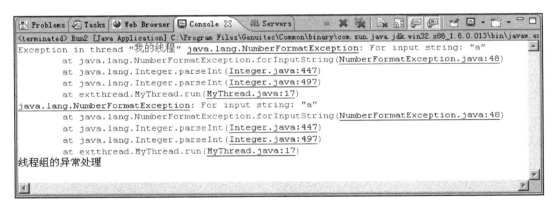

图 7-30　程序运行结果

通过前面的示例可以得出一个重要结论，如果调用过 setUncaughtExceptionHandler() 方法，则此异常处理器优先处理，其他异常处理器不再处理。

7.10　本章小结

本章弥补了前面章节遗漏的技术空白点，所给示例是对多线程技术的补充，有助于理解多线程的细节。例如，理解线程的状态后，完全可以对不同状态下的线程正在做哪些事情了如指掌；学习了线程组后，可以对线程的组织进行更有效的规划；了解 SimpleDateFormat 类在遇到多线程时也会出现意想不到的意外，以及线程在出现异常时的常用处理方式。